Lecture Notes in Mathematics 1949

Editors:
J.-M. Morel, Cachan
F. Takens, Groningen
B. Teissier, Paris

FONDAZIONE CIME
ROBERTO CONTI
CENTRO INTERNAZIONALE MATEMATICO ESTIVO
INTERNATIONAL MATHEMATICAL SUMMER CENTER

C.I.M.E. means Centro Internazionale Matematico Estivo, that is, International Mathematical Summer Center. Conceived in the early fifties, it was born in 1954 and made welcome by the world mathematical community where it remains in good health and spirit. Many mathematicians from all over the world have been involved in a way or another in C.I.M.E.'s activities during the past years.

So they already know what the C.I.M.E. is all about. For the benefit of future potential users and co-operators the main purposes and the functioning of the Centre may be summarized as follows: every year, during the summer, Sessions (three or four as a rule) on different themes from pure and applied mathematics are offered by application to mathematicians from all countries. Each session is generally based on three or four main courses (24–30 hours over a period of 6-8 working days) held from specialists of international renown, plus a certain number of seminars.

A C.I.M.E. Session, therefore, is neither a Symposium, nor just a School, but maybe a blend of both. The aim is that of bringing to the attention of younger researchers the origins, later developments, and perspectives of some branch of live mathematics.

The topics of the courses are generally of international resonance and the participation of the courses cover the expertise of different countries and continents. Such combination, gave an excellent opportunity to young participants to be acquainted with the most advance research in the topics of the courses and the possibility of an interchange with the world famous specialists. The full immersion atmosphere of the courses and the daily exchange among participants are a first building brick in the edifice of international collaboration in mathematical research.

C.I.M.E. Director
Pietro ZECCA
Dipartimento di Energetica "S. Stecco"
Università di Firenze
Via S. Marta, 3
50139 Florence
Italy
e-mail: zecca@unifi.it

C.I.M.E. Secretary
Elvira MASCOLO
Dipartimento di Matematica
Università di Firenze
viale G.B. Morgagni 67/A
50134 Florence
Italy
e-mail: mascolo@math.unifi.it

For more information see CIME's homepage: http://www.cime.unifi.it

CIME's activity is supported by:

– Istituto Nazionale di Alta Mathematica "F. Severi"
– Ministero dell'Istruzione, dell'Università e della Ricerca

Hans G. Feichtinger · Bernard Helffer
Michael P. Lamoureux · Nicolas Lerner
Joachim Toft

Pseudo-Differential Operators

Quantization and Signals

Lectures given at the
C.I.M.E. Summer School
held in Cetraro, Italy
June 19–24, 2006

Editors:
Luigi Rodino
M. W. Wong

 Springer

FONDAZIONE
CIME
ROBERTO CONTI

Authors and Editors

Hans G. Feichtinger
Faculty of Mathematics
University of Vienna
Nordbergstrasse 15
1090 Vienna, Austria
hans.feichtinger@univie.ac.at

Bernard Helffer
Laboratoire de Mathématique
Université Paris-Sud, Bat 425
91405 Orsay Cedex, France
bernard.helffer@math.u-psud.fr

Michael P. Lamoureux
University of Calgary
2500 University Drive NW
T2N 1N4 Calgary, Alberta, Canada
mikel@ucalgary.ca

Nicolas Lerner
Institut de Mathématiques de Jussieu
175 rue du Chevaleret
Université Paris 6, 75013 Paris, France
lerner@math.jussieu.fr

Joachim Toft
Växjö University
Vejdes plats 6,7
35195 Växjö, Sweden
joachim.toft@vxu.se

Luigi Rodino
Dipartimento di Matematica
Università di Torino
Via Carlo Alberto 10
10123 Torino, Italy
luigi.rodino@unito.it

M.W. Wong
Department of Mathematics and Statistics
York University
Keele Street 4700
M3J 1P3 Toronto, Ontario, Canada
mwwong@mathstat.yorku.ca

ISBN: 978-3-540-68266-0   e-ISBN: 978-3-540-68268-4
DOI: 10.1007/978-3-540-68268-4

Lecture Notes in Mathematics ISSN print edition: 0075-8434
ISSN electronic edition: 1617-9692

Library of Congress Control Number: 2008927361

Mathematics Subject Classification (2000): 35S05, 47G30, 41A05, 35J05

Cover design: WMXDesign GmbH

Printed on acid-free paper

9 8 7 6 5 4 3 2 1

springer.com

Preface

This volume contains the courses delivered at the CIME meeting "Pseudo-differential Operators, Quantization and Signals" held in Cetraro, Italy, from June 19, 2006 to June 24, 2006 and includes the courses by H.-G. Feichtinger presenting new results for Gabor multipliers on modulation and Wiener amalgam spaces, by B. Helffer analyzing non-self-adjoint operators using microlocal techniques, by M. Lamoureux addressing applications of pseudo-differential operators in geophysics, and by N. Lerner applying the techniques of Wick quantization to problems on subellipticity and lower bounds. The lectures by J. Toft on Schatten–von Neumann classes of Weyl pseudo-differential operators are also included.

This introduction is written for non-specialists. We first recall the basic notions and give an account of some developments of pseudo-differential operators. Our starting point is the class of pseudo-differential operators studied in the 1965 seminal paper of Kohn and Nirenberg published in "Communications on Pure and Applied Mathematics." Then we give a brief overview of several pre-eminent ancestors and successors in the study of pseudo-differential operators before and after the Kohn–Nirenberg milestone. The connections with quantization envisaged by Hermann Weyl in his classic "Group Theory and Quantum Mechanics," first observed by Grossmann, Loupias and Stein in the 1968 paper "Annales de l'Institute Fourier (Grenoble)," will then be described in the context of Wigner transforms. These connections give new insights into the role of pseudo-differential operators in the analysis of signals and images in the perspectives of Gabor transforms and wavelet transforms. From these come the Stockwell transform that has numerous applications in geophysics and medical imaging. The recently developed mathematical underpinnings of the Stockwell transform will be highlighted.

1. Pseudo-differential Operators

The starting point is the class of classical pseudo-differential operators introduced by Kohn and Nirenberg [19] and modified almost immediately by Hörmander [16] about 40 years ago. To wit, let $m \in \mathbb{R}$. Then we let $S_{1,0}^m$ or simply S^m be the set of all C^∞ functions σ on $\mathbb{R}^n \times \mathbb{R}^n$ such that for all multi-indices α and β, there exists a positive constant $C_{\alpha,\beta}$ for which

$$|(D_x^\alpha D_\xi^\beta \sigma)(x,\xi)| \leq C_{\alpha,\beta}(1+|\xi|)^{m-|\beta|}$$

for all x and ξ in $\mathbb{R}^n$. A function σ in S^m is called a symbol of order m. Let $\sigma \in S^m$. Then we define the pseudo-differential operator T_σ on the Schwartz space $\mathcal{S}(\mathbb{R}^n)$ by

$$(T_\sigma \varphi)(x) = (2\pi)^{-n/2} \int_{\mathbb{R}^n} e^{ix\cdot\xi} \sigma(x,\xi) \hat{\varphi}(\xi) \, d\xi$$

for all φ in $\mathcal{S}(\mathbb{R}^n)$ and all x in $\mathbb{R}^n$, where

$$\hat{\varphi}(\xi) = (2\pi)^{-n/2} \int_{\mathbb{R}^n} e^{-ix\cdot\xi} \varphi(x) \, dx$$

for all ξ in $\mathbb{R}^n$. It is easy to prove that T_σ maps $\mathcal{S}(\mathbb{R}^n)$ into $\mathcal{S}(\mathbb{R}^n)$ continuously. The most fundamental properties of pseudo-differential operators which are useful in the study of partial differential equations are listed as Theorems 1.1–1.3.

Theorem 1.1. *Let $\sigma \in S^0$. Then T_σ, initially defined on $\mathcal{S}(\mathbb{R}^n)$, can be uniquely extended to a bounded linear operator from $L^2(\mathbb{R}^n)$ into $L^2(\mathbb{R}^n)$.*

Theorem 1.2. *If $\sigma \in S^m$, then $T_\sigma^* = T_\tau$, where $\tau \in S^m$ and*

$$\tau \sim \sum_\mu \frac{(-i)^{|\mu|}}{\mu!} \partial_x^\mu \partial_\xi^\mu \overline{\sigma}.$$

Here, T_σ^* is the formal adjoint of T_σ.
To recall, the formal adjoint T_σ^* of T_σ is defined by

$$(T_\sigma \varphi, \psi)_{L^2(\mathbb{R}^n)} = (\varphi, T_\sigma^* \psi)_{L^2(\mathbb{R}^n)}$$

for all φ and ψ in $L^2(\mathbb{R}^n)$, where $(\, ,\,)_{L^2(\mathbb{R}^n)}$ is the inner product in $L^2(\mathbb{R}^n)$. The asymptotic expansion $\tau \sim \sum_\mu \frac{(-i)^{|\mu|}}{\mu!} \partial_x^\mu \partial_\xi^\mu \overline{\sigma}$ means that

$$\tau - \sum_{|\mu|<N} \frac{(-i)^{|\mu|}}{\mu!} \partial_x^\mu \partial_\xi^\mu \overline{\sigma} \in S^{m-N}$$

for all positive integers N.

Theorem 1.3. *If $\sigma \in S^{m_1}$ and $\tau \in S^{m_2}$, then $T_\sigma T_\tau = T_\lambda$, where $\lambda \in S^{m_1+m_2}$ and*

$$\lambda \sim \sum_\mu \frac{(-i)^{|\mu|}}{\mu!} (\partial_\xi^\mu \sigma)(\partial_x^\mu \tau).$$

The asymptotic expansion

$$\lambda \sim \sum_{\mu} \frac{(-i)^{|\mu|}}{\mu!} (\partial_\xi^\mu \sigma)(\partial_x^\mu \tau)$$

means that

$$\lambda - \sum_{|\mu|<N} \frac{(-i)^{|\mu|}}{\mu!} (\partial_\xi^\mu \sigma)(\partial_x^\mu \tau) \in S^{m_1+m_2-N}$$

for all positive integers N.

All these results are very well known and can be found in the books [17] by Hörmander [20] by Kumano-go, [23] by Rodino, [29] by Wong and many others. We can see variants of these results in other settings in this presentation.

2. Ancestors and Successors

Earliest sources of pseudo-differential operators can be traced to problems for n-dimensional singular integral equations. The first contributions to the theory of multi-dimensional singular integrals appear to be those of Tricomi [27] in 1928. To recall, let (r, θ) be the polar coordinates of a generic point $y = (y_1, y_2)$ in $\mathbb{R}^2$ and define for suitable functions φ on $\mathbb{R}^2$,

$$(P\varphi)(x) = \lim_{\varepsilon \to 0} \int_{r>\varepsilon} \frac{h(\theta)}{r^2} \varphi(x - y) \, dy, \quad x \in \mathbb{R}^2.$$

In general, the integral $\int_{\mathbb{R}^2} \frac{h(\theta)}{r^2} \varphi(x - y) \, dy$ is not absolutely convergent, but under the so-called Tricomi condition stipulating that

$$\int_0^{2\pi} h(\theta) \, d\theta = 0$$

and appropriate assumptions on h and φ, the limit exists and $(P\varphi)(x)$ is well defined for almost all x in $\mathbb{R}^2$. If we assume for simplicity that h is C^∞ on the unit circle S^1 with center at the origin, then P is a bounded linear operator from $L^2(\mathbb{R}^2)$ into $L^2(\mathbb{R}^2)$. Despite unsuccessful attempts by Tricomi in solving the equation

$$P\varphi = \psi$$

by finding another singular integral operator P^{-1} for which

$$P^{-1}P = I$$

and

$$PP^{-1} = I,$$

where I is the identity operator, we all know nowadays that this can be done using the Fourier transform. Indeed, P can be regarded as the convolution operator given by

$$P\varphi = K * \varphi,$$

where the singular kernel K given by

$$K(y) = \frac{h(\theta)}{r^2}, \quad y = (r, \theta) \in \mathbb{R}^2,$$

has to be suitably seen as a tempered distribution on $\mathbb{R}^2$. Applying the Fourier transform, we get

$$(P\varphi)^{\wedge}(\xi) = \sigma(\xi)\hat{\varphi}(\xi), \quad \xi \in \mathbb{R}^2,$$

where

$$\sigma(\xi) = 2\pi \hat{K}(\xi), \quad \xi \in \mathbb{R}^2.$$

In view of the Tricomi condition on $h \in C^{\infty}(S^1)$, σ turns out to be C^{∞} and homogeneous of degree 0 on $\mathbb{R}^2 \setminus \{0\}$. Hence, apart from the singularity at the origin, σ is a symbol in S^0 depending on ξ only, and with the notation of the preceding section,

$$P = T_{\sigma}.$$

Furthermore, if σ is elliptic in the sense that there exists a positive constant C such that

$$|\sigma(\xi)| \geq C, \quad \xi \in \mathbb{R}^2,$$

then $\sigma^{-1} \in S^0$ and we can define P^{-1} to be $T_{\sigma^{-1}}$. Such applications of the Fourier transform were not known to Tricomi and it took almost 30 years for mathematicians to come to these simple conclusions. Milestones of the developments in this direction are the works of Giraud [13] in 1934, Calderón and Zygmund [4] in 1952 and Mihlin [21] in 1965. Additional references can be found in the introduction of [21] and the survey paper [24] of Seeley. In fact, the analysis has been extended to the case when h also depends on x, i.e., the kernel K is a function of x and y given by

$$K(x, y) = \frac{h(x, \theta)}{r^2},$$

where $y = (r, \theta)$. In the final formulation of these results in the setting of $\mathbb{R}^n$, the symbol $\sigma \in S^0$ is the Fourier transform with respect to y of the kernel $K(x, y)$ in terms of the singular integral given by

$$\sigma(x, \xi) = \lim_{\varepsilon \to 0} \int_{|y| > \varepsilon} e^{-iy \cdot \xi} K(x, y) \, dy, \quad x, \xi \in \mathbb{R}^n.$$

If σ is elliptic in the sense that there exists a positive constant C such that

$$|\sigma(x, \xi)| \geq C, \quad x, \xi \in \mathbb{R}^n,$$

then we still have

$$\sigma^{-1} \in S^0.$$

However, it is important to note that $T_{\sigma^{-1}}$ is no longer the inverse of P in this case. But, as in Theorem 1.3, we obtain

$$T_{\sigma^{-1}}P = I + K_1$$

and

$$PT_{\sigma^{-1}} = I + K_2,$$

where K_1 and K_2 are pseudo-differential operators of order -1. When we transfer the definition of P to a compact manifold M, the operators K_1 and K_2 are compact and P is then a Fredholm operator on $L^2(M)$. It is remarkable to note that this very rudimentary symbolic calculus with remainders of order -1 plays an important role in the proof of the Atiyah–Singer index formula in [1].

In addition to the obvious extension to an arbitrary order $m \in \mathbb{R}$, the most novel ideas of the Kohn–Nirenberg paper [19] in the context of the theory of singular integral operators are the precise asymptotic formulas articulated in Theorems 1.2 and 1.3. Almost immediately after the appearance of the work of Kohn and Nirenberg is the far-reaching calculus of Hörmander [16] concerning symbols σ of type (ρ, δ), $0 \leq \delta < \rho \leq 1$. Let us recall that a function σ in $C^\infty(\mathbb{R}^n \times \mathbb{R}^n)$ is a symbol of order $m \in \mathbb{R}$ and type (ρ, δ) if for all multi-indices α and β, there exists a positive constant $C_{\alpha,\beta}$ such that

$$|(D_x^\alpha D_\xi^\beta \sigma)(x,\xi)| \leq C_{\alpha,\beta}(1 + |\xi|)^{m - \rho|\beta| + \delta|\alpha|}$$

for all x and ξ in $\mathbb{R}^n$. Since then, other generalizations and variants of pseudo-differential operators have appeared. Among many interesting classes is the very general class of pseudo-differential operators developed by Beals [2] in 1975 in which the Hörmander estimates are replaced by

$$|(D_x^\alpha D_\xi^\beta \sigma)(x,\xi)| \leq C_{\alpha,\beta}\lambda(x,\xi)\Psi(x,\xi)^{-|\beta|}\Phi(x,\xi)^{|\alpha|}$$

for all x and ξ in $\mathbb{R}^n$, where

$$\Psi(x,\xi) = (\Psi_1(x,\xi), \Psi_2(x,\xi), \ldots, \Psi_n(x,\xi))$$

and

$$\Phi(x,\xi) = (\Phi_1(x,\xi), \Phi_2(x,\xi), \ldots, \Phi_n(x,\xi))$$

are n-tuples of suitable weight functions, and $\lambda(x,\xi)$ is now the "order" of the corresponding pseudo-differential operator. Recasting the calculus of Beals, another achievement is due to Hörmander [16] using the Weyl expression for pseudo-differential operators. We refer the readers to [16] for a wide range of applications to linear partial differential equations. Weyl quantization is described in the next section, and for the sake of simplicity, we begin with a motivation based on symbols in S^m, i.e., Hörmander symbols with $\rho = 1$ and $\delta = 0$.

3. Weyl Transforms

Let $\sigma \in S^m$. Then we can associate to it the pseudo-differential operator T_σ, but T_σ is not the only operator that can be assigned to σ. To see what else can be done, let us note that for all φ in $\mathcal{S}(\mathbb{R}^n)$ and all x in $\mathbb{R}^n$,

$$(T_\sigma\varphi)(x) = (2\pi)^{-n/2} \int_{\mathbb{R}^n} e^{ix\cdot\xi}\sigma(x,\xi)\hat{\varphi}(\xi)\,d\xi$$

$$= (2\pi)^{-n} \int_{\mathbb{R}^n} \int_{\mathbb{R}^n} e^{i(x-y)\cdot\xi}\sigma(x,\xi)\varphi(y)\,dy\,d\xi,$$

where the last integral is to be understood as an oscillatory integral in which the integral with respect to y has to be performed first. With this formula in hand, it requires a huge amount of ingenuity (certainly not logic) to see that we can associate to σ another useful linear operator W_σ on $\mathcal{S}$ defined by the same formula with $\sigma(x,\xi)$ replaced by $\sigma\left(\frac{x+y}{2},\xi\right)$. The linear operator W_σ can be traced back to the work [28] by Hermann Weyl and hence we call W_σ the Weyl transform associated to the symbol σ. In fact, we have the following connection between Weyl transforms and pseudo-differential operators.

Theorem 3.1. *Let $\sigma \in S^m$. Then there exists a symbol τ in S^m such that*

$$T_\sigma = W_\tau$$

and there exists a symbol κ in S^m such that

$$W_\sigma = T_\kappa.$$

Thus, there is a one-to-one correspondence between pseudo-differential operators and Weyl transforms. We have the following result, which can be thought of as the fundamental Theorem of pseudo-differential operators.

Theorem 3.2. *Let $\sigma \in S^m$, $m \in \mathbb{R}$. Then for all φ and ψ in $\mathcal{S}(\mathbb{R}^n)$,*

$$(W_\sigma\varphi, \psi)_{L^2(\mathbb{R}^n)} = (2\pi)^{-n/2} \int_{\mathbb{R}^n} \int_{\mathbb{R}^n} \sigma(x,\xi)W(\varphi,\psi)(x,\xi)\,dx\,d\xi,$$

where $W(\varphi,\psi)$ is the Wigner transform of φ and ψ defined by

$$W(\varphi,\psi)(x,\xi) = (2\pi)^{-n/2} \int_{\mathbb{R}^n} e^{-i\xi\cdot p}\varphi\left(x+\frac{p}{2}\right)\overline{\psi\left(x-\frac{p}{2}\right)}\,dp$$

for all x and ξ in $\mathbb{R}^n$.

The Wigner transform is a very well-behaved bilinear form on $L^2(\mathbb{R}^n) \times L^2(\mathbb{R}^n)$ and it satisfies the so-called Moyal identity or the Plancherel formula to the effect that

$$\|W(\varphi, \psi)\|_{L^2(\mathbb{R}^{2n})} = \|\varphi\|_{L^2(\mathbb{R}^n)} \|\psi\|_{L^2(\mathbb{R}^n)}$$

for all φ and ψ in $L^2(\mathbb{R}^n)$.

A *tour de force* from Theorems 3.1 and 3.2 shows that we can now define pseudo-differential operators with nonsmooth symbols not in the Hörmander class S^m. To be specific, we look at symbols in $L^2(\mathbb{R}^n \times \mathbb{R}^n)$ only.

Let $\sigma \in L^2(\mathbb{R}^n \times \mathbb{R}^n)$. Then we define the Weyl transform W_σ on $L^2(\mathbb{R}^n)$ by

$$(W_\sigma f, g)_{L^2(\mathbb{R}^n)} = (2\pi)^{-n/2} \int_{\mathbb{R}^n} \int_{\mathbb{R}^n} \sigma(x, \xi)\, W(f, g)(x, \xi)\, dx\, d\xi$$

for all f and g in $L^2(\mathbb{R}^n)$. Then we have the following analogs of Theorems 1.1–1.3.

Theorem 3.3. *Let $\sigma \in L^2(\mathbb{R}^n \times \mathbb{R}^n)$. Then $W_\sigma : L^2(\mathbb{R}^n) \to L^2(\mathbb{R}^n)$ is a Hilbert–Schmidt operator.*

Theorem 3.4. *Let $\sigma \in L^2(\mathbb{R}^n \times \mathbb{R}^n)$. Then the adjoint W_σ^* of W_σ is given by*

$$W_\sigma^* = W_{\overline{\sigma}}.$$

Theorem 3.5. *Let σ and τ be symbols in $L^2(\mathbb{R}^n \times \mathbb{R}^n)$. Then*

$$W_\sigma W_\tau = W_\lambda,$$

where $\lambda \in L^2(\mathbb{R}^n \times \mathbb{R}^n)$ and is given by

$$\hat{\lambda} = (2\pi)^{-n}(\hat{\sigma} *_{1/4} \hat{\tau}).$$

Theorem 3.5, which is attributed to Grossmann, Loupias and Stein [15], tells us that the product of two Weyl transforms with symbols in $L^2(\mathbb{R}^n \times \mathbb{R}^n)$ is again a Weyl transform with symbol in $L^2(\mathbb{R}^n \times \mathbb{R}^n)$ and is given by a twisted convolution. Let us recall that the twisted convolution $f *_{1/4} g$ of two measurable functions f and g on $\mathbb{C}^n(= \mathbb{R}^n \times \mathbb{R}^n)$ is defined by

$$(f *_{1/4} g)(z) = \int_{\mathbb{C}^n} f(z - w) g(w) e^{i[z, w]/4} dw$$

for all z in $\mathbb{C}^n$, where $[z, w]$ is the symplectic form of z and w given by

$$[z, w] = 2\operatorname{Im}(z \cdot \overline{w}).$$

See the books [3] by Boggiatto, Buzano and Rodino, [12] by Folland, [25] by Stein and [30] by Wong for details and related topics.

4. Gabor Transforms

If we make a change of variables in the definition of the Wigner transform, then we get for all f and g in $L^2(\mathbb{R}^n)$, and all x and ξ in $\mathbb{R}^n$,

$$W(f,g)(x,\xi) = 2^n e^{2ix\cdot\xi}(G_{\tilde{g}}f)(2x, 2\xi),$$

where

$$\tilde{g}(x) = g(-x)$$

for all x in $\mathbb{R}^n$ and $G_{\tilde{g}}f$ is the well-known Gabor transform or the short-time Fourier transform of f with window $\tilde{g}$ given by

$$(G_{\tilde{g}}f)(x,\xi) = (2\pi)^{-n/2} \int_{\mathbb{R}^n} e^{-it\cdot\xi} f(t)\overline{\tilde{g}(t-x)}\, dt$$

for all x and ξ in $\mathbb{R}^n$. In image analysis, we can think of $(G_{\tilde{g}}f)(x,\xi)$ as the spectral content of the image f with frequency ξ at the point x.

Let us now fix a window φ in $L^1(\mathbb{R}^n) \cap L^2(\mathbb{R}^n)$ with $\int_{\mathbb{R}^n} \varphi(x)\, dx = 1$. Then the Gabor transform $G_\varphi f$ of f is given by

$$(G_\varphi f)(x,\xi) = (2\pi)^{-n/2}(f, M_\xi T_{-x}\varphi)_{L^2(\mathbb{R}^n)}$$

for all x and ξ in $\mathbb{R}^n$, where M_ξ and T_{-x} are the modulation operator and the translation operator given by

$$(M_\xi h)(t) = e^{it\cdot\xi} h(t)$$

and

$$(T_{-x}h)(t) = h(t-x)$$

for all measurable functions h on $\mathbb{R}^n$ and all t in $\mathbb{R}^n$. Now, for all x and ξ in $\mathbb{R}^n$, we define the function $\varphi_{x,\xi}$ on $\mathbb{R}^n$ by

$$\varphi_{x,\xi} = M_\xi T_{-x}\varphi.$$

We call the functions $\varphi_{x,\xi}$, $x, \xi \in \mathbb{R}^n$, the Gabor wavelets generated from the Gabor mother wavelet φ by translations and modulations.

The usefulness of the Gabor wavelets in signal and image analysis is enhanced by the following resolution of the identity formula, which allows the reconstruction of a signal or an image from its Gabor spectrum.

Theorem 4.1. For all f in $L^2(\mathbb{R}^n)$,

$$f = (2\pi)^{-n} \int_{\mathbb{R}^n} \int_{\mathbb{R}^n} (f, \varphi_{x,\xi})_{L^2(\mathbb{R}^n)} \varphi_{x,\xi}\, dx\, d\xi.$$

Let $\sigma \in L^2(\mathbb{R}^n \times \mathbb{R}^n)$. Then we define the Gabor multiplier $G_{\sigma,\varphi} : L^2(\mathbb{R}^n) \to L^2(\mathbb{R}^n)$ by

$$(G_{\sigma,\varphi}f, g)_{L^2(\mathbb{R}^n)} = \int_{\mathbb{R}^n} \int_{\mathbb{R}^n} \sigma(x,\xi)(G_{\varphi}f)(x,\xi)\overline{(G_{\varphi}g)(x,\xi)} \, dx \, d\xi$$

for all f and g in $L^2(\mathbb{R}^n)$. Using the Gabor wavelets, we see that $G_{\sigma,\varphi}f$ is equal to

$$(2\pi)^{-n} \int_{\mathbb{R}^n} \int_{\mathbb{R}^n} \sigma(x,\xi)(f, \varphi_{x,\xi})_{L^2(\mathbb{R}^n)} \varphi_{x,\xi} \, dx \, d\xi$$

for all f in $L^2(\mathbb{R}^n)$.

Gabor multipliers are also known as localization operators, Daubechies operators, anti-Wick quantization and Wick quantization. The following results are the analogs of Theorems 1.1–1.3 for Gabor multipliers.

Theorem 4.2. *Let* $\sigma \in L^2(\mathbb{R}^n \times \mathbb{R}^n)$. *Then the Gabor multiplier* $G_{\sigma,\varphi} : L^2(\mathbb{R}^n) \to L^2(\mathbb{R}^n)$ *is a Hilbert–Schmidt operator.*

Theorem 4.3. *Let* $\sigma \in L^2(\mathbb{R}^n \times \mathbb{R}^n)$. *Then the adjoint* $G_{\sigma,\varphi}^*$ *of* $G_{\sigma,\varphi}$ *is given by*

$$G_{\sigma,\varphi}^* = G_{\bar{\sigma},\varphi}.$$

Theorem 4.4. *Let* σ *and* τ *be functions in* $L^2(\mathbb{R}^n \times \mathbb{R}^n)$. *Then*

$$G_{\sigma,\varphi}G_{\tau,\varphi} = G_{\lambda,\varphi},$$

where

$$\hat{\lambda} = (2\pi)^{-n}(\hat{\sigma} *^{1/2} \hat{\tau}).$$

In Theorem 4.4, we have a new twisted convolution. To wit, the new twisted convolution $f *^{1/2} g$ of two measurable functions f and g on $\mathbb{C}^n$, is defined by

$$(f *^{1/2} g)(z) = \int_{\mathbb{C}^n} f(z - w) g(w) e^{(z \cdot \bar{w} - |w|^2)/2} dw$$

for all z in $\mathbb{C}^n$ provided that the integral exists. Theorem 4.4 can be found in the 2000 paper [10] by Du and Wong.

The interesting feature with Theorem 4.4 is that the new twisted convolution $f *^{1/2} g$ of two functions f and g in $L^2(\mathbb{R}^n \times \mathbb{R}^n)$ need not be in $L^2(\mathbb{R}^n \times \mathbb{R}^n)$. This phenomenon is the motivation for many interesting research papers on the product of Gabor multipliers. It suffices to mention the works [5] by Coburn, [7] by Cordero and Gröchenig and [8] by Cordero and Rodino.

What is a Gabor multiplier? Is it something already well known to us? The answer is yes.

Theorem 4.5. *Let $\sigma \in L^2(\mathbb{R}^n \times \mathbb{R}^n)$. Then*

$$G_{\sigma,\varphi} = W_{\sigma*V(\varphi,\varphi)},$$

where

$$V(\varphi,\varphi)^\wedge = W(\varphi,\varphi).$$

References for the materials in this section are the books [9] by Daubechies, [14] by Gröchenig, [31] by Wong and many others.

5. Wavelet Transforms

Let $\varphi \in L^2(\mathbb{R})$ be such that $\|\varphi\|_2 = 1$ and

$$\int_{-\infty}^{\infty} \frac{|\hat{\varphi}(\xi)|^2}{|\xi|} d\xi < \infty.$$

Then we call φ a mother wavelet and φ is said to satisfy the admissibility condition.

Let φ be a mother wavelet. Then for all b in $\mathbb{R}$ and a in $\mathbb{R}\backslash\{0\}$, we can define the wavelet $\varphi_{b,a}$ by

$$\varphi_{b,a}(x) = \frac{1}{\sqrt{|a|}} \varphi\left(\frac{x-b}{a}\right), \quad x \in \mathbb{R}.$$

We call $\varphi_{b,a}$ the affine wavelet generated from the mother wavelet φ by translation and dilation. To put things in perspective, let $b \in \mathbb{R}$ and let $a \in \mathbb{R}\backslash\{0\}$. Then we let T_b be the translation operator as before and D_a be the dilation operator defined by

$$(D_a f)(x) = \sqrt{|a|} f(ax)$$

for all x in $\mathbb{R}$ and all measurable functions f on $\mathbb{R}$. So, the wavelet $\varphi_{b,a}$ can be expressed as

$$\varphi_{b,a} = T_{-b} D_{1/a} \varphi.$$

Let φ be a mother wavelet. Then the wavelet transform $\Omega_\varphi f$ of a function f in $L^2(\mathbb{R})$ is defined to be the function on $\mathbb{R} \times \mathbb{R}\backslash\{0\}$ by

$$(\Omega_\varphi f)(b,a) = (f, \varphi_{b,a})_{L^2(\mathbb{R})}$$

for all b in $\mathbb{R}$ and a in $\mathbb{R}\backslash\{0\}$. At the heart of the analysis of the wavelet transform is the following resolution of the identity formula.

Theorem 5.1. *Let φ be a mother wavelet. Then for all functions f and g in $L^2(\mathbb{R})$,*

$$(f,g)_{L^2(\mathbb{R})} = \frac{1}{c_\varphi} \int_{-\infty}^{\infty} \int_{-\infty}^{\infty} (\Omega_\varphi f)(b,a)\overline{(\Omega_\varphi g)(b,a)} \frac{db\,da}{a^2},$$

where

$$c_\varphi = 2\pi \int_{-\infty}^{\infty} \frac{|\hat{\varphi}(\xi)|^2}{|\xi|} d\xi.$$

The resolution of the identity formula leads to the reconstruction formula which says that

$$f = \frac{1}{c_\varphi} \int_{-\infty}^{\infty} \int_{-\infty}^{\infty} (f, \varphi_{b,a})_{L^2(\mathbb{R})} \varphi_{b,a} \frac{db\, da}{a^2}$$

for all f in $L^2(\mathbb{R})$. In other words, we have a reconstruction formula for the signal f from a knowledge of its time-scale spectrum.

Let φ be a mother wavelet and let $\sigma \in L^2(\mathbb{R} \times \mathbb{R})$. Then we define the wavelet multiplier $\Omega_{\sigma,\varphi} : L^2(\mathbb{R}) \to L^2(\mathbb{R})$ by

$$\Omega_{\sigma,\varphi} f = \frac{1}{c_\varphi} \int_{-\infty}^{\infty} \int_{-\infty}^{\infty} \sigma(b, a)(f, \varphi_{b,a})_{L^2(\mathbb{R})} \varphi_{b,a} \frac{db\, da}{a^2}$$

for all f in $L^2(\mathbb{R})$.

As in the case of the Gabor multipliers, we have the following results.

Theorem 5.2. *The wavelet multiplier*

$$\Omega_{\sigma,\varphi} : L^2(\mathbb{R}) \to L^2(\mathbb{R})$$

is a Hilbert–Schmidt operator.

Theorem 5.3. *The adjoint $\Omega_{\sigma,\varphi}^*$ of the wavelet multiplier $\Omega_{\sigma,\varphi}$ is given by*

$$\Omega_{\sigma,\varphi}^* = \Omega_{\overline{\sigma},\varphi}.$$

What is the product of two wavelet multipliers? The answer is not so simple and seems to depend on the availability of a useful formula for a wavelet multiplier. Some technical information in this direction can be found in the paper [32] by Wong. If

$$\sigma(b, a) = \alpha(a)\beta(b)$$

for all b in $\mathbb{R}$ and all a in $\mathbb{R} \setminus \{0\}$, then $\Omega_{\sigma,\varphi}$ is a paracommutator in the sense of Janson and Peetre [18], and Peng and Wong [22]. If σ is a function of a only, then $\Omega_{\sigma,\varphi}$ is a paraproduct in the sense of Coifman and Meyer [6]. If σ is a function of b only, then $\Omega_{\sigma,\varphi}$ is a Fourier multiplier.

6. Stockwell Transforms

Let us recall that for a signal f in $L^2(\mathbb{R})$, the Gabor transform $(G_\varphi f)(x, \xi)$ with respect to the window φ gives the time–frequency content of f at time x and frequency ξ by using the window φ at time x. The drawback here is that a window of fixed width is used for all time x. It is more accurate if we can have an adaptive window that gives a wide window for low frequency and a narrow window for high frequency. That this can be done comes from our experiences with the wavelet transform. Indeed, we see that the window $\varphi_{b,a}$ is narrow if the scale a is small and the window is wide when the scale is big.

Now, the Stockwell transform $S_\varphi f$ with window φ of a signal f is defined by

$$(S_\varphi f)(x,\xi) = (2\pi)^{-1/2}|\xi| \int_{-\infty}^{\infty} e^{-it\xi} f(t)\overline{\varphi(\xi(t-x))}\, dt$$

for all x and ξ in $\mathbb{R}$. Formally, we note that for all f in $L^2(\mathbb{R})$, all x in $\mathbb{R}$ and all ξ in $\mathbb{R}\backslash\{0\}$,

$$(S_\varphi f)(x,\xi) = (f, \varphi^{x,\xi})_{L^2(\mathbb{R})},$$

where

$$\varphi^{x,\xi} = (2\pi)^{-1/2} M_\xi T_{-x} \tilde{D}_\xi \varphi.$$

Here, the dilation operator $\tilde{D}_\xi$ is defined by

$$(\tilde{D}_\xi f)(t) = |\xi| f(\xi t)$$

for all t in $\mathbb{R}$ and all measurable functions f on $\mathbb{R}$. Besides the modulation, a notable feature in the Stockwell transform is the normalizing factor in the dilation operator, which is $|\cdot|$ and not $|\cdot|^{1/2}$ as in the case of the wavelet transforms. These features distinguish the Stockwell transform from the wavelet transforms.

The Stockwell transform has recently been successfully used in seismic waves [26] by Stockwell, Mansinha and Lowe and in medical imaging [34] by Zhu and others. An attempt in understanding the mathematical underpinnings of the Stockwell transform is underway by Wong and Zhu. See [33] in this direction and we describe some of the results therein.

Theorem 6.1. *Let φ be a window with*

$$\int_{-\infty}^{\infty} \varphi(x)dx = 1.$$

Then for all f in $L^1(\mathbb{R}) \cap L^2(\mathbb{R})$,

$$\int_{-\infty}^{\infty} (S_\varphi f)(x,\xi)\, dx = \hat{f}(\xi)$$

for all ξ in $\mathbb{R}$.

See Fig. 1 for an illustration of Theorem 6.1. In view of Theorem 6.1, we have a reconstruction formula for a signal f in terms of its Stockwell spectrum, which says that

$$f = \mathcal{F}^{-1} A S_\varphi f,$$

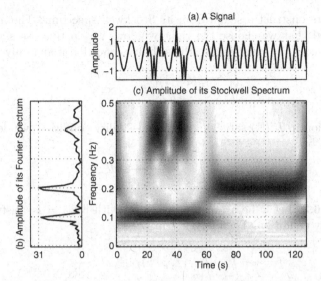

Fig. 1 Time–frequency representation of the Stockwell transform: (**a**) a signal consisting of multiple frequency components (**b**) the amplitude of the corresponding Fourier spectrum, i.e., $|(\mathcal{F}f)(k)|$ (**c**) the contour plotting the amplitude of the corresponding Stockwell transform, i.e., $|(Sf)(\tau, k)|$

where $\mathcal{F}^{-1}$ is the inverse Fourier transform and A is the time average operator given by

$$(AF)(\xi) = \int_{-\infty}^{\infty} F(x, \xi)\, dx$$

for all ξ in $\mathbb{R}$ and all measurable functions F on $\mathbb{R} \times \mathbb{R}$.

For the second result, we let M be the set of all measurable functions F on $\mathbb{R} \times \mathbb{R}$ such that

$$\int_{-\infty}^{\infty} \left| \int_{-\infty}^{\infty} F(x, \xi)\, dx \right|^2 d\xi < \infty.$$

Then M is an indefinite Hilbert space in which the indefinite inner product $(\,,\,)_M$ is given by

$$(F, G)_M = (AF, AG)_{L^2(\mathbb{R})}$$

for all F and G in M.

Then we have a characterization of the Stockwell spectra given by the following theorem.

Theorem 6.2. $\{S_\varphi f : f \in L^2(\mathbb{R})\} = M/Z$, where

$$Z = \{F : \mathbb{R} \times \mathbb{R} \to \mathbb{C} : AF = 0\}.$$

Can we reconstruct a signal from its Stockwell spectrum? The answer is yes provided that we choose the right window. To do this, we say that a function φ in $L^2(\mathbb{R})$ satisfies the admissibility condition if and only if

$$\int_{-\infty}^{\infty} \frac{|\hat{\varphi}(\xi - 1)|^2}{|\xi|} d\xi < \infty.$$

For a function in $L^2(\mathbb{R})$ satisfying the admissibility condition, we define the constant c_φ by

$$c_\varphi = \int_{-\infty}^{\infty} \frac{|\hat{\varphi}(\xi - 1)|^2}{|\xi|} d\xi.$$

Theorem 6.3. Let φ be a function in $L^2(\mathbb{R})$ with $\|\varphi\|_2 = 1$ satisfying the admissibility condition. Then for all f in $L^2(\mathbb{R})$,

$$f = \frac{1}{c_\varphi} \int_{-\infty}^{\infty} \int_{-\infty}^{\infty} (f, \varphi^{x,\xi})_{L^2(\mathbb{R})} \varphi^{x,\xi} \frac{dx\, d\xi}{|\xi|}.$$

Remark: It is important to note that an admissible wavelet φ for the Stockwell transform has to satisfy the condition

$$\hat{\varphi}(-1) = 0.$$

So, the Gaussian window that has been used exclusively for the Stockwell transform in the literature is not admissible.

This formula and its discretization can be found in the paper [11] by Du, Wong and Zhu.

L. Rodino,
M.W. Wong

References

1. M. F. Atiyah and I. M. Singer, The index of elliptic operators, I, *Ann. Math.* **87** (1968), 484–530.
2. R. Beals, A general calculus of pseudodifferential operators, *Duke Math. J.* **60** (1975), 187–220.
3. P. Boggiatto, E. Buzano and L. Rodino, *Global Hypoellipticity and Spectral Theory*, Akademie-Verlag, 1996.
4. A. P. Calderón and A. Zygmund, On the existence of certain singular integrals, *Acta Math.* **88** (1952), 85–139.
5. L. A. Coburn, The Bargmann isometry and Gabor–Daubechies wavelet localization operators, in *Systems, Approximation, Singular Operators, and Related Topics*, Birkhäuser, 2001, 169–178.
6. R. R. Coifman and Y. Meyer, *Au Delà des Opérateurs Pseudo-Différentiels*, Astérisque 57, 1978.

7. E. Cordero and K. Gröchenig, On the product of localization operators, in *Modern Trends in Pseudo-Differential Operators*, Editors: J. Toft, M. W. Wong and H. Zhu, Birkhäuser, 279–295 .

8. E. Cordero and L. Rodino, Wick calculus: a time–frequency approach, *Osaka J. Math.* **42** (2005), 43–63.

9. I. Daubechies, *Ten Lectures on Wavelets*, SIAM, 1992.

10. J. Du and M. W. Wong, A product formula for localization operators, *Bull. Korean Math. Soc.* **37** (2000), 77–84.

11. J. Du, M. W. Wong and H. Zhu, Continuous and discrete inversion formulas for the Stockwell transform, *Integral Transforms Spect. Funct.* **18** (2007), 537–543.

12. G. B. Folland, *Harmonic Analysis in Phase Space*, Princeton University Press, 1989.

13. G. Giraud, Équations à intégrales principale; étude suivie d' une application, *Ann. Sci. École Norm. Sup. Paris* **51** (1934), 251–372.

14. K. Gröchenig, *Foundations of Time–Frequency Analysis*, Birkhäuser, 2001.

15. A. Grossmann, G. Loupias and E. M. Stein, An algebra of pseudodifferential operators and quantum mechanics in phase space, *Ann. Inst. Fourier (Grenoble)* **18** (1968), 343–368.

16. L. Hörmander, Pseudo-differential operators and hypoelliptic equations, in *Singular Integrals*, AMS, 1967, 138–183.

17. L. Hörmander, *The Analysis of Linear Partial Differential Operators III*, Springer-Verlag, 1985.

18. S. Janson and J. Peetre, Paracommutators - boundedness and Schatten–von Neumann properties, *Trans. Amer. Math. Soc.* **305** (1988), 467–504.

19. J. J. Kohn and L. Nirenberg, An algebra of pseudo-differential operators, *Comm. Pure Appl. Math.* **18** (1965), 269–305.

20. H. Kumano-go, *Pseudo-Differential Operators*, MIT Press, 1981.

21. S. G. Mihlin, *Multidimensional Singular Integrals and Integral Equations*, Pergamon Press, 1965.

22. L. Peng and M. W. Wong, Compensated compactness and paracommutators, *J. London Math. Soc.* **62** (2000), 505–520.

23. L. Rodino, *Linear Partial Differential Operators in Gevrey Spaces*, World Scientific, 1993.

24. R. T. Seeley, Elliptic singular equations, in *Singular Integrals*, AMS, 1967, 308–315.

25. E. M. Stein, *Harmonic Analysis: Real-Variable Methods, Orthogonality and Oscillatory Integrals*, Princeton University Press, 1993.

26. R. G. Stockwell, L. Mansinha and R. P. Lowe, Localization of the complex spectrum: the S transform, *IEEE Trans. Signal Processing* **44** (1996), 998–1001.

27. F. G. Tricomi, Equazioni integrali contenenti il valor principale di un integrale doppio, *Math. Z.* **27** (1928), 87–133.

28. H. Weyl, *The Theory of Groups and Quantum Mechanics*, Dover, 1950.

29. M. W. Wong, *An Introduction to Pseudo-Differential Operators*, Second Edition, World Scientific, 1999.

30. M. W. Wong, *Weyl Transforms*, Springer-Verlag, 1998.

31. M. W. Wong, *Wavelet Transforms and Localization Operators*, Birkhäuser, 2002.

32. M. W. Wong, Localization operators on the affine group and paracommutators, in *Progress in Analysis*, World Scientific, 2003, 663–669.

33. M. W. Wong and H. Zhu, A characterization of the Stockwell spectrum, in *Modern Trends in Pseudo-Differential Operators*, Birkhäuser, 2007, 251–257.

34. H. Zhu, B. G. Goodyear, M. L. Lauzon, R. A. Brown, G. S. Mayer, L. Mansinha, A. G. Law and J. R. Mitchell, A new multiscale Fourier analysis for MRI, *Med. Phys.* **30** (2003), 1134–1141.

Contents

Banach Gelfand Triples for Gabor Analysis

H. Feichtinger, F. Luef, and E. Cordero

Abstract It is the purpose of this survey note to show the relevance of a Gelfand triple which is closely connected with time–frequency analysis and Gabor analysis. The Segal algebra $S_0(\mathbb{R}^d)$ and its dual can be shown to be – for a large variety of concrete cases – a convenient substitute for the Schwartz space $\mathcal{S}(\mathbb{R}^d)$ and it's dual, the space of tempered distributions $\mathcal{S}'(\mathbb{R}^d)$. This concrete pair of Banach spaces is actually a Gelfand triple, which allows to describe in a very intuitive way the properties of the classical Fourier transform and other unitary operators arising in the treatment of various mathematical questions, e.g., multipliers in harmonic analysis. We will demonstrate the usefulness of the Banach Gelfand triple $\left(S_0(\mathbb{R}^d), L^2(\mathbb{R}^d), S_0(\mathbb{R}^d)\right)$ within time–frequency analysis, with a special emphasis on questions from *time–frequency analysis* and *Gabor analysis*.

1 Introduction

Gabor analysis is often described as "the part of analysis" which in one way or the other makes use of the family of so-called *time–frequency shift operators*. Here we have to mention first the short-time Fourier transform or sliding-window Fourier transform (in short, STFT), or Dennis Gabor's

Hans Feichtinger
Universität Wien, Nordbergstrasse 15, Vienna, Austria
e-mail: hans.feichtinger@univie.ac.at

Franz Luef
Universität Wien, Nordbergstrasse 15, Vienna, Austria
e-mail: franz.luef@univie.ac.at

Elena Cordero
University of Torino, Turin, Italy
e-mail: elena.cordero@unito.it

L. Rodino, M.W. Wong (eds.) *Pseudo-Differential Operators*. Lecture Notes in Mathematics 1949.
© Springer-Verlag Berlin Heidelberg 2008

claim of 1946, that "every function" can be written as a (double) series of time–frequency shifted copies (with suitable complex amplitudes). Even if one wants to discuss the subtleties, these operations on different natural function spaces such as the standard space $\boldsymbol{L}^2(\mathbb{R}^d)$ of "signals of finite energy" (i.e., square integrable functions) the question of convergence of the Gabor series expansions, or the stable reconstruction of a signal from a densely sampled STFT cannot be answered without various extra conditions (typically conditions on the smoothness and decay of the window that is used in forming the STFT).

The correct class of function spaces for time–frequency analysis and Gabor analysis are the *modulation spaces*, since they possess an intrinsic description in terms of STFT or Gabor frames. The smallest member of this class is *Feichtinger's algebra* $S_0(\mathbb{R}^d)$. Consequently, the dual space $S_0'(\mathbb{R}^d)$ of Feichtinger's algebra serves as the largest class of functions and distributions for the discussion of operators and their properties. In between of $S_0(\mathbb{R}^d)$ and $S_0'(\mathbb{R}^d)$ sits the Hilbert space $L^2(\mathbb{R}^d)$, actually this triple of Banach spaces forms a Gelfand triple. The notion of *Gelfand triples* allows to express mapping properties of operators (such as the Fourier transform, Gabor frame operators, etc.) in a convenient way. An important consequence is the description of the mapping properties of a linear operator at three levels: at the inner level such operators may often be described via integrals (or transformations applied to ordinary functions); at the intermediate Hilbert space level one can describe unitarity properties, while to outer level one can describe the mapping at the level of distributions.

In Sect. 2 we recall basic facts about Bessel sequences, frames and Riesz basis in Hilbert spaces. In Sect. 3 we discuss Gabor frames and their main features in the setting of $L^2(\mathbb{R}^d)$. In Sect. 4 we briefly describe time–frequency representations, especially the short-time Fourier transform. After these preparations we are in the position to introduce in Sect. 5 the key players of our presentation Gelfand triples, and concretely the Gelfand triple $(S_0(\mathbb{R}^d), L^2(\mathbb{R}^d), S_0'(\mathbb{R}^d))$. In Sect. 6 we give an overview of the main results of Feichtinger and Kozek on time–frequency quantization, pseudo-differential operators and their spreading function. Additionally we give the reader a flavor of the usefulness of Banach Gelfand triples for Gabor frame operators. In Sect. 7 we conclude our survey note with some results about Gabor multipliers and its relation to localization operators. The topic developed in this survey note can be found mainly in [16, 18, 19].

Notation. We define $t^2 = t \cdot t$, for $t \in \mathbb{R}^d$, and $xy = x \cdot y$ is the scalar product on $\mathbb{R}^d$.

The Schwartz class is denoted by $\mathcal{S}(\mathbb{R}^d)$, the space of tempered distributions by $\mathcal{S}'(\mathbb{R}^d)$. We use the brackets $\langle f, g \rangle$ to denote the extension to $\mathcal{S}(\mathbb{R}^d) \times \mathcal{S}'(\mathbb{R}^d)$ of the inner product $\langle f, g \rangle = \int f(t)\overline{g(t)}dt$ on $\boldsymbol{L}^2(\mathbb{R}^d)$. The Fourier transform is normalized to be $\hat{f}(\omega) = \mathcal{F}f(\omega) = \int f(t)e^{-2\pi it\omega}dt$. We recall the space of p-summable sequences

$$\ell^p(J) = \left\{ (a_n)_{n \in J} \ : \ \|(a_n)_{n \in J}\|_{\ell^p} := \left(\sum_{n \in J} |a_n|^p \right)^{1/p} < \infty \right\}.$$

2 Preliminaries

Let $\{g_k\}_{k \in J}$ be a family of an infinite dimensional (separable) Hilbert space $\mathscr{H}$ with inner product $\langle \cdot, \cdot \rangle$. The classical examples are the Hilbert space $L^2(\mathbb{R}^d)$ of (equivalence classes of measurable) functions of finite energy ($L^2(\mathbb{R}^d)$-norm) and the sequence space $\ell^2(\mathbb{Z}^d)$ consisting of square summable complex-valued sequences. Similar to the finite dimensional case, we want to represent a signal f in $\mathscr{H}$ as a (now possibly infinite) linear combination of the form

$$f \sim \sum_{k \in J} c_k g_k \,.$$

At this point there a several questions that arise naturally when considering an infinite sum. First, by convention, we want the sum to converge in the (prescribed) Hilbert-norm, i.e., $\lim_{K \to \infty} \|f - \sum_{k=1}^{K} c_k g_k\| \longrightarrow 0$. Secondly, the sum should converge to the same limit (preferably f) regardless of the summation order we choose (known as *unconditional convergence*). A more subtle point is that we would like to have a continuous linear dependency between the signal f and the coefficients c_k in order to avoid pathological cases in which small alterations in the signal result in uncontrollable changes in the corresponding coefficient sequence and vice-versa. This technical detail accounts for numerical stability.

Obviously, all these requirements are trivially fulfilled in the finite dimensional case. For an infinite family $\{g_k\}$, however, these assumptions have to be ensured before dealing with decomposition and reconstruction issues. Fortunately, there exists concepts in functional analysis that do exactly fit these kind of requirements. For a precise description we need the following definitions.

Definition 2.1. A family $\{g_k\}$ of a Hilbert space $\mathscr{H}$ is complete in $\mathscr{H}$ if the set of finite linear combination of $\{g_k\}$, write $\mathrm{span}(g_k)$, is dense in $\mathscr{H}$, i.e., every f in $\mathscr{H}$ can be arbitrarily well approximated by elements in $\mathrm{span}(g_k)$ with respect to the $\mathscr{H}$-norm.

In the mathematical literature complete systems are often called "total". The definition makes no claim about the "cost" of approximation. In other words, it is allowed to use more and more complicated coefficient sequences as the approximation quality is increased. In particular, total families do not necessarily allow a series expansion of arbitrary elements from the given Hilbert space.

Definition 2.2. The family $\{g_k\}_{k \in J}$ of a Hilbert space $\mathcal{H}$ is a basis for $\mathcal{H}$ if for all $f \in \mathcal{H}$ there exists unique scalars $c_k(f)$ such that

$$f = \sum_{k \in J} c_k(f) g_k .$$

Definition 2.3. A family $\{g_k\}_{k \in J}$ of a Hilbert space $\mathcal{H}$ is a Riesz sequence if there exist bounds $A, B > 0$ such that

$$A \|c\|^2_{\ell^2(J)} \leq \left\| \sum_{k \in J} c_k g_k \right\|^2 \leq B \|c\|^2_{\ell^2(J)} , \qquad c \in \ell^2(J) .$$

A Riesz sequence which generates all $\mathcal{H}$ is called a *Riesz basis* for $\mathcal{H}$.

Riesz bases are somehow "distorted" orthonormal bases as described in the following lemma which reveals all useful properties of a Riesz basis [28].

Lemma 2.1. *Let $\{g_k\}$ be a sequence in a Hilbert space $\mathcal{H}$. The following are equivalent:*

1. *$\{g_k\}_{k \in J}$ is a Riesz basis for $\mathcal{H}$.*
2. *$\{g_k\}_{k \in J}$ is an unconditional basis for $\mathcal{H}$ and g_k are uniformly bounded.*
3. *$\{g_k\}_{k \in J}$ is a basis for $\mathcal{H}$, and $\sum_{k \in J} c_k g_k$ converges if and only if $\sum_{k \in J} |c_k|^2$ converges.*
4. *There is an equivalent inner product on $\mathcal{H}$ for which $\{g_k\}_{k \in J}$ is an orthonormal basis for $\mathcal{H}$.*
5. *$\{g_k\}_{k \in J}$ is a complete Bessel sequence and possesses a bi-orthogonal system $\{h_k\}_{k \in J}$ that is also a complete Bessel sequence.*

The last item of the lemma says that there exists a unique sequence $\{h_k\}_{k \in J}$ such that $\langle g_k, h_j \rangle = \delta_{kj}$ which, combined with the second statement, induces the representation

$$f = \sum_{k \in J} \langle f, h_k \rangle g_k = \sum_{k \in J} \langle f, g_k \rangle h_k , \qquad f \in \mathcal{H} .$$

Hence, Riesz bases are potential candidates for our purpose of signal representation. We point out that the coefficient sequence is always square summable which is an important stability criterion.

A basis allows only unique expansions with respect to the coefficients. In applications it is sometimes more useful to weaken this property. This can be obtained by looking for overcomplete (linearly dependent) sets which is implemented in the concept of frames introduced by Duffin and Schaeffer in 1952 [12].

Definition 2.4. The sequence $\{g_k\}_{k \in J}$ in a Hilbert space $\mathcal{H}$ is called a Bessel sequence if

$$\sum_{k \in J} |\langle f, g_k \rangle|^2 < \infty, \qquad f \in \mathcal{H}.$$

Definition 2.5. A family $\{g_k\}_{k \in J}$ of a Hilbert space $\mathcal{H}$ is a frame of $\mathcal{H}$ if there exist bounds $A, B > 0$ such that

$$A\|f\|^2 \le \sum_{k \in J} |\langle f, g_k \rangle|^2 \le B\|f\|^2, \qquad f \in \mathcal{H}. \tag{1}$$

If $A = B$, then $\{g_k\}_{k \in J}$ is called a tight frame.

The synthesis map $D : \ell^2(J) \to \mathcal{H}$ of a frame $\{g_k\}_{k \in J}$ is defined by

$$D : (c_k) \to \sum_{k \in J} c_k g_k.$$

Its adjoint D^* is the analysis operator $D^* f = (\langle f, g_k \rangle)$. The *frame operator* S is defined by

$$Sf = DD^* f = \sum_{k \in J} \langle f, g_k \rangle g_k, \qquad f \in \mathcal{H}.$$

By (1), the frame operator satisfies

$$A\langle f, f \rangle \le \langle Sf, f \rangle \le B\langle f, f \rangle, \qquad f \in \mathcal{H},$$

and is, therefore, bounded, positive, and invertible. The inverse operator S^{-1} is obviously also positive and has therefore a square root $S^{-1/2}$ (self-adjoint), [37]. The sequence $\{S^{-1/2} g_k\}$ is a tight frame with $A = B = 1$. Indeed,

$$\sum_{k \in J} \langle f, S^{-\frac{1}{2}} g_k \rangle S^{-\frac{1}{2}} g_k = S^{-\frac{1}{2}} \sum_{k \in J} \langle f, S^{-\frac{1}{2}} g_k \rangle g_k = S^{-\frac{1}{2}} S (S^{-\frac{1}{2}} f)$$

$$= S^{-\frac{1}{2}} S^{-\frac{1}{2}} S f = If, \quad \forall f \in \mathcal{H}.$$

Every orthonormal basis of $\mathcal{H}$ is a Riesz basis of $\mathcal{H}$ and every Riesz basis of $\mathcal{H}$ is also a frame. The important difference between a Riesz basis and a frame is that the null space $N(G)$ of the synthesis map D of a frame $\{g_k\}_{k \in J}$ is in general non-trivial which is equivalent to the statement that the range of the analysis map D^* is a (closed) proper subspace of $\ell^2(J)$.

The sequence $\{\tilde{g}_{k \in J}\}$ with $\tilde{g}_k = S^{-1} g_k$, is also a frame with frame bounds $1/B$ and $1/A$. It is a dual frame for $\{g_k\}$ in the sense that

$$f = \sum_{k \in J} \langle f, \tilde{g}_k \rangle g_k = \sum_{k \in J} \langle f, g_k \rangle \tilde{g}_k, \qquad f \in \mathcal{H}.$$

Again we see, that frames do indeed fit our purpose for signal analysis and signal recovery. In contrast to Riesz bases, frames have, in general, no bi-orthogonal relation. Moreover, the dual frame is not unique. The canonical dual $\{S^{-1}g_k\}$ is the one that is producing minimal ℓ^2 coefficients as already shown in [12]. For alternative dual frames there exist constructive approaches that rely on the canonical dual. In [4, 33], it is shown that any dual frame of $\{g_k\}$ can be written as

$$S^{-1}g_k + h_k - \sum_{j \in J} \langle S^{-1}g_k, g_j \rangle h_j, \tag{2}$$

where $\{h_k\}$ is a Bessel sequence.

The lack of uniqueness has the advantage that if one coefficient is missing out of the sequence $\langle f, g_k \rangle$, the whole signal can still be completely recovered as long as $\{g_k\}$ is a frame but no Riesz basis. Similarly, any frame that is not a Riesz basis is still a frame when discarding single frame elements. Studies about the conservation of the frame property when discarding frame elements are known as *excesses of frames* [1, 2].

3 Gabor Analysis on L^2

We define the Fourier transform of an integrable function by $\hat{f}(\omega) = \int_{\mathbb{R}^d} f(t)e^{-2\pi it\omega}dt$. The translation operator T_x and the modulation operator M_ω are given by

$$T_x f = f(\cdot - x), \qquad M_\omega f = e^{2\pi i\omega \cdot}f(\cdot), \quad x, \omega \in \mathbb{R}^d. \tag{3}$$

Combined together they give rise to the so-called time–frequency shift $\pi(\lambda)$:

$$\pi(\lambda) = M_\omega T_x, \quad (x, \omega) \in \mathbb{R}^{2d}. \tag{4}$$

Note that

$$\pi(\lambda_2)\pi(\lambda_1) = e^{2\pi i(x_1\omega_2 - x_2\omega_1)}\pi(\lambda_1)\pi(\lambda_2)$$

for $\lambda_1 = (x_1, \omega_1), \lambda_2 = (x_2, \omega_2) \in \mathbb{R}^{2d}$.

A time–frequency lattice Λ is a discrete subgroup of $\mathbb{R}^{2d}$ ($= \mathbb{R}^d \times \hat{\mathbb{R}}^d$) with compact quotient. Its redundancy $|\Lambda|$ is the reciprocal value of the measure of a fundamental domain for the quotient $\mathbb{R}^{2d}/\Lambda$.

For a lattice Λ in $\mathbb{R}^{2d}$ and a so-called *Gabor atom* $g \in \boldsymbol{L}^2$ we define the associated Gabor family by

$$\mathcal{G}(g, \Lambda) = \{\pi(\lambda)g\}_{\lambda \in \Lambda}.$$

If $\mathcal{G}(g, \Lambda)$ is a frame for $\boldsymbol{L}^2$, we call it a *Gabor frame*. Since Λ has a group structure, the frame operator

$$Sf = \sum_{\lambda \in \Lambda} \langle f, \pi(\lambda)g \rangle \pi(\lambda)g$$

has the property that it commutes with all time–frequency shifts of the form $\pi(\lambda)$ for $\lambda \in \Lambda$. Therefore, the canonical dual frame of $\mathcal{G}(g, \Lambda)$ is simply given by $\mathcal{G}(h, \Lambda)$ with $h = S^{-1}g$. The fact, that a canonical dual frame of a Gabor frame is again a Gabor frame, i.e., generated by a single function, is the key property in many applications. It reduces computational issues to solving the linear system $Sh = g$.

A special and widely studied case are separable lattices of the form $\alpha\mathbb{Z} \times \beta\mathbb{Z}$ for some positive lattice parameters α and β, whose redundancy is simply $(\alpha\beta)^{-1}$. The prototype of a function generating Gabor frames for such separable lattices is the Gaussian

$$\psi(x) = e^{-\pi x^2 \sigma^2} . \tag{5}$$

for some real $\sigma > 0$. The Gaussian generates a Gabor frame if and only if $\alpha\beta < 1$ [31, 35, 38, 39]. We emphasize that for $\alpha\beta = 1$ the Gaussian generates a *unstable* generating system for $L^2(\mathbb{R}^d)$, i.e., the resulting Gabor family is complete but coefficient sequences must not be bounded. In this context we mention a central result, the so-called *density theorem* and refer to [26] for detailed discussions. An elegant elementary proof of the density theorem has been provided by Janssen [30].

Theorem 3.1. *Assume that $\mathcal{G}(g, \alpha, \beta)$ is a frame. Then, $\alpha\beta \leq 1$. Moreover, $\mathcal{G}(g, \alpha, \beta)$ is a Riesz basis for $L^2(\mathbb{R}^d)$ if and only if $\alpha\beta = 1$.*

In his seminal paper [22], Gabor chose the integer lattice $a = b = 1$ in $\mathbb{R}^2$ and used the Gaussian in order to define a Gabor system with maximal time–frequency localization. However, as mentioned above, this system is no longer stable though complete, and, indeed, the celebrated Balian–Low Theorem [3, 34] states that good time–frequency localization and Gabor Riesz bases are not compatible:

Theorem 3.2 (Balian–Low). *If $\mathcal{G}(g, 1, 1)$ constitutes a Riesz basis for $L^2(\mathbb{R})$, then*

$$\int_{\mathbb{R}} |g(t)|^2 t^2 dt \int_{\mathbb{R}} |\hat{g}(\omega)|^2 \omega^2 d\omega = \infty .$$

The Balian–Low Theorem reveals a form of uncertainty principle and has inspired fundamental research, see [26] and references therein.

In the sequel we state some fundamental results of Gabor frames and the Gabor frame operator (Gabor frame-type operator). To this end we need the notion of the *adjoint lattice* Λ° of Λ which is the set of all elements in $\mathbb{R}^{2d}$ that satisfy the commutation property

$$\pi(\lambda^\circ)\pi(\lambda) = \pi(\lambda)\pi(\lambda^\circ) \quad \text{for all} \quad \lambda \in \Lambda .$$

Note that Λ° is again a lattice of $\mathbb{R}^{2d}$ (and that $\Lambda^{\circ\circ} = \Lambda$). Instead of the frame operator we will use the more general notion of a frame-type operator

$S_{g,\gamma,\Lambda}$ associated to the pair (g,γ), where γ takes the role of an "analyzing" and g the role of a "synthesizing" window:

$$S_{g,\gamma,\Lambda}f = \sum_{\lambda \in \Lambda} \langle f, \pi(\lambda)\gamma \rangle \, \pi(\lambda)g \,, \qquad f \in \boldsymbol{L}^2(\mathbb{R}^d)\,.$$

This sum converges in $\boldsymbol{L}^2(\mathbb{R}^d)$ for all $f \in \boldsymbol{L}^2(\mathbb{R}^d)$ as long both functions g, γ are Bessel atoms for Λ, that is, $\mathcal{G}(g,\Lambda)$ and $\mathcal{G}(\gamma,\Lambda)$ are Gabor Bessel families. For the fundamental results to hold with respect to norm convergence we need a little bit more than Bessel sequences. It is that both atoms g (and analogously γ) satisfy

$$\sum_{\lambda^\circ \in \Lambda^\circ} |\langle g, g_{\lambda^\circ} \rangle| < \infty \,, \tag{A$'$}$$

also known as the *Tolimieri–Orr's condition*. This somewhat technical property is used for controlling convergence problems (by altering the convergence definition Condition (A$'$) can be weakened). Condition (A$'$) is in general not easy to verify. In particular, if Condition (A$'$) holds for one lattice, there is, in general, no guarantee that it holds also for a different lattice. This problem, however, is overcome by the Feichtinger algebra $\boldsymbol{S}_0$ (Sect. 5) which defines a class of functions for which Condition (A$'$) is satisfied for any lattice in $\mathbb{R}^{2d}$.

We summarize the fundamental results of Gabor analysis in the following theorem that is given in [20] in a slightly more general context. The statements go back to the seminal papers [10, 29, 40]. They are, however, all consequences of the fundamental identity of Gabor analysis extensively studied in [17].

Theorem 3.3. *Let Λ be a lattice in $\mathbb{R}^{2d}$ with adjoint lattice Λ°. Then, for g, h satisfying* (A$'$), *the following holds.*

1. (Fundamental Identity of Gabor Analysis)

$$\sum_{\lambda \in \Lambda} \langle f, \pi(\lambda)\gamma \rangle \langle \pi(\lambda)g, h \rangle = |\Lambda| \sum_{\lambda^\circ \in \Lambda^\circ} \langle g, \pi(\lambda^\circ)\gamma \rangle \langle \pi(\lambda^\circ)f, h \rangle \tag{7}$$

for all $f, h \in \boldsymbol{L}^2(\mathbb{R}^d)$, where both sides converge absolutely.
2. (Wexler–Raz Identity)

$$S_{g,\gamma,\Lambda}f = |\Lambda| \cdot S_{f,\gamma,\Lambda^\circ}g \tag{8}$$

for all $f \in \boldsymbol{L}^2(\mathbb{R}^d)$.
3. (Janssen Representation)

$$S_{g,\gamma,\Lambda} = |\Lambda| \sum_{\lambda^\circ \in \Lambda^\circ} \langle \gamma, \pi(\lambda^\circ)g \rangle \, \pi(\lambda^\circ) \tag{9}$$

where the series converges unconditionally in the strong operator sense.

In Sect. 6 we explicitly derive the Janssen representation of the Gabor frame operator from advanced concepts in harmonic analysis and provide a much deeper insight into this topic.

Another important result is the Ron–Shen Duality Principle which is often referred to [36] although it appeared already in [29] and [10].

Theorem 3.4. *Let $g \in \boldsymbol{L}^2(\mathbb{R}^d)$ and Λ be a lattice in $\mathbb{R}^{2d}$ with adjoint Λ°. Then the Gabor system $\mathcal{G}(g, \Lambda)$ is a frame for $\boldsymbol{L}^2(\mathbb{R}^d)$ if and only if $\mathcal{G}(g, \Lambda^\circ)$ is a Riesz basis for its closed linear span. In this case, the quotient of the two frame bounds and quotient of the Riesz bounds (alternatively the condition number of the corresponding frame operator and the Gramian matrix, respectively) coincide.*

The last important identity in Gabor Analysis that we want to present in this section is the Wexler–Raz Biorthogonality Relation which basically says that g and γ are dual Gabor windows if and only if $S_{g,\gamma,\Lambda} = Id$. That is, according to Janssen Representation, exactly the case when

$$\langle \gamma, \pi(\lambda^\circ)g \rangle = |\Lambda|^{-1}\delta_{0,\lambda^\circ}.$$

Alternatively this relation can be described by what is a true biorthogonality (using again Kronecker's Delta):

$$\langle \pi(\lambda^{\circ\prime})\gamma, \pi(\lambda^\circ)g \rangle = |\Lambda|^{-1}\delta_{\lambda^{\circ\prime},\lambda^\circ}.$$

In the next section we describe basic and more advanced studies in harmonic analysis that contribute to a better understanding of the Gabor frame operator.

4 Time–Frequency Representations

Traditionally we extract the frequency information of a signal f by means of the Fourier transform $\hat{f}(\omega) = \int_{\mathbb{R}^d} f(t)e^{-2\pi it\omega}dt$. If we know $\hat{f}(\omega)$ for all frequencies ω, then our signal f can be reconstructed by the inversion formula $f(t) = \int_{\widehat{\mathbb{R}^d}} \hat{f}(\omega)e^{2\pi it\omega}d\omega$ (valid pointwise or in the quadratic mean).

However, in many situations it is of relevance to know, *how long* each frequency appears in the signal f, e.g., for a pianist playing a piece of music. Mathematically this leads to the study of functions $S(f)(t,\omega)$ of the signal f, which describe the time–frequency content of f over "time" t. In the following we mention the most prominent time–frequency representations.

In the last century researchers such as E. Wigner, Kirkwood, and Rihaczek had invented different time–frequency representations [32, 41]. The work of Wigner and Kirkwood was motivated by the description of a particle in quantum mechanics by a joint probability distribution of position and momentum of the particle. More concretely, in 1932 Wigner introduced the first time–frequency representation of a function $f \in L^2(\mathbb{R}^d)$ by

$$W(f)(x, \omega) = \int_{\mathbb{R}^d} f(x + \frac{t}{2})\overline{f(x - \frac{t}{2})}e^{-2\pi i\omega t}dt, \tag{10}$$

the so called *Wigner distribution* of f. Later Kirkwood proposed another time–frequency representation, which was in a different context rediscovered by Rihaczek. Both researchers associated to a function $f \in \boldsymbol{L}^2(\mathbb{R}^d)$ the following expression

$$R(f)(x, \omega) = f(x)\overline{\hat{f}(\omega)}e^{-2\pi i x\omega},$$

the *Kirkwood–Rihaczek distribution* of f.

Nowadays, the *short-time Fourier transform* (STFT) has become the standard tool for (linear) time–frequency analysis. It is used as a measure of the time–frequency content of a signal f (energy distribution), but it also establishes a connection to the Heisenberg group.

The STFT provides information about local (smoothness) properties of the signal f. This is achieved by localization of f near t through multiplication with some *window function* g and a subsequent Fourier transform providing information about the frequency content of f in this segment. Typically g is concentrated around the origin. If g is compactly supported only a segment of f in some interval or ball around t is relevant, but g can be any non-zero Schwartz function such as the Gaussian. Overall we have:

$$V_g f(x, \omega) = \int_{\mathbb{R}^d} f(t)\overline{g(t - x)}e^{-2\pi i t\omega}dt, \quad \text{for} \quad (x, \omega) \in \mathbb{R}^{2d}, \tag{11}$$

In 1927 Weyl pointed out that the translation and modulation operator satisfy the following commutation relation

$$T_x M_\omega = e^{-2\pi i x\omega} M_\omega T_x, \quad (x, \omega) \in \mathbb{R}^{2d}. \tag{12}$$

$\{T_x : x \in \mathbb{R}^d\}$ and $\{M_\omega : \omega \in \mathbb{R}^d\}$ are Abelian groups of unitary operators, with the infinitesimal generators given by differentiation and multiplication operator, respectively. Therefore the commutation relation (12) is the analogue of Heisenberg's commutation relation for the differentiation and multiplication operator.

The time–frequency shifts $M_\omega T_x$ for $(x, \omega) \in \mathbb{R}^{2d}$ satisfy the following composition law:

$$\pi(x, \omega)\pi(y, \eta) = e^{-2\pi i x \cdot \eta}\pi(x + y, \omega + \eta), \tag{13}$$

for $(x, \omega), (y, \eta)$ in the time–frequency plane $\mathbb{R}^d \times \widehat{\mathbb{R}^d}$, i.e., the mapping $(x, \omega) \mapsto \pi(x, \omega)$ defines (only) a *projective representation* of the time–frequency plane (viewed as an Abelian group) $\mathbb{R}^d \times \widehat{\mathbb{R}^d}$. By adding a toral component, i.e., $\tau \in \mathbb{C}$ with $|\tau| = 1$ one can augment the phase space $\mathbb{R}^d \times \widehat{\mathbb{R}^d}$ to the so-called *Heisenberg group* $\mathbb{R}^d \times \widehat{\mathbb{R}^d} \times T$ and the mapping $(x, \omega, \tau) \mapsto \tau M_\omega T_x$ defines a (true) unitary representation of the Heisenberg group [21], the so-called *Schrödinger representation*. From this point of view the definition of $V_g f$ can be interpreted as representation coefficients:

$$V_g f(x, \omega) = \langle f, M_\omega T_x g \rangle, \qquad f, g \in L^2(\mathbb{R}^d).$$

The STFT is linear in f and conjugate linear in g. The choice of the window function g influences the properties of the STFT remarkably. One example of a good window class is the Schwartz space of rapidly decreasing functions. Later we will discuss another function space, which is perfectly suited as a good class of windows, *Feichtinger's algebra*.

Furthermore, for $f, g \in L^2(\mathbb{R}^d)$ the STFT $V_g f$ is uniformly continuous on $\mathbb{R}^{2d}$, i.e., we can sample the $V_g f$ without problem. This fact is of great relevance in the discussion of Gabor frames.

By Parseval's theorem and an application of the commutation relations (12) we derive the following relation

$$V_g f(x, \omega) = e^{-2\pi i x \omega} V_{\hat{g}} \hat{f}(\omega, -x), \tag{14}$$

which is sometimes called the *fundamental identity of time–frequency analysis* [26]. Equation (14) expresses the fact that the STFT is a joint time–frequency representation and that the Fourier transform amounts to a rotation of the time–frequency plane $\mathbb{R}^d \times \widehat{\mathbb{R}^d}$ by an angle of $\frac{\pi}{2}$ whenever the window g is Fourier invariant. Another important consequence of the definition of STFT (11) and the commutation relations (12) is the *covariance property* of the STFT:

$$V_g (T_u M_\eta f)(x, \omega) = e^{-2\pi i u \omega} V_g f(x - u, \omega - \eta). \tag{15}$$

Later we will draw an important conclusion of the basic identity of time–frequency analysis (14) and the covariance property of the STFT (15): isometric Fourier invariance and the invariance under TF-shifts of Feichtinger's algebra.

As for the Fourier transform there is also a Parseval's equation for the STFT which is referred to as *Moyal's formula*.

Lemma 4.1 (Moyal's formula). *Let* $f_1, f_2, g_1, g_2 \in L^2(\mathbb{R}^d)$ *then* $V_{g_1} f_1$ *and* $V_{g_2} f_2$ *are in* $L^2(\mathbb{R}^{2d})$ *and the following identity holds:*

$$\langle V_{g_1} f_1, V_{g_2} f_2 \rangle_{L^2(\mathbb{R}^{2d})} = \langle f_1, f_2 \rangle \overline{\langle g_1, g_2 \rangle}. \tag{16}$$

Moyal's formula implies that orthogonality of windows g_1, g_2 resp. of signals f_1, f_2 implies orthogonality of their STFT's. Most importantly we observe that one has for normalized $g \in L^2(\mathbb{R}^d)$ (i.e., with $\|g\|_2 = 1$):

$$\|V_g f\|_{L^2(\mathbb{R}^{2d})} = \|f\|_{L^2(\mathbb{R}^d)},$$

for all $f \in L^2(\mathbb{R}^d)$, i.e., the STFT is an isometry from $L^2(\mathbb{R}^d)$ to $L^2(\mathbb{R}^{2d})$.

Another consequence of Moyal's formula is an inversion formula for the STFT. Assume that the analysis window $g \in L^2(\mathbb{R}^d)$ and the synthesis window $\gamma \in L^2(\mathbb{R}^d)$ satisfy $\langle g, \gamma \rangle \neq 0$. Then for $f \in L^2(\mathbb{R}^d)$

$$f = \frac{1}{\langle g, \gamma \rangle} \iint_{\mathbb{R}^{2d}} \langle f, \pi(x,\omega)g \rangle \, \pi(x,\omega)\gamma \, dxd\gamma. \tag{17}$$

We observe that in contrast to the Fourier inversion the building blocks of the STFT inversion formula are just time–frequency shifts of a square-integrable function. Therefore also the Riemannian sums corresponding to this inversion integral are functions in $L^2(\mathbb{R}^d)$ and are even norm convergent in $L^2(\mathbb{R}^d)$ for nice windows (from Feichtinger's algebra, see later).

5 The Gelfand Triple $(S_0, L^2, S_0')(\mathbb{R}^d)$

Since Feichtinger's discovery of the Segal algebra $S_0(\mathbb{R}^d)$ in 1979 many results have shown that $S_0(\mathbb{R}^d)$ is a good substitute of Schwartz's space $\mathcal{S}(\mathbb{R}^d)$ of test functions (except if one is interested in a discussion of partial differential equations). Furthermore $S_0(\mathbb{R}^d)$ has turned out as the appropriate setting for the treatment of questions in harmonic analysis on $\mathbb{R}^d$ (actually on a general locally compact Abelian group G, even without using their structure theory). In this section we recall well-known properties of $S_0(\mathbb{R}^d)$ which we will need later in our discussion of Gabor frame operators. Nowadays the space $S_0(\mathbb{R}^d)$ is called *Feichtinger's algebra* since it is a Banach algebra with respect to pointwise multiplication and convolution.

Definition 5.1. A function in $f \in L^2(\mathbb{R}^d)$ is in the subspace $S_0(\mathbb{R}^d)$ if, for some non-zero g (called the "window") in the Schwartz space $\mathcal{S}(\mathbb{R}^d)$,

$$\|f\|_{S_0} := \|V_g f\|_{L^1} = \iint_{\mathbb{R}^d \times \widehat{\mathbb{R}^d}} |V_g f(x,\omega)| dxd\omega < \infty.$$

The space $(S_0(\mathbb{R}^d), \|\cdot\|_{S_0})$ is a Banach space, for any fixed, non-zero $g \in S_0(\mathbb{R}^d)$, and different windows g define the same space and equivalent norms. Since $S_0(\mathbb{R}^d)$ contains the Schwartz space $\mathcal{S}(\mathbb{R}^d)$ any Schwartz function is suitable, but also compactly supported functions having an integrable Fourier transform (such as a trapezoidal or triangular function) are suitable windows. Often it is convenient to use the Gaussian as a window.

The above definition of $S_0(\mathbb{R}^d)$ [26] (different from the original one [13]) allows for an easy derivation of the basic properties of Feichtinger's algebra in the following lemma.

Lemma 5.1. *Let $f \in S_0(\mathbb{R}^d)$, then the following holds:*

(1) $\pi(u, \eta)f \in S_0(\mathbb{R}^d)$ *for* $(u, \eta) \in \mathbb{R}^d \times \widehat{\mathbb{R}}^d$, *and* $\|\pi(u, \eta)f\|_{S_0} = \|f\|_{S_0}$.
(2) $\hat{f} \in S_0(\mathbb{R}^d)$, *and* $\|\hat{f}\|_{S_0} = \|f\|_{S_0}$.

Proof. 1. For $z = (u, \eta)$ in the time–frequency plane $\mathbb{R}^d \times \widehat{\mathbb{R}}^d$ one has:

$$\|\pi(u, \eta)f\|_{S_0} = \iint_{\mathbb{R}^{2d}} |V_g f(x - u, \omega - \eta)| \, dx d\omega =$$

$$= \iint_{\mathbb{R}^{2d}} |V_g f(x, \omega)| \, dx d\omega = C \|f\|_{S_0}.$$

2. The key of the argument is an application of the fundamental identity of time–frequency analysis (14) to a Fourier invariant window g and the independence of the definition of $S_0(\mathbb{R}^d)$ for $g \in \mathcal{S}(\mathbb{R}^d)$. For simplicity we choose g the (Fourier invariant) Gaussian $g_0(x) = 2^{d/4} e^{-\pi x^2}$:

$$\|\hat{f}\|_{S_0} = \iint_{\mathbb{R}^{2d}} |V_{g_0} \hat{f}(x, \omega)| \, dx d\omega = \iint_{\mathbb{R}^{2d}} |V_{\widehat{g_0}} \hat{f}(x, \omega)| \, dx d\omega =$$

$$\iint_{\mathbb{R}^{2d}} |V_{g_0} f(-\omega, x)| \, dx d\omega = \iint_{\mathbb{R}^{2d}} |V_{g_0} f(x, \omega)| \, dx d\omega = \|f\|_{S_0}.$$

$\square$

Later we will need that $S_0(\mathbb{R}^d)$ is *dense* and *continuously embedded* into $L^p(\mathbb{R}^d)$ for any $p \in [1, \infty)$. The original motivation for Feichtinger's introduction of $S_0(\mathbb{R}^d)$ was the search for a *smallest* member in the family of all time–frequency homogenous Banach spaces. For a proof of all these assertions we refer the reader to the original paper of Feichtinger or Gröchenig's book on time–frequency analysis [26].

Another reason for usefulness of $S_0(\mathbb{R}^d)$ is the fact that $S_0(\mathbb{R}^d)$ is a natural domain for the application of Poisson summation [25].

Lemma 5.2. *Let Λ be a lattice in $\mathbb{R}^d$ and $f \in S_0(\mathbb{R}^d)$ then*

$$\sum_{\lambda \in \Lambda} f(\lambda) = |\Lambda|^{-1} \sum_{\lambda^\perp \in \Lambda^\perp} \hat{f}(\lambda^\perp) \tag{18}$$

holds pointwise and with absolute convergence.

Here $\Lambda^\perp$ is the orthogonal lattice for Λ, e.g., $\Lambda^\perp = (A^{-1})^t \mathbb{Z}^d$ for $\Lambda = A\mathbb{Z}^d$, where A is a non-singular matrix describing Λ.

In 1958 I. M. Gelfand and A. G. Kostyuchenko introduced Gelfand triples in their study of the spectral theory of self-adjoint operators [23]. They were motivated by the work of Dirac on the foundations of quantum mechanics and Schwartz's theory of distributions.

An important result of linear algebra is the theorem on the existence of eigenvectors for any self-adjoint linear operator A on $\mathbb{R}^d$. The situation changes drastically when one passes from the finite to the infinite-dimensional case, since it can happen that a unitary operators does not have any (non-zero) eigenvector. Particular examples of such operators are the translation operator T_x and the modulation operator M_ω on $L^2(\mathbb{R}^d)$. Let us present an easy argument showing that the translation operator $T_x, x \neq 0$, has no eigenvectors in $L^2(\mathbb{R}^d)$. Assume that $f \in L^2(\mathbb{R}^d)$ satisfies

$$T_x f(t) = a f(t), \tag{19}$$

which by the Fourier transform is equivalent to

$$M_{-x}\hat{f}(\omega) = a\hat{f}(\omega) \quad \text{a.e..} \tag{20}$$

But this is only possible if the function $\hat{f}$ equals zero a.e., up to the points with $e^{2\pi i \omega x} \neq a$, i.e., it differs from zero only on a set of measure zero, hence $\hat{f} = 0$ and finally $f = 0 \in L^2(\mathbb{R}^d)$. In other words, the translation operator T_x does not have eigenvectors in the space $L^2(\mathbb{R}^d)$. On the other hand we are not too far off with the claim that T_x has the eigenvectors $e^{-2\pi i t \omega}$ corresponding to the eigenvalue $e^{2\pi i x \omega}$, and the claim that any function f in $L^2(\mathbb{R}^d)$ can be (kind of) expanded in terms of the eigenvectors $e^{-2\pi i t \omega}$, by suitable interpretation of the inversion formula for the Fourier transform (valid pointwise for $f \in S_0(\mathbb{R}^d)$):

$$f(t) = \int_{\mathbb{R}^d} \hat{f}(\omega) e^{2\pi i t \omega} d\omega. \tag{21}$$

Furthermore, the action of the translation operator is given by

$$T_x f(t) = \int_{\mathbb{R}^d} e^{2\pi i x \omega} \hat{f}(\omega) e^{2\pi i t \omega} d\omega,$$

which is a continuous analog of the spectral decomposition of a self-adjoint operator in $\mathbb{R}^d$.

More concretely, the system of eigenfunctions $\{e^{-2\pi i t \omega} : \omega \in \widehat{\mathbb{R}^d}\}$ is complete in the sense that for any function f in $L^2(\mathbb{R}^d)$ Parseval's equality holds

$$\int_{\mathbb{R}^d} |f(t)|^2 dt = \int_{\mathbb{R}^d} |\hat{f}(\omega)|^2 d\omega.$$

The obvious problem is the fact that $L^2(\mathbb{R}^d)$ does not contain the system of eigenvectors of the translation operator T_x. But they can be considered as linear functionals on $S_0(\mathbb{R}^d)$. This as well es several similar observations suggests to study operators on a Hilbert space via a dense subspace and its associated dual space. In our example it is actually possible to start from $S_0(\mathbb{R}^d)$ and construct $L^2(\mathbb{R}^d)$ as completion of $S_0(\mathbb{R}^d)$ with respect to norm corresponding to the usual scalar product $\langle f, g \rangle = \int_{\mathbb{R}^d} f(t)\overline{g(t)}dt$.

In this context it turns out that $S_0(\mathbb{R}^d)$ has the important additional property that both δ-distributions and the pure frequencies $\chi_\omega(x) = e^{-2\pi i x \omega}$ (for all $\omega \in \mathbb{R}^d$) are in a natural way elements of $S_0'(\mathbb{R}^d)$, i.e., define bounded linear functionals on $S_0(\mathbb{R}^d)$. This dual space can be defined via STFT as follows [26]:

$$S_0'(\mathbb{R}^d) = \left\{ f \in \mathcal{S}'(\mathbb{R}^d) : \|f\|_{S_0'(\mathbb{R}^d)} = \|V_g f\|_{L^\infty} = \sup_{\mathbb{R}^d \times \widehat{\mathbb{R}^d}} |V_g f(x,\omega)| < \infty. \right\}$$

It is now easy to verify that $\delta \in S_0'(\mathbb{R}^d)$. Indeed, for $g \in \mathcal{S}(\mathbb{R}^d)$, we have

$$\sup_{(x,\omega) \in \mathbb{R}^d \times \widehat{\mathbb{R}^d}} |V_g \delta(x,\omega))| = \sup_{(x,\omega) \in \mathbb{R}^d \times \widehat{\mathbb{R}^d}} |\langle \delta, M_\omega T_x g \rangle| =$$

$$= \sup_{(x,\omega) \in \mathbb{R}^d \times \widehat{\mathbb{R}^d}} |g(-x)| = \|g\|_{L^\infty} < \infty.$$

We are now in a situation similar to the one inspiring Gelfand to introduce what is nowadays called a Gelfand triple. The main idea being the observation, that a triple of spaces – consisting of the Hilbert space itself, a small (topological vector) space contained in the Hilbert space, and its dual – allows a much better description of the situation. The advantage in our case is the fact that we can even take a Banach space, namely $S_0(\mathbb{R}^d)$. Hence we can work with the following formal definition:

Definition 5.2. A (Banach) *Gelfand triple* consists of some Banach space $(B, \|\cdot\|_B)$ which is continuously and densely embedded into some Hilbert space $\mathcal{H}$, which in turn is w^*-continuously and densely embedded into the dual Banach space $(B', \|\cdot\|_{B'})$.

We shall use the symbol $(B, \mathcal{H}, B')$ for such a triple of spaces. In this setting the inner product on $\mathcal{H}$ extends in a natural way to a pairing between B and B' (producing an anti-linear functional of the same norm).

As another consequence we mention an extension of an eigenvector of a bounded operator on a Hilbert space $\mathcal{H}$. Let A be a linear operator on a Banach space B then a linear functional F is a *generalized eigenvector* of A to the eigenvalue λ if

$$F(Af) = \lambda F(f), \qquad \text{for all } f \in B.$$

This notion allows to interpret the characters $\chi_\omega(x) = e^{-2\pi i \omega x}$ as generalized eigenvectors for the translation operator T_x on $S_0(\mathbb{R}^d)$. Furthermore the set of generalized eigenvectors $\{\chi_\omega : \omega \in \mathbb{R}^d\}$ is complete by Plancherel's theorem, i.e., if $\hat{f}(\omega) = \langle \chi_\omega, f \rangle = 0$ for all $\omega \in \mathbb{R}^d$ implies $f \equiv 0$. This suggests to think of the Fourier transform of f at frequency ω as the evaluation of the linear functional $\langle \chi_\omega, f \rangle$.

The treatment of the translation operator T_x on $L^2(\mathbb{R}^d)$ is a particular case of a general theorem by Gelfand that for any self-adjoint operator A

on a Hilbert space $\mathcal{H}$ there exists a nuclear space and a complete system of generalized eigenvectors, see [24]. The advantage of the approach presented here is that instead of a (may be complicated) nuclear topological vector space a relatively simple-minded Banach space can be used.

The introduction of Gelfand triples does not only offer a better description of a self-adjoint operator but it also allows to a simplification of proofs. For example, in the discussion of the Fourier transform $\mathcal{F}$ we consider it as an object on $S_0(\mathbb{R}^d)$ where everything is well-defined and Parseval's formula and taking the inverse Fourier transform is justified by the nice properties of $S_0(\mathbb{R}^d)$. By a density argument we get all properties of the Fourier transform on the level of $L^2(\mathbb{R}^d)$. And we obtain an extension of the Fourier transform to $S_0'(\mathbb{R}^d)$ by duality, the so-called *generalized Fourier transform*.

The preceding discussion suggests the following lemma which says that assertions for an operator on the S_0-level are actually statements for $L^2(\mathbb{R}^d)$ and S_0', respectively.

Lemma 5.3. *The Fourier transform $\mathcal{F}$ on $\mathbb{R}^d$ has the following properties:*

1. *$\mathcal{F}$ is an isomorphism from $S_0(\mathbb{R}^d)$ to $S_0(\widehat{\mathbb{R}^d})$,*
2. *$\mathcal{F}$ is a unitary map between $L^2(\mathbb{R}^d)$ and $L^2(\widehat{\mathbb{R}^d})$,*
3. *$\mathcal{F}$ is a weak* (as well as a norm-to-norm) continuous bijection from $S_0'(\mathbb{R}^d)$ to $S_0'(\widehat{\mathbb{R}^d})$.*

Furthermore we have that Parseval's formula

$$\langle f, g \rangle = \langle \hat{f}, \hat{g} \rangle \tag{22}$$

is valid for $(f, g) \in S_0(\mathbb{R}^d) \times S_0'(\mathbb{R}^d)$ and therefore on each level of the Gelfand triple $(S_0, L^2, S_0')(\mathbb{R}^d)$.

The properties of Fourier transform are expressed by the *Gelfand bracket*

$$\langle f, g \rangle_{(S_0, L^2, S_0')(\mathbb{R}^d)} = \langle \hat{f}, \hat{g} \rangle_{(S_0, L^2, S_0')(\widehat{\mathbb{R}^d})} \tag{23}$$

which combines the functional brackets of Banach spaces and that of the inner-product for the Hilbert space.

The Fourier transform is a prototype for the notion of a Gelfand triple isomorphism.

Definition 5.3. If $(B_1, \mathcal{H}_1, B'_1)$ and $(B_2, \mathcal{H}_2, B'_2)$ are Gelfand triples then an operator A is called a *[unitary] Gelfand triple isomorphism* if

1. A is an isomorphism between B_1 and B_2.
2. A is a [unitary operator resp.] isomorphism from $\mathcal{H}_1$ to $\mathcal{H}_2$.
3. A extends to a weak* isomorphism as well as a norm-to-norm continuous isomorphism between B'_1 and B'_2.

In this terminology the Fourier transform is a unitary Gelfand triple isomorphism of the Gelfand triple automorphism on $(S_0, L^2, S_0')(\mathbb{R}^d)$ (isomorphism with itself). In the following lemma we give conditions for the extension of a linear mapping given on $S_0(\mathbb{R}^d)$ to a unitary mapping on $L^2(\mathbb{R}^d)$.

Lemma 5.4 (cf. [20]). *Let U be a unitary mapping from $L^2(\mathbb{R}^d)$ to $L^2(\mathbb{R}^d)$. The mapping U extends to a Gelfand triple isomorphism between $(S_0, L^2, S_0')(\mathbb{R}^d)$ and $(S_0, L^2, S_0')(\mathbb{R}^d)$ if and only if the restrictions of U to $S_0(\mathbb{R}^d)$ defines a bounded bijective linear mapping from $S_0(\mathbb{R}^d)$ onto itself.*

Due to this lemma we only have to check the properties of U at the S_0-level, i.e., to verify the existence of some $C > 0$ such that

$$\|Uf\|_{S_0(\mathbb{R}^d)} \leq C\|f\|_{S_0(\mathbb{R}^d)}. \tag{24}$$

The discussion of the Fourier transform $\mathcal{F}$ on the Gelfand triple $(S_0, L^2, S_0')(\mathbb{R}^d)$ allows to think of $\mathcal{F}$ as a bounded operator between $S_0(\mathbb{R}^d)$ and $S_0'(\mathbb{R}^d)$ with a distributional kernel $k(t, \omega) = e^{-2\pi i t \omega}$. The existence of a distributional kernel holds for any bounded operator between $S_0(\mathbb{R}^d)$ and $S_0'(\mathbb{R}^d)$. This important fact is the so-called *kernel theorem* for $S_0(\mathbb{R}^d)$ (cf. [16], Theorem 7.4.2). Before we give a precise description of this important fact we recall the notion of a Wilson basis. With the help of a Wilson basis we can adapt a linear algebra reasoning to the infinite-dimensional setting.

In 1991 Daubechies, Jaffard and Journé [9] followed an idea of Wilson in their construction of an orthonormal basis from a Gabor system $\mathcal{G}(g, \Lambda)$ of $L^2(\mathbb{R}^d)$. Wilson suggested that the building blocks $\pi(x, \omega)g$ of an orthonormal basis of $L^2(\mathbb{R}^d)$ should be symmetric in ω and should be concentrated at ω and $-\omega$.

Definition 5.4. For $g \in L^2$ the associated Wilson system $\mathcal{W}(g)$ consists of functions

$$\psi_{k,n} = c_n T_{\frac{k}{2}}(M_n + (-1)^{k+n}M_{-n})g, \quad (k, n) \in \mathbb{Z}^d \times \mathbb{N}_0,$$

where $c_0 = \frac{1}{2}$ and $c_n = \frac{1}{\sqrt{2}}$ for $n \geq 1$, $\psi_{k,0} = T_k g$ and $\psi_{2k+1,0} = 0$ for $k \in \mathbb{Z}$.

They proved the following theorem which shows a method for the construction of a Wilson basis from a Gabor system $\mathcal{G}(g, \frac{1}{2}\mathbb{Z} \times \mathbb{Z})$. Later Feichtinger, Gröchenig, and Walnut [14] showed that Wilson systems provide an unconditional basis for $S_0(\mathbb{R}^d)$ and $S_0'(\mathbb{R}^d)$ endowed with the w^*-topology. Therefore Wilson systems provide us with a natural class of bases for time–frequency analysis. The existence of an unconditional basis for $S_0(\mathbb{R}^d)$ will be very helpful in our discussion of the kernel theorem for $S_0(\mathbb{R}^d)$ and their construction relies heavily on the functorial properties of S_0, cf. [16].

Theorem 5.1. *Let $\mathcal{G}(g, \frac{1}{2}\mathbb{Z} \times \mathbb{Z})$ be a tight frame for $L^2(\mathbb{R})$ with $\|g\| = 1$ and $g(x) = \overline{g(-x)}$. Then the Wilson system $\mathcal{W}(g)$ is an orthonormal basis of $L^2(\mathbb{R})$.*

As a corollary we get Wilson bases for $L^2(\mathbb{R}^d)$ by taking tensor products.

Corollary 5.1. *Let $\mathcal{W}(g)$ be a Wilson basis for $L^2(\mathbb{R})$ and define $\mathbf{\Psi}_{k,n} = \prod_{j=1}^{d} \psi_{r_j,s_j}$ for $(r,s) \in \mathbb{Z}^d \times \mathbb{N}_0$. Then $\mathbf{\Psi}_{k,n}$ is an orthonormal basis for $L^2(\mathbb{R}^d)$.*

In applications of mathematics one often has to deal with linear systems. In the discrete and finite case each linear system is a linear mapping from the input space $\mathbb{R}^n$ into the output space $\mathbb{R}^m$ of our system and its action is given by matrix multiplication after a choice of bases in $\mathbb{R}^n$ and $\mathbb{R}^m$, respectively (similarly from $\mathbb{C}^n$ to $\mathbb{C}^m$ using complex matrices).

A linear system in infinite dimensions may be considered as a continuous analog of matrix multiplication (replacing summation by integration), i.e.,

$$g(x) = Kf(x) = \int_{\mathbb{R}^d} k(x,y)f(y)dy.$$

We can think of the input values $f(y)$ as being listed in an infinite column vector and $k(x,y)$ as an infinite matrix, the so-called *kernel* of K, and the integral $\int_{\mathbb{R}^d} k(x,y)f(y)dy$ providing the entries of the output vector in the expected way. In signal processing, such a model is known as a *linear time-variant system*.

For a wide range of function spaces (covering practically all cases relevant for applications) and by means of the use of generalized functions this analogy can be given a precise mathematical meaning. The natural way of describing this context is via so-called *kernel theorems*. Although only Hilbert Schmidt operators can be described as integral operators with L^2-kernels, every bounded linear system A on $L^2(\mathbb{R}^d)$ can be uniquely described by some distributional kernel $K \in S_0'(\mathbb{R}^{2d})$.

Suppose we have an integral operator K with distributional kernel k on $S_0(\mathbb{R}^d)$, i.e., we think of K in a weak sense

$$\langle Kf, g \rangle = \langle k, g \otimes \bar{f} \rangle, \quad f, g \in S_0(\mathbb{R}^d),$$

where $g \otimes f$ denotes the tensor product $g(x)f(y)$, then K is a bounded operator between $S_0(\mathbb{R}^d)$ and $S_0'(\mathbb{R}^d)$. Since by duality we deduce that

$$|\langle Kf, g \rangle| = |\langle k, g \otimes \bar{f} \rangle| \le \|k\|_{S_0'} \|g \otimes \bar{f}\|_{S_0} = \|k\|_{S_0'} \|f\|_{S_0} \|g\|_{S_0}$$

is true for all $g \in S_0(\mathbb{R}^d)$, we have that $Kf \in S_0'(\mathbb{R}^d)$. Therefore the operator K is bounded between $S_0(\mathbb{R}^d)$ and $S_0'(\mathbb{R}^d)$ with the following estimate for the operator norm of K:

$$\|K\|_{op} \le \|k\|_{S_0'}.$$

The non-trivial aspect of the kernel theorem is that the converse is true.

Theorem 5.2. *If K is a bounded operator from $S_0(\mathbb{R}^d)$ to $S_0'(\mathbb{R}^d)$, then there exists a unique kernel $k \in S_0'(\mathbb{R}^{2d})$ such that $\langle Kf, g \rangle = \langle k, g \otimes \bar{f} \rangle$ for $f, g \in S_0(\mathbb{R}^d)$.*

We only sketch a proof and refer the interested reader to the book of Gröchenig [26] for the technical details.

We define the infinite matrix $\mathbf{a} = \big(a_{(l,m),(r,s)}\big)$ of the operator K with respect to a multivariate Wilson basis $\mathcal{W}(g)$ by

$$a_{(l,m),(r,s)} = \langle K\boldsymbol{\Psi}_{r,s}, \boldsymbol{\Psi}_{l,m} \rangle. \tag{25}$$

Then the matrix $(\mathbf{a})$ is bounded from $\ell^1(\mathbb{Z}^d \times \mathbb{N}_0)$ to $\ell^\infty(\mathbb{Z}^d \times \mathbb{N}_0^d)$. We therefore can define a kernel k for K as in linear algebra by

$$k = \sum_{l,m,r,s} a_{(l,m),(r,s)}\, \boldsymbol{\Psi}_{l,m} \otimes \boldsymbol{\Psi}_{r,s}. \tag{26}$$

Now, we know that $\{\boldsymbol{\Psi}_{l,m} \otimes \boldsymbol{\Psi}_{r,s}\}$ is an orthonormal basis for $\boldsymbol{L}^2(\mathbb{R}^{2d})$ which yields that $k \in S_0'(\mathbb{R}^{2d})$ with weak*-convergence of the sum.

An important corollary of the preceding discussion is the following observation.

Corollary 5.2. *Let $(\boldsymbol{\Psi}_{\mathbf{k},\mathbf{n}})$ be an orthonormal Wilson basis for $\boldsymbol{L}^2(\mathbb{R}^d)$ then the coefficient mapping $D : f \mapsto \langle f, \boldsymbol{\Psi}_{\mathbf{k},\mathbf{n}} \rangle$ induces a Gelfand triple isomorphism between $(\boldsymbol{S_0}, \boldsymbol{L}^2, \boldsymbol{S_0'})(\mathbb{R}^d)$ and $(\ell^1, \ell^2, \ell^\infty)(\mathbb{Z}^d \times \mathbb{N}^d)$.*

Proof. Since $(\boldsymbol{\Psi}_{\mathbf{k},\mathbf{n}})$ is an orthonormal basis of $\boldsymbol{L}^2(\mathbb{R}^d)$ the analysis operator $f \mapsto \langle f, \boldsymbol{\Psi}_{\mathbf{k},\mathbf{n}} \rangle$ is an isomorphism between $\boldsymbol{L}^2(\mathbb{R}^d)$ and $\ell^2(\mathbb{Z}^d \times \mathbb{N}^d)$. The Wilson system $(\boldsymbol{\Psi}_{\mathbf{k},\mathbf{n}})$ is an unconditional basis for $S_0(\mathbb{R}^d)$ and therefore the analysis operator gives an isomorphism between $S_0(\mathbb{R}^d)$ and $\ell^1(\mathbb{Z}^d \times \mathbb{N}^d)$. By duality we obtain that $S_0'(\mathbb{R}^d)$ is isomorph to $\ell^\infty(\mathbb{Z}^d \times \mathbb{N}^d)$. $\square$

6 The Spreading Function and Pseudo-Differential Operators

The notion of a Gelfand triple has turned out to be a very fruitful concept for investigations in Gabor analysis, see [16], [8], [11]. In this section we present some results of Feichtinger and Kozek on Gelfand triples for time–frequency analysis. All these results have their origin in the search of a mathematical framework for problems in signal analysis. Many problems in applications are modelled as linear time-variant systems (LTV). In the last section we learned that a LTV is just an integral operator K acting on signals with finite energy,

$$Kf(x) = \int_{\mathbb{R}^d} k(x,y)f(y)dy, \qquad f \in \boldsymbol{L}^2(\mathbb{R}^d). \tag{27}$$

The quality of an integral operator K on $L^2(\mathbb{R}^d)$ relies on properties of its kernel k. For example integrability conditions on k yield to classes of nice operators. The most prominent class of operators, the *Hilbert–Schmidt* operators $\mathcal{HS}$ are defined in terms of integrability conditions. Namely, an integral operator K on $L^2(\mathbb{R}^d)$ is a *Hilbert–Schmidt* operator if $k \in L^2(\mathbb{R}^d \times \mathbb{R}^d)$.

The class of Hilbert–Schmidt operators $\mathcal{HS}$ has a natural inner product. Let $K_1, K_2 \in \mathcal{HS}$ with kernels k_1, k_2, respectively. Then

$$\langle K_1, K_2 \rangle_{\mathcal{HS}} := \langle k_1, k_2 \rangle_{L^2(\mathbb{R}^d \times \mathbb{R}^d)} \tag{28}$$

defines an inner product on $\mathcal{HS}$. The associated *Hilbert–Schmidt norm* $\|K\|_{\mathcal{HS}} := (\langle K_1, K_2 \rangle_{\mathcal{HS}})^{1/2}$ gives $\mathcal{HS}$ the structure of a Hilbert space [37]. Furthermore we recall that every Hilbert–Schmidt operator on $\mathcal{HS}$ is a compact operator on $L^2(\mathbb{R}^d)$. Recall that a compact operator K on $L^2(\mathbb{R}^d)$ is of Hilbert–Schmidt type if and only if there exists an orthonormal basis $(e_n)_{n \in \mathbb{N}}$ in $L^2(\mathbb{R}^d)$ and a sequence of scalars $(\lambda_n)_{n \in \mathbb{N}} \in \ell^2(\mathbb{N})$ such that

$$Kf = \sum_{n \in \mathbb{N}} \lambda_n \langle e_n, f \rangle e_n. \tag{29}$$

The sequence of scalars $(\lambda_n)_{n \in \mathbb{N}}$ are actually the eigenvalues of K and $\|K\|_{\mathcal{HS}} = \left(\sum_{n \in \mathbb{N}} |\lambda_n|^2 \right)^{1/2}$. The space of Hilbert–Schmidt operators $\mathcal{HS}$ is not closed in the C^*-algebra $\mathcal{K}$ of compact operators on $L^2(\mathbb{R}^d)$ with respect to the operator norm and there exist compact operators which are not of Hilbert–Schmidt type. But $\mathcal{HS}$ is a two-sided ideal in $\mathcal{K}$.

If we choose as orthonormal basis of $L^2(\mathbb{R}^d)$ a Wilson basis $(\Psi_{\mathbf{k},\mathbf{n}})$ then the preceding observations led to an isomorphism between $\mathcal{HS}$ and $\ell L^2(\mathbb{Z}^d \times \mathbb{N}^d)$. Now we can make use of the concept of Gelfand triples, but this time we take the Hilbert–Schmidt operators as Hilbert space of an *Operator Gelfand triple*. We observe that the kernel theorem for $S_0(\mathbb{R}^d)$ provides us with another class of operators with "smooth kernels". We write $\mathcal{L}$ for the space of bounded linear operators on a Banach space B. One finds that $K \in \mathcal{L}(S_0'(\mathbb{R}^d), S_0(\mathbb{R}^d))$ can be identified with kernels $k \in S_0(\mathbb{R}^{2d})$ and is dense in $\mathcal{HS}$. But the class of Hilbert–Schmidt operators $\mathcal{HS}$ is dense in $\mathcal{L}(S_0(\mathbb{R}^d), S_0'(\mathbb{R}^d))$ and therefore $(\mathcal{L}(S_0'(\mathbb{R}^d), S_0(\mathbb{R}^d)), \mathcal{HS}, \mathcal{L}(S_0(\mathbb{R}^d), S_0'(\mathbb{R}^d)))$ is indeed a Gelfand triple. In this setting the kernel theorem can be interpreted as a unitary Gelfand triple isomorphism between this triple and their kernels in $(S_0, L^2, S_0')(\mathbb{R}^d \times \mathbb{R}^d)$. There is another Gelfand triple isomorphism that associates the $\mathcal{HS}$ Gelfand triple with the Gelfand triple $(S_0, L^2, S_0')(\mathbb{R}^d \times \widehat{\mathbb{R}}^d)$: the so-called *spreading symbol* of operators.

As a motivation we discuss a problem of great practical interest: communication with cellar phones. In modern communication cellar phones play a crucial rule in everyday life. How do engineers solve the problem of transmitting a signal f from a sender A to a receiver B? In the most general situation sender A and receiver B move in different directions with certain velocities which yields to a variation of the path lengths of the transmitted signal f

and due to the Doppler effect to a change of frequencies. Therefore B receives a signal of the following form

$$\tilde{f} = \iint_{\mathbb{R}^2} \eta(K)(x,\omega)M_\omega T_x f \, dx \, d\omega, \tag{30}$$

where the function $\eta(K)$ models the effect of the channel by the amount of time–frequency shifts arising as just described, applied the signal f. The receiver B is not interested in the signal $\tilde{f}$ but in the original signal f. From a mathematical point of view $\tilde{f}$ is just the action of an operator K on the signal f, i.e., $\tilde{f} = Kf$. In this picture B has to invert the operator K to get the information contained in the signal f. Operators of this form are called *pseudo-differential operators* and arise naturally in many problems of physics, engineering and mathematics. The function $\eta(K)$ is the so-called *spreading function* of the operator K. In the following we look for conditions on the spreading function $\eta(K)$ which allow an inversion of our pseudo-differential operator K.

First, (30) suggests a decomposition of a general operator K on $L^2(\mathbb{R}^d)$ as a continuous superposition of time–frequency shifts.

$$K = \iint_{\mathbb{R}^{2d}} \eta(K)(x,\omega)M_\omega T_x \, dx \, d\omega. \tag{31}$$

We already know such a decomposition of the identity operator on $L^2(\mathbb{R}^d)$ since this is the inversion formula for the STFT:

$$I_{L^2(\mathbb{R}^d)} = \frac{1}{\langle g, \gamma \rangle} \iint_{\mathbb{R}^{2d}} V_g f(x,\omega) M_\omega T_x \, dx \, d\omega \tag{32}$$

for $g, \gamma \in L^2(\mathbb{R}^d)$ with $\langle g, \gamma \rangle \neq 0$.

The non-commutativity of translation and modulation operators on $L^2(\mathbb{R}^d)$ leads to a twisted convolution of the spreading functions of two operators K and L. Let $K, L \in \mathcal{L}(S_0, S_0')$ and $\eta(K), \eta(L)$ their spreading functions respectively. Then the spreading function of the composition KL is given by *twisted convolution* of $\eta(K)$ and $\eta(L)$:

$$\eta(KL)(x,\omega) = \iint_{\mathbb{R}^2} \eta(K)(x',\omega')\eta(L)(x-x',\omega-\omega')e^{-2\pi ix'(\omega-\omega')}d\omega'. \tag{33}$$

The spreading function of the adjoint operator K^* is given by

$$\eta(K^*)(x,\omega) = \overline{\eta(K)(-x,-\omega)} \cdot e^{-2\pi ix\omega} \tag{34}$$

and therefore leads to a noncommutative involution. Later we will return to this topic in the context of Gröchenig/Leinert's resolution of the "irrational case"-conjecture [27].

The relation between the kernel k of an operator K from the Gelfand triple

$(\mathcal{L}(\boldsymbol{S}_0, \boldsymbol{S}_0'), \mathcal{H}S, \mathcal{L}(\boldsymbol{S}_0', \boldsymbol{S}_0))$ and its spreading function $\eta(K)$ is given by the following mapping from $\mathbb{R}^d \times \mathbb{R}^d$ to $\mathbb{R}^d \times \widehat{\mathbb{R}}^d$

$$\eta(K)(x, \omega) = \int_{\mathbb{R}^d} k(y, y - x) e^{-2\pi i y \omega} dy, \tag{35}$$

which is very useful in the calculation of the spreading function of an operator K. It can be interpreted literally at the lowest level (integrals, etc., exist), and extend by continuity to the "upper levels". Moreover it can be described by the fact that it is the unique Gelfand triple isomorphism which maps TF-shift operators onto the corresponding Dirac measures in the TF-plane (hence reproducing exactly the situation we had in the finite case).

The spreading function of an operator K is an object living on the time–frequency plane $\mathbb{R}^d \times \widehat{\mathbb{R}}^d$. Therefore a further understanding of its properties should be done according to the structure of $\mathbb{R}^d \times \widehat{\mathbb{R}}^d$ which is closely related to the structure of the Euclidean plane $\mathbb{R}^d \times \mathbb{R}^d$. Namely, the time–frequency plane is a symplectic manifold, i.e., there exists a non-degenerate 2-form $\Omega(X, Y) = y \cdot \omega - x \cdot \eta$ for two points $X = (x, \omega), Y = (y, \eta)$ in $\mathbb{R}^d \times \widehat{\mathbb{R}}^d$. Since Ω is non-degenerate there is a unique invertible skew-symmetric linear operator $\mathcal{J}$ on $\mathbb{R}^d \times \widehat{\mathbb{R}}^d$ such that the symplectic form Ω and the Euclidian inner product are related as follows: $\Omega(X, Y) = \langle \mathcal{J}X, Y \rangle$ for all $X, Y \in \mathbb{R}^d \times \widehat{\mathbb{R}}^d$. This implies an important fact about the characters of $\mathbb{R}^d \times \widehat{\mathbb{R}}^d$. Namely, the characters are given by $\{\chi_s(X, Y) = e^{2\pi i \Omega(X, Y)} | X \in \mathbb{R}^d \times \widehat{\mathbb{R}}^d\}$ for a fixed $Y \in \mathbb{R}^d \times \widehat{\mathbb{R}}^d$. Therefore it is natural to analyse a function F on $\mathbb{R}^d \times \widehat{\mathbb{R}}^d$ with the *symplectic Fourier Transform*

$$\mathcal{F}_s F(X) = \iint_{\mathbb{R}^d \times \widehat{\mathbb{R}}^d} F(Y) e^{2\pi i \Omega(X, Y)} dY \tag{36}$$

instead of the Fourier transform $\mathcal{F}$ induced by the standard inner-product $\langle \cdot, \cdot \rangle$ on $\mathbb{R}^d \times \mathbb{R}^d$. From the relation between symplectic form and inner-product we obtain that the symplectic Fourier transform $\mathcal{F}_s$ is just a Fourier transform followed by a rotation by $\frac{\pi}{2}$ since $\mathcal{J}$ describes a rotation by $\frac{\pi}{2}$ around the origin of $\mathbb{R}^d \times \mathbb{R}^d$. This fact allows us to derive similar statements for the symplectic Fourier transform as for the Euclidian Fourier transform:

1. $\mathcal{F}_s$ is a unitary mapping from $\boldsymbol{L}^2(\mathbb{R}^d \times \widehat{\mathbb{R}}^d)$ onto $\boldsymbol{L}^2(\widehat{\mathbb{R}}^d \times \mathbb{R}^d)$.
2. $\mathcal{F}_s^{-1} = \mathcal{F}_s$ (involutive property).
3. $\mathcal{F}_s \left(\boldsymbol{S}_0(\mathbb{R}^d \times \widehat{\mathbb{R}}^d) \right) = \boldsymbol{S}_0(\mathbb{R}^d \times \widehat{\mathbb{R}}^d)$.

By duality we obtain that

Proposition 6.1. *The symplectic Fourier transform $\mathcal{F}_s$ defines a unitary Gelfand triple automorphism on $(\boldsymbol{S}_0, \boldsymbol{L}^2, \boldsymbol{S}_0')(\mathbb{R}^d \times \widehat{\mathbb{R}}^d)$.*

Another reason for our choice of $\boldsymbol{S}_0(\mathbb{R}^{2d})$ as space of test functions is that the Poisson summation formula for symplectic Fourier transform holds pointwise and with absolute convergence. Recently, we have shown that the Fundamental Identity of Gabor Analysis can be derived by an application of Poisson summation to a product of two STFT's:

Theorem 6.1. *Let Λ a lattice in $\mathbb{R}^d \times \widehat{\mathbb{R}}^d$ with adjoint lattice Λ° and $F \in S_0(\mathbb{R}^{2d})$. Then*

$$\sum_{\lambda \in \Lambda} F(\lambda) = \frac{1}{|\Lambda|} \sum_{\lambda^\circ \in \Lambda^\circ} \mathcal{F}_s F(\lambda^\circ) \qquad (37)$$

holds pointwise and with absolute convergence on both sides.

The spreading function is an important tool for the description of (slowly) time-variant channels in communication theory, but it is not the only symbol of associated with a linear operator. In the theory of pseudo-differential operators the *Kohn–Nirenberg symbol* (KN), denoted by $\sigma(K)$, is used for an operator $K \in (S_0, L^2, S_0')(\mathbb{R}^d \times \mathbb{R}^d)$. It is defined as the symplectic Fourier transform of the spreading function $\eta(K)$:

$$\sigma(x, \omega) = \mathcal{F}_s \eta(K) = \iint_{\mathbb{R}^d \times \widehat{\mathbb{R}}^d} \eta(K) e^{2\pi i (y \cdot \omega - x \cdot \eta)} dy d\eta, \quad (x, \omega) \in \mathbb{R}^d \times \widehat{\mathbb{R}}^d. \tag{38}$$

If $Kf(x) = \int_{\mathbb{R}^d} k(x, y) f(y) dy$ then $\sigma(K) = \int_{\mathbb{R}^d} k(x, x - y) e^{-2\pi i y \cdot \omega} dy$. In signal analysis $\sigma(K)$ was introduced by Zadeh and is called the *time-varying transfer function* of a system modelled by K. As an example we mention the KN symbol of a rank-one operator $f \otimes \bar{g}$, which describes the mapping $h \mapsto \langle h, g \rangle f$, is equal to

$$\sigma(f \otimes \bar{g})(x, \omega) = f(x) \overline{\hat{g}(\omega)} e^{-2\pi i x \cdot \omega}, \quad (x, \omega) \in \mathbb{R}^d \times \widehat{\mathbb{R}}^d, \tag{39}$$

the Rihaczek distribution of f against g. For $f, g \in S_0(\mathbb{R}^d)$ we have that the KN-symbol $\sigma(f \otimes \bar{g}) \in S_0(\mathbb{R}^d \times \widehat{\mathbb{R}}^d)$ which in turn implies (using the last equation) that $(x, \omega) \mapsto e^{2\pi i x \cdot \omega}$ is a pointwise multiplier on $S_0(\mathbb{R}^d \times \widehat{\mathbb{R}}^d)$.

After these preparations we can state one of our main results:

Theorem 6.2. *The spreading function $K \mapsto \eta(K)$ is a unitary Gelfand triple isomorphism from $(\mathcal{L}(S_0', S_0), \mathcal{HS}, \mathcal{L}(S_0, S_0'))$ to $(S_0, L^2, S_0')(\mathbb{R}^d \times \widehat{\mathbb{R}}^d)$.*

Corollary 6.1. *The KN symbol of K induces a unitary Gelfand triple isomorphism between $(\mathcal{L}(S_0', S_0), \mathcal{HS}, \mathcal{L}(S_0, S_0'))$ and $(S_0, L^2, S_0')(\mathbb{R}^d \times \widehat{\mathbb{R}}^d)$.*

Another consequence of the preceding theorem is the following Gelfand-bracket identities for $K_1, K_2 \in (\mathcal{L}(S_0, S_0'), \mathcal{HS}, \mathcal{L}(S_0', S_0))$:

$$\langle K_1, K_2 \rangle_{(\mathcal{B}, \mathcal{HS}, \mathcal{B}')} = \langle \eta(k_1), \eta(k_2) \rangle_{(S_0, L^2, S_0')(\mathbb{R}^d \times \widehat{\mathbb{R}}^d)} \tag{40}$$

$$= \langle \sigma(k_1), \sigma(k_2) \rangle_{(S_0, L^2, S_0')(\mathbb{R}^d \times \widehat{\mathbb{R}}^d)}, \tag{41}$$

with $\mathcal{B} = \mathcal{L}(S_0, S_0')$ and $\mathcal{B}' = \mathcal{L}(S_0', S_0)$ respectively.

The KN symbol of a rank-one operator $f \otimes \bar{g}$, which is the mapping $h \mapsto \langle h, g \rangle f$, is the Rihaczek distribution and by an application of the (inverse) symplectic Fourier transform we get another time–frequency distribution: the STFT!

Lemma 6.1. *For $f, g \in S_0(\mathbb{R}^d)$ the rank-one operator $f \otimes \bar{g}$ has a kernel in $S_0(\mathbb{R}^d)$. Moreover the corresponding spreading function is*

$$\eta(f \otimes \bar{g})(x, \omega) = \int_{\mathbb{R}^d} f(x)\overline{g(y - x)}e^{-2\pi i y \cdot \omega} dy \tag{42}$$

and hence coincides with $V_g f \in S_0(\mathbb{R}^d \times \widehat{\mathbb{R}}^d)$.

In the light of this result the inversion formula for the STFT is a superposition of time–frequency shifts with the spreading function of the rank-one operator $g \otimes \bar{\gamma}$ for $g, \gamma \in L^2(\mathbb{R}^d)$ with $\langle g, \gamma \rangle \neq 0$:

$$f = \frac{1}{\langle g, \gamma \rangle} \iint_{\mathbb{R}^d \times \widehat{\mathbb{R}}^d} \eta(f \otimes \bar{g})(x, \omega) T_x M_\omega \gamma \, dx d\omega. \tag{43}$$

Recall that in analogy with the characters $\{\chi_\omega : \omega \in \widehat{\mathbb{R}}^d\}$ the time–frequency shifts $\{\pi(X) : X = (x, \omega) \in \mathbb{R}^d \times \widehat{\mathbb{R}}^d\}$ would be an orthonormal set with respect to the Hilbert–Schmidt inner product $\langle \cdot, \cdot \rangle_{\mathcal{HS}}$ and $\eta(f \otimes \bar{g})(x, \omega) = \langle f \otimes \bar{g}, \pi(x, \omega) \rangle_{\mathcal{HS}}$ but as in the case of Fourier transform the building blocks $\pi(X)$ for $X \in \mathbb{R}^d$ of our orthonormal system $\{\pi(X) : X = (x, \omega) \in \mathbb{R}^d \times \widehat{\mathbb{R}}^d\}$ are not Hilbert–Schmidt. As in our treatment of the Fourier transform it is not so important that the building blocks are elements of our Hilbert space but that they allow us to get expressions as they would be an orthonormal set of elements in our Hilbert space.

As a first example we state a generalization of the inversion formula for the STFT from $L^2(\mathbb{R}^d)$ to the Gelfand triple $(S_0, L^2, S_0')(\mathbb{R}^d)$, where for $f \in S_0'(\mathbb{R}^d)$ the formula is interpreted in a weak sense.

Proposition 6.2. *Let $g, \gamma \in S_0(\mathbb{R}^d)$ with $\langle g, \gamma \rangle \neq 0$. Then*

$$f = \frac{1}{\langle g, \gamma \rangle} \iint_{\mathbb{R}^d \times \widehat{\mathbb{R}}^d} \eta(f \otimes \bar{g})(x, \omega) T_x M_\omega \gamma \, dx d\omega. \tag{44}$$

holds for $f \in (S_0, L^2, S_0')(\mathbb{R}^d)$.

That is a special case of a general statement about the spreading function.

Theorem 6.3. *Any $K \in (\mathcal{L}(S_0, S_0'), \mathcal{HS}, \mathcal{L}(S_0', S_0))$ has a representation*

$$K = \iint_{\mathbb{R}^d \times \widehat{\mathbb{R}}^d} \langle K, \pi(x, \omega) \rangle_{\mathcal{L}(S_0, S_0')} \pi(x, \omega) \, dx d\omega \tag{45}$$

convergent in the strong resp. weak*-sense. The (complex-valued) amplitude function arising in this context, i.e., $\eta(K)(x,\omega) = \langle K, \pi(x,\omega)\rangle_{\mathcal{L}(S_0, S_0')}$, is called the spreading distribution of the operator K.

The basic tool in the proof is the fact that the spreading representation maps a time–frequency shift $\pi(x,\omega)$ for $(x,\omega) \in \mathbb{R}^d \times \widehat{\mathbb{R}}^d$ on the Dirac measure δ_X at $X = (x,\omega)$, i.e., $\eta(\pi(X)) = \delta_X$ and the relation between the spreading function and the kernel of an operator K.

The preceding theorem is the mathematical justification of a widely used statement that the spreading function of an operator K is a measure for the time–frequency content of K.

In our intuition we move an operator K over $\mathbb{R}^d \times \widehat{\mathbb{R}}^d$ and want there is a simply relation between the original symbol of K and the symbol after a movement to $(x,\omega) \in \mathbb{R}^d \times \widehat{\mathbb{R}}^d$. The KN-symbol of an operator K is shifted by $T_{x,\omega}$ in the time–frequency plane.

Lemma 6.2. *Let K belong to one of the spaces $(\mathcal{L}(S_0', S_0), \mathcal{H}S, \mathcal{L}(S_0', S_0))$, then*
$\pi(x,\omega)K\pi(x,\omega)^*$, the conjugation of K by $\pi(x,\omega) \in \mathbb{R}^d \times \widehat{\mathbb{R}}^d$ corresponds to translation of the KN symbol $\sigma(K)$,

$$\sigma(\pi(x,\omega)K\pi(x,\omega)^*) = T_{(x,\omega)}(\sigma(K)). \tag{46}$$

This property of the KN symbol is of central importance in our study of the Gabor frame operator to which we devote the final part of this section. Let $\mathcal{G} = (g, \Lambda)$ be a Gabor system for a lattice $\Lambda \in \mathbb{R}^d \times \widehat{\mathbb{R}}^d$. Then the Gabor frame operator $S_{g,\Lambda}$ commutes with all time–frequency shifts of the lattice Λ, i.e.,

$$\pi(\lambda)S_{g,\Lambda}\pi(\lambda)^* = S_{g,\Lambda}, \quad \text{for all} \quad \lambda \in \Lambda. \tag{47}$$

This fact was the motivation for Feichtinger and Kozek to introduce the class of Λ-invariant operators [16].

Definition 6.1. Let Λ a lattice in $\mathbb{R}^d \times \widehat{\mathbb{R}}^d$ and K an operator concentrated on Λ. Then K is called Λ-*invariant* if $\pi(\lambda)K = K\pi(\lambda)$ for all $\lambda \in \Lambda$.

In the following we want to find the support of the spreading function $\eta(K)$ of an Λ-invariant operator $K \in (\mathcal{L}(S_0, S_0'), \mathcal{H}S, \mathcal{L}(S_0', S_0))$. As a first step towards this result we study spreading representations of K on $\mathbb{R}^d \times \widehat{\mathbb{R}}^d$.

Lemma 6.3. *Let $K_1, K_2 \in \mathcal{L}(S_0, S_0')$ with spreading function $\eta(K_1), \eta(K_2)$, respectively. Then*

1. $\eta(K_1 K_2)(\lambda) = \iint_{\mathbb{R}^d \times \widehat{\mathbb{R}}^d} \eta(K_1)(\mu)\eta(K_2)(\lambda-\mu)\rho(\lambda-\mu,\mu)d\mu$ with $\rho(X,Y) = e^{2\pi i(y\cdot\omega - x\cdot\eta)}$ for $X = (x,\omega), Y = (y,\eta) \in \mathbb{R}^d \times \widehat{\mathbb{R}}^d$.
2. $supp(\eta(K_1)\eta(K_2)) \subset supp(K_1) + supp(K_2)$.
3. $|\eta(K_1 K_2)| = |\eta(K_1)| * |\eta(K_2)|$ for $\eta(K_1), \eta(K_2) \in L^1_{loc}(\mathbb{R}^d \times \widehat{\mathbb{R}}^d)$.

The proof of (1) is a consequence of the commutation relation for time–frequency shifts and the fact that for $K_1 \in \mathcal{L}(S_0, S_0')$ and $K_2 \in \mathcal{L}(S_0, S_0')$ also $K_1 K_2 \in \mathcal{L}(S_0', S_0)$. Now each operator K in $\mathcal{L}(S_0, S_0')$ has an absolutely convergent spreading representation and therefore our result holds pointwise. The support condition follows from the analogous result for the ordinary convolution.

By abstract reasons each Λ-invariant operator K has a representation in the set of all operators concentrated on $\Lambda^\circ = \{\lambda^\circ \in \mathbb{R}^d \times \widehat{\mathbb{R}}^d | \pi(\lambda)\pi(\lambda^\circ)\} = \pi(\lambda^\circ)\pi(\lambda)\}$ since K lies in the commutant of the $(C^*$, von Neumann) algebra generated by $\{\pi(\lambda) : \lambda \in \Lambda\}$. The set Λ° is the so-called *adjoint lattice* since it is the annihilator subgroup of Λ for the symplectic Fourier transform $\mathcal{F}_s$ and if $\Lambda^\perp$ is the annihilator subgroup of Λ with respect to $\mathcal{F}$ then $\Lambda^\circ = \mathcal{J}\Lambda^\perp$.

The time–frequency invariance of $S_0(\mathbb{R}^d)$ implies that K and $\pi(\lambda)K$ are in the Gelfand triple $(\mathcal{L}(S_0, S_0'), \mathcal{H}S, \mathcal{L}(S_0', S_0))$, too. Therefore, the Λ-invariance of T translates into a periodicity condition for the symbol $\sigma(K)$

$$\sigma(K) = T_\lambda(\sigma(K)), \quad \lambda \in \Lambda. \tag{48}$$

This periodicity condition corresponds to a support condition for the spreading function since $\eta(K)(\lambda) = \eta(K)(\lambda)e^{-2\pi\Omega(\lambda,\mu)}$. But $\{e^{-2\pi\Omega(\lambda,\mu)} | \lambda \in \Lambda\}$ for a fixed $\mu \in \mathbb{R}^d \times \widehat{\mathbb{R}}^d$ is a group of characters on $\mathbb{R}^d \times \widehat{\mathbb{R}}^d$ yields that

$$\mathrm{supp}(\eta(K)) \subset \mathcal{J}\Lambda^\perp = \Lambda^\circ. \tag{49}$$

The fact that distributions in $S_0'(\mathbb{R}^d)$ with support in a discrete subgroup is a sum of Dirac measures with a bounded sequence of coefficients implies that for some bounded sequence (c_{λ°) over Λ°

$$\eta(K) = \sum_{\lambda^\circ \in \Lambda^\circ} c_{\lambda^\circ} \delta_{\lambda^\circ} \tag{50}$$

with $c_{\lambda^\circ} = (K)_{\lambda^\circ} = \iint_{\mathbb{R}^d \times \widehat{\mathbb{R}}^d / \Lambda^\circ} \sigma(K)(\mu)e^{2\pi i \Omega(\lambda,\mu)} d\mu$.

Returning to the description in the operator domain we arrive at the following characterization

Theorem 6.4. *Let $K \in (\mathcal{L}(S_0, S_0'), \mathcal{H}S, \mathcal{L}(S_0', S_0))$ and $\sigma(K)$ the KN symbol. Then $\sigma(K)$ is a Λ-periodic distribution with a symplectic Fourier transform supported on Λ^0. Furthermore*

$$K = \sum_{\lambda^\circ \in \Lambda^\circ} (K)_{\lambda^\circ} \pi(\lambda^\circ). \tag{51}$$

Corollary 6.2. *The mapping $\sigma(K) \mapsto (K)_{\lambda^\circ}$ is a unitary Gelfand triple isomorphism between $(S_0, L^2, S_0')(\mathbb{R}^d \times \widehat{\mathbb{R}}^d / \Lambda)$ and $(\ell^1, \ell^2, \ell^\infty)(\Lambda^\circ)$.*

Note that the time–frequency invariance of $S_0(\mathbb{R}^d)$ implies the boundedness of K on $S_0(\mathbb{R}^d)$ since

$$\|K\|_{\mathcal{L}(S_0)} = \| \sum_{\lambda^\circ \in \Lambda^\circ} (K)_{\lambda^\circ} \pi(\lambda^\circ)\|_{\mathcal{L}(S_0)} \leq \sum_{\lambda^\circ \in \Lambda^\circ} |(K)_{\lambda^\circ}|. \qquad (52)$$

The next theorem shows that for any Λ-invariant operator K with $\sigma(K) \in$ $S_0'((\mathbb{R}^d \times \widehat{\mathbb{R}}^d)/\Lambda)$ there exists a prototype operator $P \in \mathcal{L}(S_0, S_0')$ such that periodization of P in the time–frequency plane is corresponds to sampling of the spreading function $\eta(P)$ on Λ^0.

Theorem 6.5. *Let K be a Λ-invariant operator with $\sigma(K) \in S_0'((\mathbb{R}^d \times \widehat{\mathbb{R}}^d)/\Lambda)$. Then there exists some $P \in \mathcal{L}(S_0, S_0')$ such that its periodization is exactly K*

$$K = \sum_{\lambda \in \Lambda} \pi(\lambda) \, P \, \pi(\lambda)^* = \frac{1}{|\Lambda|} \sum_{\lambda^\circ \in \Lambda^\circ} \langle P, \pi(\lambda^\circ) \rangle_{\mathcal{L}(S_0, S_0')} \, \pi(\lambda^\circ). \qquad (53)$$

Remark 6.1. The preceding result is a discrete analog of our spreading representation for operators in $\mathcal{L}(S_0, S_0')$ which in the context of Gabor analysis leads to the so-called *Janssen representation* of the Gabor frame operator.

The proof of the theorem is based on two important features of the time–frequency plane $\mathbb{R}^d \times \widehat{\mathbb{R}}^d$:

1. $\{U \mapsto \pi(\lambda) \, U \, \pi(\lambda)^* | \lambda \in \Lambda\}$ defines unitary representation of Λ which gives the Λ-invariance of K.
2. An application of the Poisson summation formula for the symplectic Fourier transform to $\sigma(P)$ with respect to the lattice Λ maps the periodization of

$$\sigma(K) = \sum_{\lambda \in \Lambda} T_\lambda(\sigma(P)) \qquad (54)$$

to the sampling of the spreading function $\eta(P)$ on the lattice Λ°.

As an application we state that the Gabor frame operator $S_{g,\Lambda}$ of a Gabor system $\mathcal{G}(g, \Lambda)$ with $g \in S_0(\mathbb{R}^d)$ is generated by shifting a rank-one operator along the lattice Λ. In addition we use the fact the spreading function of a rank-one operator is the STFT. Altogether we therefore have

$$S_{g,\Lambda} = \frac{1}{|\Lambda|} \sum_{\lambda^\circ \in \Lambda^\circ} \langle g, \pi(\lambda^\circ)\gamma \rangle \pi(\lambda^\circ) \qquad (55)$$

with $\gamma \in S_0(\mathbb{R}^d)$. Equation (55) is the so-called Jannsen representation of $S_{g,\Lambda}$ which decomposes $S_{g,\Lambda}$ into an *absolutely convergent* series of time–frequency shifts. In (55) we used implicitly another pleasant property of $S_0(\mathbb{R}^d)$.

Lemma 6.4. *Let $g, \gamma \in S_0(\mathbb{R}^d)$ and Λ a lattice in $\mathbb{R}^d \times \widehat{\mathbb{R}}^d$. Then (g, γ) satisfy Tolimieri–Orr's condition (A'):*

$$\sum_{\lambda \in \Lambda} |\langle g, \gamma_\lambda \rangle| < \infty, \qquad (A').. \qquad (A')$$

This stability of Condition (A′) for $g, \gamma \in S_0(\mathbb{R}^d)$ with respect to lattice changes makes Feichtinger's algebra such an important object in Gabor analysis. In a recent work Feichtinger and Kaiblinger have drawn some deep consequences from this fact. Roughly speaking, they proved that the set of functions in $S_0(\mathbb{R}^d)$ which generate a Gabor frame is *"open"* [15].

We close our discussion of the Gabor frame operator with a striking result of Gröchenig/Leinert on the quality of the canonical dual of a Gabor system $\mathcal{G}(g, \Lambda)$ generated by a window $g \in S_0(\mathbb{R}^d)$.

Theorem 6.6. *Let* $g \in S_0(\mathbb{R}^d)$ *and* $\mathcal{G}(g, \Lambda)$ *a Gabor frame of* $L^2(\mathbb{R}^d)$. *Then* $\gamma_0 = S_{g,\Lambda}^{-1} g$ *is in* $S_0(\mathbb{R}^d)$.

Their proof is based on a noncommutative version of Wiener's lemma for the Banach algebra $\ell^1(\Lambda)$ with twisted convolution $\natural$ as product and noncommutative involution $*$ as described above for the spreading function of a product of two operators in $\mathcal{L}(S_0, S_0')$ and the spreading function of the adjoint of an operator in $\mathcal{L}(S_0, S_0')$. A special case of their main result is that $(\ell^1, \natural, *)$ is a *symmetric* Banach algebra. In this context their Wiener lemma is expressed as the inverse-closedness of the Banach algebra

$$\mathcal{A}(\Lambda) = \{A \in \mathcal{B}(L^2(\mathbb{R}^d)) \,|\, A = \sum_{\lambda \in \Lambda} a_\lambda \pi(\lambda), \ (a_\lambda) \in \ell^1(\Lambda)\}$$

of absolutely convergent time–frequency series in the C^*-algebra $C^*(\Lambda)$ generated by the time–frequency shifts $\{\pi(\lambda) : \lambda \in \Lambda\}$. In other words, the argument is based on the highly non-trivial fact that a element of $\mathcal{A}(\Lambda)$ which is invertible in $C^*(\Lambda)$ has its inverse already in $\mathcal{A}(\Lambda)$.

We end up this section by recalling another way of representing an integral operator: the Weyl form of a pseudo-differential operator. First, the Wigner distribution defined in (10) can be generalized to a pair of functions f, g as follows:

$$W(f, g)(x, \omega) = \int f(x + \frac{t}{2}) \overline{g(x - \frac{t}{2})} e^{-2\pi i \omega t} \, dt. \tag{56}$$

Then the Weyl operator L_{σ^w} of symbol $\sigma^w \in \mathcal{S}'(\mathbb{R}^{2d})$ is defined by

$$\langle L_{\sigma^w} f, g \rangle = \langle \sigma^w, W(g, f) \rangle, \qquad f, g \in \mathcal{S}(\mathbb{R}^d). \tag{57}$$

An easy computation shows that

$$L_{\sigma^w} = \iint_{\mathbb{R}^{2d}} \widehat{\sigma^w}(\omega, -x) M_\omega T_x \, dx \, d\omega,$$

i.e., L_{σ^w} is the operator K defined in (31), with symbol $\eta(K)$ given by

$$\eta(K)(x, \omega) = \widehat{\sigma^w}(\omega, -x).$$

As a consequence,

Corollary 6.3. *The Weyl symbol σ^w of L_{σ^w} induces a unitary Gelfand triple isomorphism between $(\mathcal{L}(\boldsymbol{S}_0', \boldsymbol{S}_0), \mathcal{H}S, \mathcal{L}(\boldsymbol{S}_0, \boldsymbol{S}_0'))$ and $(\boldsymbol{S}_0, \boldsymbol{L}^2, \boldsymbol{S}_0')(\mathbb{R}^d \times \widehat{\mathbb{R}}^d)$.*

7 Gabor Multipliers

In this section we study the interplay between *Gabor multipliers* and suitable Gelfand triples. A number of basic results can be obtained as a combination of known facts about both the *analysis* and the *synthesis* mapping associated with a Gabor or Weyl–Heisenberg family, and the standard properties of multiplication operators, acting between Banach sequence spaces, based for example, on Hölder's inequality. For a detailed treatment of this subject we refer the reader to [19].

Since the atoms used to build Gabor multipliers should generate Bessel families with respect to general TF-lattices Λ, windows g will be most often taken from the Segal algebra $\boldsymbol{S}_0(\mathbb{R}^d)$. In particular, such windows will generate Bessel families for all the lattices $a\mathbb{Z}^d \times b\mathbb{Z}^d \lhd \mathbb{R}^d \times \widehat{\mathbb{R}}^d, a > 0, b > 0$.

Definition 7.1. Let g_1, g_2 be two $\boldsymbol{L}^2$-functions, Λ a TF-lattice for $\mathbb{R}^d$, i.e., a discrete subgroup of the phase space $\Lambda \lhd \mathbb{R}^d \times \widehat{\mathbb{R}}^d$. Furthermore let $\mathbf{m} = (\mathbf{m}(\lambda))_{\lambda \in \Lambda}$ be a complex-valued sequence on Λ. Then the *Gabor multiplier* associated to the triple (g_1, g_2, Λ) with (*strong* or) *upper symbol* $\mathbf{m}$ is given by

$$G_{\mathbf{m}}(f) = G_{g_1, g_2, \Lambda, \mathbf{m}}(f) = \sum_{\lambda \in \Lambda} m(\lambda) \langle f, \pi(\lambda)g_1 \rangle \, \pi(\lambda)g_2.$$

We simply write $G_{g, \Lambda, \mathbf{m}}$ for the case $g_1 = g = g_2$.

It is obvious from this definition that Gabor multipliers are essentially (infinite) linear combinations of rank-one operators $f \mapsto \langle f, \pi(\lambda)g_1 \rangle \, \pi(\lambda)g_2$, with coefficients m_λ. Whenever $g_1 = g = g_2$ and $\|g\|_2 = 1$ these building blocks are just the orthogonal projections onto the $1D$-subspaces of $\boldsymbol{L}^2$ generated by the elements of the WH-family $(\pi(\lambda)g)_{\lambda \in \Lambda}$. Depending on the properties of the *analysis window* g_1, the *synthesis window* g_2 and the *multiplier sequence* $\mathbf{m} = (m_\lambda)_{\lambda \in \Lambda}$ the overall operator $G_{g_1, g_2, \Lambda, \mathbf{m}}$ is bounded between various spaces. Typically one would require that both g_1 and g_2 are *Bessel atoms* with respect to the given lattice Λ, and that $\mathbf{m}$ is bounded. In this case the coefficient mapping using g_1, mapping f to the sequence of sampling values of the STFT $V_{g_1} f(\Lambda)$ maps $\boldsymbol{L}^2(\mathbb{R}^d)$ into $\ell^2(\Lambda)$ (by definition), and also the synthesis mapping $\mathbf{c} \mapsto \sum_{\lambda \in \Lambda} c_\lambda \pi(\lambda)g_2$ is bounded from $\ell^2(\Lambda)$ to $\boldsymbol{L}^2(\mathbb{R}^d)$, and thus the overall operator is bounded on $\boldsymbol{L}^2(\mathbb{R}^d)$.

There are many good reasons to assume that the windows g_1 and g_2 should be chosen from $\boldsymbol{S}_0(\mathbb{R}^d)$. Among them notice that $\boldsymbol{S}_0(\mathbb{R}^d)$ is much larger than the Schwartz space $\boldsymbol{S}(\mathbb{R}^d)$, used often in such a context *just for convenience*. On the other hand, $\boldsymbol{L}^2(\mathbb{R}^d)$ is a too large reservoir, since some of the more interesting results described below are not valid for all windows in $\boldsymbol{L}^2(\mathbb{R}^d)$.

In order to concentrate on the essential properties we shall state some of our results only for the case $g_1 = g_2 = g$, assuming that (g, Λ) generates a *tight* Gabor frame. In this particular case a minimal *symbolic calculus* is valid, in the sense that the constant multiplier $\mathbf{m} \equiv 1$ yields a multiple of the identity operator. Summarizing these basic facts we have:

Theorem 7.1. *Assume that* $g \in S_0(\mathbb{R}^d)$. *Then one has:*

1. *If* $\mathbf{m} \in \ell^\infty(\Lambda)$, *then* $G_\mathbf{m} = G_{g,\Lambda,\mathbf{m}}$ *defines a bounded operator on* (S_0, L^2, S_0'), *and the operator norm of* $G_\mathbf{m}$ *can be estimated (up to some constant) by* $\|\mathbf{m}\|_\infty$.
2. *The Gabor multiplier generated by* $\mathbf{m}(\lambda) \equiv 1$ *is a multiple of the identity operator if and only if* (g, Λ) *generates a tight Gabor frame.*
3. $G_\mathbf{m}$ *is a compact operator on* $L^2(\mathbb{R}^d)$ *and on* $S_0(\mathbb{R}^d)$, *if* $\mathbf{m} \in c_o(\Lambda)$, *i.e., if* $m(\lambda) \to 0$ *for* $\lambda \to \infty$ *(in the sense of* Λ).
4. *If* $\mathbf{m} \in \ell^2(\Lambda)$, *then* $G_\mathbf{m} : S_0'(\mathbb{R}^d) \to L^2(\mathbb{R}^d)$ *and* $L^2(\mathbb{R}^d) \to S_0(\mathbb{R}^d)$.
5. *For* $\mathbf{m} \in \ell^1(\Lambda)$ *the operator* $G_\mathbf{m}$ *operator on* $L^2(\mathbb{R}^d)$, *maps* $S_0'(\mathbb{R}^d)$ *into* $S_0(\mathbb{R}^d)$.

Proof. These statements follow from the boundedness properties of the coefficient resp. synthesis mappings (for fixed lattice Λ), as described in some detail in Sect. 3.3.3 of [20].   □

Of course it would be possible to make similar statements for other classes of windows. For example, any $g \in S_0'(\mathbb{R}^d)$ in combination with an ℓ^1 multiplier sequence yields still a (compact) linear operator from $S_0(\mathbb{R}^d)$ into $S_0'(\mathbb{R}^d)$, to mention a rather extreme possible variant. A more traditional approach to TF-analysis making use of Schwartz functions and tempered distributions would probably make use of $\mathcal{S}(\mathbb{R}^d)$ and $\mathcal{S}'(\mathbb{R}^d)$ (instead of $S_0(\mathbb{R}^d)$ and $S_0'(\mathbb{R}^d)$) in the above context.

For general pairs (g_1, g_2) from $S_0(\mathbb{R}^d)$ an even more compact formulation of the above theorem using the terminology of *Gelfand triples* can be given:

Theorem 7.2. *For every pair* (g_1, g_2) *in* $S_0(\mathbb{R}^d)$, *and any TF-lattice* Λ, *the mapping from the strong symbol (multiplier)* $(\mathbf{m}(\lambda))_{\lambda \in \Lambda}$ *to the corresponding Gabor multiplier* $G_{g_1,g_2,\Lambda,\mathbf{m}}$ *maps the Gelfand triple* $(\ell^1(\Lambda), \ell^2(\Lambda), \ell^\infty(\Lambda))$ *into the bounded operators with kernel in the corresponding Gelfand triple* $\left(S_0(\mathbb{R}^d \times \widehat{\mathbb{R}}^d), L^2(\mathbb{R}^d \times \widehat{\mathbb{R}}^d), S_0'(\mathbb{R}^d \times \widehat{\mathbb{R}}^d)\right)$, *i.e., into* $(\mathcal{B}, \mathcal{HS}, \mathcal{B}')$.

In the last part of this section we summarize the mapping properties between the space of symbols and the membership of the resulting Gabor multiplier in one of the typical operator ideals within the bounded operators on the Hilbert space $L^2(\mathbb{R}^d)$. Again we fix a pair (g_1, g_2) in $S_0(\mathbb{R}^d)$, and the TF-lattice Λ.

Theorem 7.3. *Assume that* g, g_1, g_2 *are in* $S_0(\mathbb{R}^d)$. *Then one has:*

1. *If* $\mathbf{m}$ *is bounded, then* $G_{g_1,g_2,\Lambda,\mathbf{m}}$ *is a bounded operator on* $L^2(\mathbb{R}^d)$.
2. *If* $\mathbf{m}$ *is real-valued, then* $G_{g,\Lambda,\mathbf{m}}$ *is a self-adjoint operator on* $L^2(\mathbb{R}^d)$.

3. If $\mathbf{m} \in c_o(\Lambda)$, then $G_{g_1,g_2,\Lambda,\mathbf{m}}$ is a compact operator on $L^2(\mathbb{R}^d)$.

4. If $\mathbf{m} \in \ell^2(\Lambda)$, then $G_{g_1,g_2,\Lambda,\mathbf{m}}$ is a Hilbert–Schmidt operator on $L^2(\mathbb{R}^d)$.

5. If $\mathbf{m} \in \ell^1(\Lambda)$, then $G_{g_1,g_2,\Lambda,\mathbf{m}}$ is a trace-class operator on $L^2(\mathbb{R}^d)$.

Proof. Most of these statements follow from general facts about operator ideal properties of linear operators on $L^2(\mathbb{R}^d)$ with kernels in the Gelfand triple $(\mathcal{B}, \mathcal{HS}, \mathcal{B}')$. Obviously L^2-kernels correspond (exactly) to Hilbert–Schmidt operators. On the other hand the operators in $\mathcal{B}$, i.e., with S_0-kernels, are absolutely convergent sums of rank-one operators, and hence they are trace-class. Since the sequences with a finite number of non-zero coefficients generate finite rank operators, the density of such sequences in $c_o(\Lambda)$ implies (3). Relation (2) is easily verified directly and the main application of the symmetry assumption between analysis and synthesis, i.e., the choice $g_1 = g_2 = g$, is the investigation of the eigenvalue behaviour of operators with real symbols. □

Remark 7.1. The main statements of the above theorem can be summarized in the terminology of Gelfand triples by saying that for atoms $g_1, g_2 \in S_0(\mathbb{R}^d)$ the mapping $(m_\lambda)_{\lambda \in \Lambda} \mapsto G_{g_1,g_2,\Lambda,\mathbf{m}}$ maps the Gelfand triple of sequence spaces $(\ell^1(\Lambda), \ell^2(\Lambda), \ell^\infty(\Lambda))$ into the Gelfand triple of operator ideals, consisting of trace-class operators, $\mathcal{HS}$ and the class of all bounded linear operators on $L^2(\mathbb{R}^d)$.

Remark 7.2. The Gabor multipliers are special cases of the so-called *localization operators*. They have been studied by many authors, we refer the reader to [5–7] and references therein.

References

1. R. Balan, P. Casazza, C. Heil, and Z. Landau. Deficits and excesses of frames. *Adv. Comput. Math.*, 18(2-4):93–116, 2003.
2. R. Balan, P. Casazza, C. Heil, and Z. Landau. Excesses of Gabor frames. *Appl. Comput. Harmon. Anal.*, 14(2):87–106, 2003.
3. R. Balian. Un principe d'incertitude fort en théorie du signal on en mécanique quantique. *C. R. Acad. Sci. Paris*, 292:1357–1362, 1981.
4. O. Christensen and Y. C. Eldar. Oblique dual frames and shift-invariant spaces. *Appl. and Comp. Harm. Anal.*, 17(1):48–68, 2004.
5. E. Cordero and K. Gröchenig. Time-frequency analysis of Localization operators. *J. Funct. Anal.*, 205(1):107–131, 2003.
6. E. Cordero and K. Gröchenig. *Necessary conditions for Schatten class localization operators, Proc. Amer. Math. Soc.*, 133:3573–3579, 2005.
7. E. Cordero and K. Gröchenig. Symbolic Calculus and Fredholm Property for Localization Operators. *J. Fourier Anal. Appl.*, 2006.
8. S. Dahlke, M. Fornasier, and T. Raasch. Adaptive frame methods for elliptic operator equations. to appear in Adv. Comp. Math., 2005.
9. I. Daubechies, S. Jaffard, and J.L. Journé. A simple Wilson orthonormal basis with exponential decay. *SIAM J. Math. Anal.*, 22(2):554–572, 1991.

10. I. Daubechies, H. J. Landau, and Z. Landau. Gabor time–frequency lattices and the Wexler-Raz Identity. *J. Four. Anal. and Appl.*, 1(4):437–478, 1995.
11. M. Dörfler, H. G. Feichtinger, and K. Gröchenig. Time–frequency partitions for the Gelfand triple (S_0, L^2, S_0'). Submitted, 2004.
12. R. Duffin and A. Schaeffer. A class of nonharmonic Fourier series. *Trans. Amer. Math. Soc.*, 72:341–366, 1952.
13. H. G. Feichtinger, Modulation spaces on locally compact abelian groups, *Technical Report, University Vienna, 1983.* and also in *Wavelets and Their Applications*, M. Krishna, R. Radha, S. Thangavelu, editors, Allied Publishers, 99–140, 2003.
14. H. G. Feichtinger, K. Gröchenig, and D. Walnut. Wilson bases and modulation spaces. *Math. Nachrichten*, 155:7–17, 1992.
15. H. G. Feichtinger and N. Kaiblinger. Varying the time–frequency lattice of Gabor frames. *Trans. Am. Math. Soc.*, 356(5):2001–2023, 2004.
16. H. G. Feichtinger and W. Kozek. Quantization of TF–lattice invariant operators on elementary LCA groups. In H.G. Feichtinger and T. Strohmer, editors, *Gabor Analysis and Algorithms: Theory and Applications*, chapter 7, pages 233–266. Birkhäuser, Boston, 1998.
17. H. G. Feichtinger and F. Luef. Wiener amalgam spaces for the fundamental identity of Gabor analysis. In *Proc. Conf. El-Escorial, summer 2004*, El-Escorial, 2005. To appear.
18. H. G. Feichtinger, F. Luef, and T. Werther. A Guided Tour from Linear Algebra to the Foundations of Gabor Analysis. 2005.
19. H. G. Feichtinger and K. Nowak. A first survey of Gabor multipliers. In H.G. Feichtinger and T. Strohmer, editors, *Advances in Gabor Analysis*, Appl. Numer. Harmon. Anal., pages 99–128. Birkhäuser, 2003.
20. H. G. Feichtinger and G. Zimmermann. A Banach space of test functions for Gabor analysis. In H. G. Feichtinger and T. Strohmer, editors, *Gabor Analysis and Algorithms: Theory and Applications*, pages 123–170. Birkhäuser, Boston, 1998.
21. G. B. Folland. *A course in abstract harmonic analysis*. Studies in Advanced Mathematics. Boca Raton, FL: CRC Press, 1995.
22. D. Gabor. Theory of communication. *Proc. IEE (London)*, 93(III):429–457, November 1946.
23. I. M. Gel'fand and A. G. Kostyuchenko. Entwicklung nach Eigenfunktionen von Differentialoperatoren und anderen Operatoren. *Dokl. Akad. Nauk SSSR*, 103:349–352, 1955.
24. I. M. Gelfand and N. Ya. Vilenkin. *Applications of harmonic analysis*. New York and London: Academic Press. XIV, 1964.
25. K. Gröchenig. An uncertainty principle related to the Poisson summation formula. *Stud. Math.*, 121(1):87–104, 1996.
26. K. Gröchenig. *Foundations of Time–Frequency Analysis*. Birkhäuser, 2001.
27. K. Gröchenig and M. Leinert. Wiener's Lemma for twisted convolution and Gabor frames. *J. Amer. Math. Soc.*, 17(1):1–18, 2003.
28. C. Heil. A basis theory primer. Unpublished Manuscript, July 1997.
29. A. J. E. M. Janssen. Duality and biorthogonality for Weyl-Heisenberg frames. *J. Four. Anal. and Appl.*, 1(4):403–436, 1995.
30. A. J. E. M. Janssen. Classroom proof of the density theorem for Gabor systems. Unpublished Manuscript, 2005.
31. N. Kaiblinger. Approximation of the Fourier transform and the dual Gabor window. *J. Fourier Anal. Appl.*, 11(1):25–42, 2005.
32. J. G. Kirkwood. Quantum statistics of almost classical assemblies. *Phys. Rev.*, II. Ser. 44:31–37, 1933.
33. S. Li and H. Ogawa. Pseudoframes for subspaces with applications. *J. Four. Anal. and Appl.*, 10(4):409–431, 2004.
34. F. Low. Complete sets of wave packets. In C. DeTar, editor, *A Passion for Physics - Essay in Honor of Geoffrey Chew*, pages 17–22. World Scientific, Singapore, 1985.

35. Y. I. Lyubarskii. Frames in the Bargmann space of entire functions. *Adv. Soviet Math.*, 429:107–113, 1992.

36. A. Ron and Z. Shen. Weyl-Heisenberg frames and Riesz basis on $L^2(\mathbb{R}^d)$. *Duke Math. J.*, 89(2):273–282, 1997.

37. W. Rudin. *Functional Analysis*. Mc Graw-Hill, 2nd edition, 1991.

38. K. Seip and R. Wallsten. Density theorems for sampling and interpolation in the Bargmann-Fock space II. *J. reine angewandte Mathematik*, 429:107–113, 1992.

39. T. Strohmer. Numerical algorithms for discrete Gabor expansions. In H. G. Feichtinger and T. Strohmer, editors, *Gabor Analysis and Algorithms: Theory and Applications*, pages 267–294. Birkhäuser, Boston, 1998.

40. J. Wexler and S. Raz. Discrete Gabor expansions. *Signal Processing*, 21(3):207–220, 1990.

41. E. P. Wigner. On the quantum correction for thermo-dynamic equilibrium. *Phys. Rerv. Letters*, 40:749–759, 1932.

Four Lectures in Semiclassical Analysis for Non Self-Adjoint Problems with Applications to Hydrodynamic Instability

B. Helffer

Abstract Our aim is to show how semi-classical analysis can be useful in questions of stability appearing in hydrodynamics. We will emphasize on the motivating examples and see how these problems can be solved or by harmonic approximation techniques used in the semi-classical analysis of the Schrödinger operator or by recently obtained semi-classical versions of estimates for operators of principal type (mainly subelliptic estimates). These notes correspond to an extended version of the course given at the school in Cetraro. We have in particularly kept the structure of these lectures with an alternance between the motivating examples and the presentation of the theory. Many of the results which are presented have been obtained in collaboration with Olivier Lafitte.

1 General Introduction

In Hydrodynamics an important question is to analyze the stability or the instability of the solutions. This question appears at least at the first stage (analysis of the linearized problem) to be a question of spectral analysis. This question appears to depend strongly on the various physical parameters. In some asymptotics regime, this question can be analyzed by techniques coming from semi-classical analysis: this means that there is a small parameter h which plays in the analysis the role of the Planck constant in an analogous way to the Quantum Mechanics.

We will emphasize on the motivating examples and see how these problems can be solved or by harmonic approximation techniques used in the semi-classical analysis of the Schrödinger operator or by recently obtained

Bernard Helffer
Laboratoire de Mathématiques, Université Paris-Sud, 91405 Orsay Cedex, France
e-mail: bernard.helffer@math.u-psud.fr

L. Rodino, M.W. Wong (eds.) *Pseudo-Differential Operators*. Lecture Notes in Mathematics 1949.
© Springer-Verlag Berlin Heidelberg 2008

35

semi-classical versions of estimates for operators of principal type (mainly subelliptic estimates). In this way, we hope to show that these recent results are much more than academic transpositions of former theorems developed more than thirty years ago when analyzing the main properties of Partial Differential Equations: local solvability, hypoellipticity, propagation of singularities... (see Egorov [Eg], Trèves [Trev], the treatise by Hörmander [Ho3] and references therein).

Actually, we will not need at the moment the most sophisticated theorems of this theory (see the lectures by N. Lerner [Le]) but the most generic. We will give explicit proofs for the simple examples we have. They are based mainly on two tools: the semiclassical elliptic theory for h-pseudodifferential operators and the construction of WKB solutions.

We consider four different models coming from different modelizations appearing in hydrodynamics. The first one is the *Rayleigh–Taylor model*. Although the subject has a long story starting with [St] (see also [Cha]), the semi-classical analysis appears in [La1, La2, HelLaf1]. The problem we meet in this case is self-adjoint and related to the analysis of the bottom of the spectrum for a Schrödinger operator. The three other examples are not selfadjoint. We will see that we meet problems related to the notion of pseudospectrum. The second one extends the previous one by introducing some velocity at the surface between the two fluids. This is an extension of the *Kelvin–Helmholtz* classical model which is analyzed in [CCLa]. The third one, the *Rayleigh with convection model* was studied in [CCLaRa] and is a natural generalization with a convective velocity of the classical Rayleigh problem for a transition region. The fourth one is called the *Kull–Anisimov ablation front model*. It has been analyzed by many physicists and more recently in the PhD theses of L. Masse [Mas] and V. Goncharov [Go].

Finally as other relevant references, we quote [Ag], [BeSh], [Bo], [BudkoL], [ChLa], [Col], [Da1], [Da3], [GH], [He1], [HeRo1], [HeRo2], [HelSj2], [KeSu], [Kull], [Si2], [Tay].

Organization of the Course

This course is divided in four (unequal) lectures.

Lecture 1 is devoted to the analysis of the Rayleigh–Taylor model. We show how the initial problem of analyzing the possible instability of the model leads to a spectral problem for a compact selfadjoint operator which appears to be an h-pseudodifferential operator.

When needed, we will recall various basic things on the h-pseudodifferential operators.

We are let to the analysis of the largest eigenvalue of a compact operator. We show that either harmonic analysis or WKB solutions permit to have a good asymptotic of this eigenvalue.

Lecture 2 is devoted to the presentation of some mathematical tools adapted to the analysis of non-selfadjoint problems. We first start by presenting a new example (Kelvin–Helmholtz) as a motivation. We then give the main definitions related to the pseudo-spectrum. Here we will emphasize on the "elliptic" h-pseudodifferential theory and on what can be done by WKB constructions. We then apply the techniques for analyzing our Kelvin–Helmholtz model.

Lecture 3 is devoted to the presentation of the results on subellipticity in the semi-classical context. We will see how the question of the subellipticity of h-pseudodifferential operators can appear naturally. In comparison with what was done in the course of N. Lerner [Le], this will illustrate the most simple examples which were presented!

Lecture 4 explains the origin of two other models. We will show that they lead to similar questions for some suitable regimes of parameters. Again, we arrive to the analysis of a system, which can be reduced to a high order non symmetric differential operator. We then sketch the mathematical treatment of these two models. This gives us also a good opportunity for presenting other results in subellipticity mainly obtained by Dencker–Sjöstrand–Zworski.

2 Lecture 1: The Rayleigh–Taylor Model

2.1 The Rayleigh–Taylor Model: Physical Origin

The starting point for this model is the analysis of the following differential system in $\mathbb{R}^4 = \mathbb{R}^3_x \times \mathbb{R}_t$. With $x = (x_1, x_2, x_3)$ this system reads:

$$\begin{aligned} \partial_t \varrho + \mathrm{div}\,(\varrho \mathbf{u}) &= 0 \\ \partial_t(\varrho \mathbf{u}) + \nabla \cdot (\varrho \mathbf{u} \otimes \mathbf{u}) + \nabla p &= \varrho \mathbf{g} \,. \end{aligned} \tag{1}$$

The unknowns are $\mathbf{u} = (u_1, u_2, u_3)$, the density ϱ and the pressure p. We assume that $\mathbf{g} = (0, 0, 1)g$. The second line in (1) corresponds to three equations and reads more explicitly:

$$\begin{aligned} \partial_t(\varrho u_1) + \mathrm{div}\,(\varrho\, u_1\, \mathbf{u}) + \partial_1 p &= 0 \,; \\ \partial_t(\varrho u_2) + \mathrm{div}\,(\varrho\, u_2\, \mathbf{u}) + \partial_2 p &= 0 \,; \\ \partial_t(\varrho u_3) + \mathrm{div}\,(\varrho\, u_3\, \mathbf{u}) + \partial_3 p &= \varrho g \,. \end{aligned} \tag{2}$$

Here we have used the short notations:

$$\partial_t = \frac{\partial}{\partial t} \,, \ \partial_i = \frac{\partial}{\partial x_i} \text{ for } i = 1, 2, 3 \,.$$

The reader can look in the first pages of the book by P.L. Lions [Li] for the way to get these equations from the principles of conservation of mass (for the first line of (1)) and of momentum (for the second line of (1)).

This system models the so-called Rayleigh–Taylor instability, which occurs when a heavy fluid is above a light fluid in a gravity field directed from the heavy to the light fluid. We refer to Chap. X in Chandrasekhar's book [Cha] for a presentation of the theory. Here we intend to study the linear growth rate of this instability in a situation where there is a mixing region. This linear growth rate will corresponds to γ in (16) below.

We would like to analyze the linearized problem around a stationary solution (i.e. t-independent):

$$\varrho = \rho^0 \, , \ \mathbf{u} = \mathbf{u}^0 = 0 \, , \ p = p^0 \, , \tag{3}$$

where ρ^0 is assumed to depend only on x_3 and p^0 and ρ^0 are related, as imposed by the second line in (1), by:

$$\nabla p^0 = \rho^0 \mathbf{g} \, . \tag{4}$$

We assume that the perturbation $(\hat{\mathbf{u}}, \hat{p}, \hat{\rho})$ is incompressible that is satisfying:

$$\operatorname{div} \hat{\mathbf{u}} = 0 \, . \tag{5}$$

The linearized system takes the form:

$$\partial_t \hat{\rho} + (\rho^0)' \hat{u}_3 = 0 \, ; \tag{6}$$

$$\rho^0 \partial_t \hat{u}_1 + \partial_1 \hat{p} = 0 \, ; \tag{7}$$

$$\rho^0 \partial_t \hat{u}_2 + \partial_2 \hat{p} = 0 \, ; \tag{8}$$

$$\rho^0 \partial_t \hat{u}_3 + \partial_3 \hat{p} = g\hat{\rho} \, . \tag{9}$$

In order to analyze (at least formally this system) we extract from the system an equation involving only $\hat{u}_3$ (by eliminating the other unknowns). This is done along the following lines.
We first differentiate with respect to t (9). This leads to:

$$\rho^0 \partial_t^2 \hat{u}_3 + \partial_t \partial_3 \hat{p} = g \frac{\partial \hat{\rho}}{\partial t} \, . \tag{10}$$

We now use (6) in order to eliminate $\frac{\partial \hat{\rho}}{\partial t}$ and get:

$$\rho^0 \partial_t^2 \hat{u}_3 + \partial_t \partial_3 \hat{p} + g(\rho^0)'(x_3)\hat{u}_3 = 0 \, . \tag{11}$$

We now differentiate (7) and (8) respectively with respect to x_1 and x_2. This gives:

$$\rho^0 \partial_t \partial_1 \hat{u}_1 + \partial_1^2 \hat{p} = 0 \, , \tag{12}$$

and

$$\rho^0 \partial_t \partial_2 \hat{u}_2 + \partial_2^2 \hat{p} = 0 \, . \tag{13}$$

Differentiating (5) with respect to t and using (12) and (13), we get:

$$\Delta_{12}\,\hat{p} = \rho^0 \partial_t \partial_3 \hat{u}_3 \,, \tag{14}$$

where Δ_{12} is the Laplacian with respect to the two first variables (x_1, x_2):

$$\Delta_{12} = \partial_1^2 + \partial_2^2 \,.$$

It remains to eliminate $\hat{p}$ between (11) and (14):

$$\Delta_{12}\left(\rho^0 \partial_t^2 \hat{u}_3 + (\rho^0)' g \hat{u}_3\right) + \partial_3 \rho^0 \partial_3 \partial_t^2 \hat{u}_3 = 0 \,. \tag{15}$$

We now look for a solution $\hat{u}_3$ in the form:

$$\mathbb{R}^3 \times \mathbb{R} \ni (x, t) \mapsto \hat{u}_3(x_1, x_2, x_3, t) = v(x_3) \exp(\gamma t + ik_1 x_1 + ik_2 x_2) \,, \tag{16}$$

where:

- v is an unknown real function in $L^2(\mathbb{R})$.
- γ is a real parameter.
- (k_1, k_2) is in $\mathbb{R}^2$ and corresponds to the momentum variables dual to (x_1, x_2).

This is what is called in the physical literature the analysis in normal modes. The reader can for example look in the introductory chapter of [Cha] for a more heuristic explanation. This leads to an ordinary differential equation (in the x_3-variable) for v:

$$-(k_1^2 + k_2^2)(\rho^0 \gamma^2 v + (\rho^0)' g v) + \gamma^2 \frac{d}{dx_3} \rho^0 \frac{d}{dx_3} v = 0 \,. \tag{17}$$

Replacing x_3 by x $(x \in \mathbb{R})$ and dividing by $\gamma^2 k^2$ with

$$k^2 = k_1^2 + k_2^2 \,,$$

we get:

$$\left[-\frac{1}{k^2} \frac{d}{dx} \rho^0 \frac{d}{dx} + \rho^0 + (\rho^0)' \frac{g}{\gamma^2} \right] v = 0 \,. \tag{18}$$

So we are interested in analyzing for which value of (γ, k) (with $\gamma > 0$) there exists a non trivial v satisfying (17).

The choice of $\gamma > 0$ corresponds to our interest for instability. Actually, we could have started by looking at possibly complex γ's but one immediately get as a necessary condition that γ^2 should be real and the pure imaginary γ's are not interesting for the problem.

2.2 Rayleigh–Taylor Mathematically

In the case of the Rayleigh–Taylor model, as we have seen in (18), the main point is to analyze as a function of $\delta \in \mathbb{R}$ the kernel in $L^2(\mathbb{R})$ of:

$$P(h, \delta) := -h^2 \frac{d}{dx} \varrho(x) \frac{d}{dx} + \varrho(x) + \delta \varrho'(x) . \tag{19}$$

Here $h > 0$ and $\varrho(x) \in C^\infty(\mathbb{R})$ satisfies:

$$\begin{aligned} \lim_{x \to -\infty} \varrho(x) = \rho_- > 0 , \\ \lim_{x \to +\infty} \varrho(x) = \rho_+ > 0 , \end{aligned} \tag{20}$$

$$\varrho(x) > 0 , \ \forall x \in \mathbb{R} , \tag{21}$$

$$\rho_- \neq \rho_+ , \tag{22}$$

$$\lim_{|x| \to +\infty} \varrho'(x) = 0 . \tag{23}$$

We look at $h \to 0$ (see[1] [HelLaf1] for the case $h \to +\infty$). The problem comes from the analysis of the Euler equations in a gravity field. The physical parameters are the intensity g of the gravity, a wave number $k > 0$ and a parameter γ which measures the large time behavior of the solution. The mathematical problem is to determine a pair (u, γ) such that

$$P\left(\frac{1}{k}, \frac{g}{\gamma^2}\right) u = 0 \qquad \text{with } u \text{ non trivial}$$

This means that the link between the physical parameters (g, k, γ) and the mathematical parameters is:

$$\delta = \frac{g}{\gamma^2} , \ h = \frac{1}{k} . \tag{24}$$

The physical situation leads to analyze the case $\delta g > 0$. This implies $\gamma^2 > 0$, and we choose $\gamma > 0$.

Note that the instability is only analyzed when

$$\rho_+ \neq \rho_- .$$

This implies that $\varrho'(x)$ is not identically 0.

The most physical case corresponds to:

$$\rho_- > \rho_+ , \ g > 0 ,$$

so δ is positive and ϱ' is negative somewhere.

---

[1] In this case the limiting model corresponds to $\rho = \rho_-$ for $x < 0$ and $\rho = \rho_+$ for $x > 0$.

Generally ϱ is assumed monotone but the semi-classical techniques are not limited to this case.

2.3 Elementary Spectral Theory

First we observe that there is no problem for defining the selfadjoint extension of $P(h, \delta)$ in $L^2(\mathbb{R})$ (which is unique starting from $C_0^\infty(\mathbb{R})$) and it is immediate that $P(h, 0)$ is injective. More precisely, the bottom of its spectrum is strictly positive.

Definition 2.1. We call generalized spectrum of the family $P(h, \delta)$ the set of the δ's in $\mathbb{R}$ such that $P(h, \delta)$ is non injective.

The standard analysis of the solution at ∞ for ordinary differential equations shows that, for all δ, the dimension of $\ker P(h, \delta)$ is zero or one.

The next result is relatively well known (connected to the Birman–Schwinger principle [Si1]).

Proposition 2.1. *Under the previous assumptions and assuming in addition that ϱ' is not identically 0, then the generalized spectrum $P(h, \delta)$ is the union of two sequences (possibly empty or finite) δ_n^+ et δ_n^- s.t.:*

$$0 < \delta_n^+ < \delta_{n+1}^+ ,$$
$$\lim_{n \to +\infty} \delta_n^+ = +\infty , \tag{25}$$

$$0 < -\delta_n^- < -\delta_{n+1}^- ,$$
$$\lim_{n \to +\infty} \delta_n^- = -\infty . \tag{26}$$

Proof. If we observe that:

$$\ker P(h, \delta) \neq \{0\} \text{ iff } \ker(K(h) - \frac{1}{\delta}) \neq \{0\} , \tag{27}$$

where

$$K(h) = -P(h, 0)^{-\frac{1}{2}} \varrho'(x) P(h, 0)^{-\frac{1}{2}} . \tag{28}$$

the proof is immediately reduced to the standard result for $K(h)$, which is a compact selfadjoint operator.

For the compactness of $K(h)$, we can for example observe that the operator $P(h, 0)^{-\frac{1}{2}}$ belongs to $\mathcal{L}(L^2(\mathbb{R}); H^1(\mathbb{R}))$ and that, under Assumption (23), the operator of multiplication by ρ' is compact from $H^1(\mathbb{R})$ in $L^2(\mathbb{R})$.

Note that when $\varrho' < 0$, which is the simplest natural physical case, the operator $K(h)$ is positive.

Let us also mention an a priori "universal" estimate of [CCLaRa]. If u is, for some $\delta \neq 0$, in the kernel of $P(h, \delta)$, we get by taking the scalar product in L^2 by u:

$$\int_{-\infty}^{+\infty} \varrho(h^2 u'(x)^2 + u(x)^2)\,dx = -\delta \int_{-\infty}^{+\infty} \varrho'(x)u(x)^2\,dx$$
$$= 2\delta \int_{-\infty}^{+\infty} \varrho u(x)u'(x)\,dx\,. \tag{29}$$

Using Cauchy–Schwarz, we get:

$$\int_{-\infty}^{+\infty} \varrho(x)(1 - \frac{|\delta|}{h})(u'(x)^2 + u(x)^2)\,dx \le 0\,. \tag{30}$$

This implies

$$\ker P(h,\delta) = \{0\}\,,\ \forall \delta \in\,] - h, h[\,. \tag{31}$$

Universal upper bound
We could have started from the operator:

$$-h^2 \varrho^{-\frac{1}{2}} \frac{d}{dx} \varrho \frac{d}{dx} \varrho^{-\frac{1}{2}} + 1 + \delta \frac{\varrho'(x)}{\varrho(x)}\,,$$

which shows more clearly the role of the function ϱ'/ϱ.
One way is to change of functions introducing

$$u = \varrho(x)^{-\frac{1}{2}} v\,.$$

This shows also that if:

$$1 + \delta \frac{\varrho'(x)}{\varrho(x)} > 0\,,\ \forall x \in \mathbb{R}\,, \tag{32}$$

then δ is not in the generalized spectrum.

Remark 2.1. The theory can be extended to the cases $\rho_+ = 0$ or $\rho_- = 0$, under Condition (34).

2.4 A Crash Course on h-Pseudodifferential Operators

At least if the profile ϱ is regular, the h-pseudodifferential calculus gives an easy way for getting the extremal eigenvalues of $K(h)$ in the semi-classical limit. Let us briefly describe this tool.

A family $(h \in\,]0, h_0])$ of h-pseudodifferential operators

$$A_h = \mathrm{Op}_h(a)\,,$$

associated to a symbol $(x, \xi) \mapsto a(x, \xi; h)$ is defined for $u \in \mathcal{S}(\mathbb{R}^m)$ by:

$$(\mathrm{Op}_h(a)u)(x) =$$

$$(2\pi h)^{-m} \int_{\mathbb{R}^m \times \mathbb{R}^m} \exp(\tfrac{i}{h}(x - y) \cdot \xi)\, a(\tfrac{x+y}{2}, \xi; h)\, u(y)\,dy d\xi\,. \tag{33}$$

The function a is called the Weyl symbol (or h-Weyl symbol if we want to recall the dependence on h) of A_h. We refer to the book of D. Robert [Rob] for a course on this theory which is specifically semi-classical (and to the course of N. Lerner [Le] in this volume[2]) and the assumptions which can be done on the symbols.

Here it is enough to consider as symbols C^∞ (with respect to the variables $(x, \xi) \in \mathbb{R}^n \times \mathbb{R}^n$) functions a s.t., for some given p, p', q and $h_0 > 0$, there exists, for all α and β in $\mathbb{N}^m$, constants $C_{\alpha,\beta}$ s.t., for all $h \in]0, h_0]$,

$$|D_x^\alpha D_\xi^\beta a(x, \xi; h)| \leq C_{\alpha,\beta} \, h^q \, \langle x \rangle^{p-|\alpha|} \langle \xi \rangle^{p'-|\beta|} \, .$$

When the symbol satisfies this condition, we write simply $a \in S^{(q,p,p')}$, and the corresponding operator $\mathrm{Op}_h(a)$ is said to belong to $\mathrm{Op}_h S^{(q,p,p')}$.

This class is an algebra by composition and the composition is just a multiplication for the principal symbols. Typically, if $a \in S^{(q,p,p')}$ and $b \in S^{(q_1,p_1,p_1')}$, then there exists c in $S^{(q+q_1,p+p_1,p'+p_1')}$ s.t.:

$$\mathrm{Op}_h(a) \circ \mathrm{Op}_h(b) = \mathrm{Op}_h(c) \, ,$$

and

$$c - ab \in S^{(q+q_1+1,p+p_1-1,p'+p_1'-1,)} \, .$$

This leads to the natural definition of "principal symbol". In the current situation, the symbol $a \in S^{q,p,p'}$ has more properties. It admits the formal expansion:

$$a(x, \xi; h) \sim h^q \sum_{j \geq 0} h^j a_j(x, \xi) \, ,$$

with:

$$a_j(x, \xi) \in S^{0,-j,-j} \, ,$$

and one has, for any $N > 0$, a good control of the remainders

$$r^N(x, \xi, h) := a(x, \xi; h) - h^q \sum_{0 \leq j \leq N} h^j a_j(x, \xi) \, ,$$

in $S^{(q-N,-N,-N)}$.

The symbol $a_0(x, \xi)$ is called the principal symbol. The symbol $a_1(x, \xi)$ is called the subprincipal symbol. We note that the principal symbol is independent of the quantization (this is not the case for the subprincipal symbol).

We have natural continuity theorems (based on the Calderon–Vaillancourt Theorem) in $H^s(\mathbb{R}^m)$, where moreover the constants are controlled with respect to h.

In addition compact operators on $L^2(\mathbb{R}^n)$ can be recognized as the operators whose symbol in $S^{(0,0,0)}$ tends to 0 as $|x| + |\xi| \to +\infty$.

---

[2] N. Lerner has a slightly different convention for the quantization. But taking $h = \frac{1}{2\pi}$ in (33) leads to this convention.

Typically, an operator in $\operatorname{Op}_h S^{(q,p,p')}$ with $p < 0$ and $p' < 0$ is compact. The role of q is to give the size of the norm of the operator with respect to h.

Finally, let us briefly discuss invertibility. As the principal symbol of an operator (sat in $\operatorname{Op}_h(S^{0,0,0})$), is invertible (=elliptic), one can inverse the operator for h small enough. This is indeed very simple. If B_h is the operator of h-Weyl symbol $\frac{1}{a_0}$, then the calculus gives that:

$$B_h \, A_h = I + h R_h$$

with $R_h \in \operatorname{Op}_h(S^{(0,-1,-1)})$.

Then the uniform control in $\mathcal{L}(L^2)$ of R_h gives the invertibility of $(I+hR_h)$ in $\mathcal{L}(L^2)$ and hence the invertibility of A_h. For the invertibility, modulo $\mathcal{O}(h^\infty)$, one can also inverse $(I + hR_h)$ by using the Neumann series:

$$(I + h R_h)^{-1} \sim \sum_{j \geq 0} (-1)^j h^j (R_h)^j \, .$$

2.5 Application for Rayleigh–Taylor: Semi-Classical Analysis for $K(h)$

Under strong assumptions on ϱ, one can use the previous h- pseudodifferential calculus. We assume:

$$|D_x^\alpha \varrho(x)| \leq C_\alpha \varrho(x) \langle x \rangle^{-|\alpha|} \, . \tag{34}$$

This assumption permits to see that:

$$K(h) = -(-h^2 \frac{d}{dx} \varrho \frac{d}{dx} + \varrho)^{-\frac{1}{2}} \varrho'(x)(-h^2 \frac{d}{dx} \varrho \frac{d}{dx} + \varrho)^{-\frac{1}{2}} \tag{35}$$

is an h-pseudodifferential operator. More precisely it belongs to $\operatorname{Op}_h S^{(0,0,0)}$. The operator $K(h)$ appears indeed as the composition of three h-pseudo-differential operators $(-h^2 \frac{d}{dx} \rho \frac{d}{dx} + \rho)^{-\frac{1}{2}}$, $-\rho'(x)$ and again $(-h^2 \frac{d}{dx} \rho \frac{d}{dx} + \rho)^{-\frac{1}{2}}$.

So the h-pseudodifferential calculus gives that it is an h-pseudodifferential operator.

The principal symbol of $K(h)$ is

$$(x,\xi) \mapsto p(x,\xi) = -(\xi^2 + 1)^{-1} \frac{\varrho'(x)}{\varrho(x)} \, . \tag{36}$$

For the analysis of the extremal eigenvalues, we have first to determine the extrema of this symbol. If these extrema are non degenerate then we can apply the harmonic approximation as in [HelSj1]. The tunneling effect together with the decay of the eigenfunctions can also be analyzed (see [BrHe], [HePa]). There is indeed a natural extension of Agmon Estimates for h-pseudodifferential operators whose symbol admit an holomorphic extension in suitable bands $|\operatorname{Im}\xi| \leq R$ in the ξ variable.

This leads to the following computations. We get

$$\frac{\partial p}{\partial \xi}(x,\xi) = 2\xi \frac{\varrho'(x)}{\varrho(x)}(\xi^2 + 1)^{-2} ,$$

$$\frac{\partial p}{\partial x}(x,\xi) = -(\varrho''(x)\varrho(x) - \varrho'(x)^2)(\varrho(x))^{-2}(\xi^2 + 1)^{-1} .$$

The condition $\varrho'(x) = 0$ should be excluded because it does not correspond to an extremum of $p(x,\xi)$. So we get:

$$\xi = 0 \; ; \; \varrho''(x)\varrho(x) - \varrho'(x)^2 = 0 .$$

This corresponds to the condition that x_0 is a critical point of the map $x \mapsto -\varrho'(x)/\varrho(x)$.

It remains to verify that the extrema are non degenerate. We obtain at a critical point $(x_0, 0)$:

$$\frac{\partial^2 p}{\partial \xi^2}(x_0, 0) = +2\varrho'(x_0)/\varrho(x_0)$$

$$\frac{\partial^2 p}{\partial \xi \partial x}(x_0, 0) = 0$$

$$\frac{\partial^2 p}{\partial x^2}(x_0, 0) = -\frac{\varrho'''(x_0)\varrho(x_0) - \varrho'(x_0)\varrho''(x_0)}{\varrho(x_0)^2}$$

It is then easy to determine if $(x_0, 0)$ corresponds to:

- A minimum of p,
  if $\varrho'(x_0)/\varrho(x_0) > 0$
  and $\varrho'''(x_0)\varrho(x_0) - \varrho'(x_0)\varrho''(x_0)) < 0$.
- A maximum of p,
  if $\varrho'(x_0)/\varrho(x_0) < 0$
  and $\varrho'''(x_0)\varrho(x_0) - \varrho'(x_0)\varrho''(x_0) > 0$.

When $\rho' < 0$ and $\rho > 0$, then the maxima of the symbol correspond to $\xi = 0$ and to the x's such that $-\frac{\varrho'}{\rho}$ is maximal.

We recall that the simplest physical situation corresponds to $\varrho'(x) < 0$. In this case we have only maxima, which actually are the points of interest if looking for largest eigenvalue.

2.6 Harmonic Approximation

If we are interested in the largest eigenvalue of $K(h)$ a very general theory has been developed (of course for Schrödinger, but also for more general h-pseudodifferential operators).

We just sketch what corresponds to the first approximation. We have just to consider the following harmonic operator associated to a point $(x_0, 0)$ corresponding to a maximum of p, and to consider the spectrum of

$$p(x_0, 0) + h \left(\frac{1}{2} \frac{\partial^2 p}{\partial \xi^2}(x_0, 0) D_y^2 + \frac{1}{2} \frac{\partial^2 p}{\partial x^2}(x_0, 0) y^2 \right) + h p_1(x_0, 0) \,,$$

where p_1 is the subprincipal Weyl symbol of $K(h)$, which actually is 0.

This operator is consequently

$$-\frac{\varrho'(x_0)}{\varrho(x_0}(1 - h D_y^2) - h \frac{\varrho'''(x_0)\varrho(x_0) - \varrho'(x_0)\varrho''(x_0)}{2\varrho(x_0)^2} y^2$$

The largest eigenvalue of this operator (which is semi-bounded from above!) is explicitly known and gives the existence of an eigenvalue for $K(h)$ (with some error $\mathcal{O}(h^{\frac{3}{2}})$).

If there are more than one critical maximum point for p, the largest eigenvalue of $K(h)$ is well approximated by the largest (over the maxima of p) of the largest eigenvalue of the approximating harmonic oscillators.

2.7 Instability of Rayleigh–Taylor: An Elementary Approach via WKB Constructions

We present here what simple constructions of WKB solutions can give for the model of Rayleigh–Taylor. A very detailed analysis have been given in [HelLaf1] extending previous works by Cherfils, Lafitte, Raviart [CCLaRa]. Here we present a simpler analysis but this will only give conditions under which one can construct approximate solutions in the kernel of $P(h, \delta)$.

In the semi-classical situation, we look for a solution in the form

$$u(x, h) = a(x, h) \exp -\frac{\varphi(x)}{h} \tag{37}$$

near some point x_0 (to be determined!) with

$$a(x, h) \sim \sum_{j \geq 0} h^j a_j(x) \,, \tag{38}$$

$$\delta(h) \sim \sum_j h^j \delta_j \tag{39}$$

such that

$$\exp \frac{\varphi}{h} \cdot P(h, \delta(h)) \cdot u(h) \sim 0 \,. \tag{40}$$

Here "~ 0" means that the right-hand side should be $\mathcal{O}(h^\infty)$.

Concretely, we expand $\exp \frac{\varphi}{h} \cdot P(h, \delta(h)) \cdot u(h)$ in powers of h and express the cancellation of each coefficient of h^j.

We get as first eikonal equation

$$-\varrho(x)\varphi'(x)^2 + \varrho(x) + \delta_0\varrho'(x) = 0 . \tag{41}$$

In order to have an (exponentially) localized (as $h \to 0$) in a neighborhood of x_0, it is natural to impose the condition that φ admits a minimum at x_0. So the first condition is:

$$\varphi'(x_0) = 0 . \tag{42}$$

This leads as a first necessary condition to

$$\varrho(x_0) + \delta_0\varrho'(x_0) = 0 . \tag{43}$$

A second necessary condition is obtained by differentiating the eikonal equation:

$$-\varrho'(x)\varphi'(x)^2 - 2\varrho(x)\varphi'(x)\varphi''(x) + \varrho'(x) + \delta_0\varrho''(x) = 0 .$$

This gives at x_0:

$$\varrho'(x_0) + \delta_0\varrho''(x_0) = 0 . \tag{44}$$

We are asking for a non-degenerate minimum of φ at x_0. Differentiating two times the eikonal equation, we obtain:

$$-2\varrho(x_0)(\varphi''(x_0))^2 + \varrho''(x_0) + \delta_0\varrho'''(x_0) = 0 \tag{45}$$

which implies

$$\varrho''(x_0) + \delta_0\varrho'''(x_0) > 0 . \tag{46}$$

We recover the condition obtained in the previous analysis.

Till now, we just looked for a phase. The next step is to determine the amplitude. The coefficient δ_1 will be determined by looking at the first transport equation:

$$\begin{aligned} 2\varrho(x)\varphi'(x)a_0'(x) + \varrho'(x)\varphi'(x)a_0(x) \\ +\varrho(x)\varphi''(x)a_0(x) + \delta_1\varrho'(x)a_0(x) = 0 . \end{aligned} \tag{47}$$

If we impose the condition[3]

$$a_0(x_0) = 1 ,$$

a necessary (and actually sufficient) condition for solving is:

$$\varrho(x_0)\varphi''(x_0) + \delta_1\varrho'(x_0) = 0 . \tag{48}$$

We then obtain a_0 by simple integration:

$$a_0'(x)/a_0(x) = (\varrho'(x)\varphi'(x) + \varrho\varphi''(x) + \delta_1\varrho'(x)) / (2\varrho(x)\varphi'(x)) .$$

---

[3] This condition corresponds to the idea that we look for the ground state, hence non vanishing.

The condition (48) permits indeed to extend the right-hand side as a C^∞ function and we get explicitly:

$$a_0(x) = \exp \int_{x_0}^{x} (\varrho'(\tau)\varphi'(\tau) + \varrho\varphi''(\tau) + \delta_1\varrho'(\tau)) \,/\, (2\varrho(\tau)\varphi'(\tau)) \; d\tau \;. \quad (49)$$

It is then not difficult to iterate at any order the construction: At each step the cancellation of the coefficient of h^j in the expansion of $\exp \frac{\varphi}{h} \cdot P(h, \delta(h)) \cdot u(h)$ permits to determine δ_j and to find $a_{j-1}(x)$, with, for $j \geq 2$, the initial condition

$$a_{j-1}(x_0) = 0 \;.$$

We have now constructed a formal solution. Let us recall now how one can associate to this formal expansion an explicit realization. The first idea is to consider a finite sum. We let $\delta^N(h) = \sum_{j=0}^{N} \delta_j h^j$ and introduce $a^N(x, h) = \sum_{j=0}^{N} h^j a_j(x)$ which is well defined in the neighborhood of x_0.

We then introduce a cut-off χ which localizes in a neighborhood of x_0. We then let

$$u_\chi^N(x, h) = \chi(x)a^N(x, h) \exp -(\varphi(x)/h) \;.$$

Computing $P(h, \delta^N(h))u_\chi^N(x, h)$, we find:

$$\begin{aligned}
&P(h, \delta^N(h))u_\chi^N(x, h) \\
&= (\chi h^N r_N(x, h) + \tilde{\chi}(x)b_0(x, h)) \exp -\tfrac{\varphi(x)}{h} \;,
\end{aligned} \quad (50)$$

where $\tilde{\chi}$ is C^∞, with a support disjoint of x_0. Here it is important to observe that $\exp -\frac{\varphi(x)}{h}$ is exponentially small on the support of $\tilde{\chi}$ (here we have used that φ has a local minimum at x_0).

What can we deduce from this construction? Under the previous assumptions, $P(h, \delta)$ is selfadjoint and we can deduce that, in an interval $]{-}Ch^N, +Ch^N[$, the spectrum of $P(h, \delta^N(h))$ is not empty for h small enough. Assumption (20) permits also to say that near 0 the spectrum is discrete.

This is not the complete answer to our question. But this strongly suggests the existence, close to $\delta^N(h))$ (modulo $\mathcal{O}(h^N)$) of an effective $\delta(h)$ such that $P(h, \delta(h)$ has a non zero kernel. Note that the answer to this last question is easier when ϱ is strictly monotone. Note that the question is more delicate as for example $\rho_- = 0$. The essential spectrum of $P(h, \delta)$ contains indeed 0. The previous analysis (see [HelLaf1]) based on the h-pseudodifferential calculus avoids this difficulty (finally artificial) if $\frac{\varrho'}{\varrho} \to 0$ at ∞ and if ϱ is regular.

Three remarks for ending this first lecture:

- One can take $N = +\infty$ by using a summation procedure à la Borel.
  The Borel Lemma says that for a given sequence of reals α_n $(n \in \mathbb{N}))$ one can always find a C^∞ function $h \mapsto f(h)$ admitting $\sum_n \alpha_n h^n$ as Taylor expansion at 0.

Here we need a version with parameters, but we can define some realizations of $\sum_{j=0}^{\infty} \delta_j h^j$ and $\sum_{j=0}^{\infty} h^j a_j(x)$, permitting to replace the remainder $\mathcal{O}(h^N)$ by $\mathcal{O}(h^\infty)$.

- With more work, one can also hope a result in the analytic category by using the notion of analytic symbol introduced by J. Sjöstrand.

  We should assume in this case that the function $x \mapsto \varrho(x)$ is analytic.

  We warn the reader that this does not mean that the above formal sums become convergent. This simply means that one can prove that, in a fixed complex neighborhood of x_0, $|a_j(x)|$ is bounded by $C^{j+1}j!$ and that we have similar estimates for the sequence $(\delta_j)_{j \in \mathbb{N}}$ (cf. the works by J. Sjöstrand [Sj1], Helffer–Sjöstrand [HelSj1], Klein–Schwarz [KlSc90]). This simply means that, by a "finite" tricky summation ($N(h) = \frac{C_0}{h}$ depending on h), one gets the existence of $\epsilon_0 > 0$, such that:

$$P(h, \delta^{N(h)}(h))u_\chi^N(x, h) = \mathcal{O}(\exp -\frac{\epsilon_0}{h}) \exp -\frac{\varphi(x)}{h} . \qquad (51)$$

- Here we have used the self-adjointness property for getting information on the spectrum. We will now see in the next lecture that for more complicate models the selfadjoint character of the problem disappears.

3 Lecture 2: Towards Non Self-Adjoint Models

3.1 Instability for Kelvin–Helmholtz I: Physical Origin

As a motivation, we will start with a generalization of the Kelvin–Helmholtz model. We refer to Chap. XI in Chandrasekhar's book [Cha] for a complete exposition of the origin of the model. This is a generalization [CCLa] of the classical Kelvin–Helmholtz instability which appears when two fluids move with different parallel velocities on each side of an interface.

When linearizing along the stationary solution $(\varrho_0, \mathbf{u_0}, p_0)$ for a given density ϱ_0 and a given (this time not zero) velocity u_0 (see (3)), where u_0 is the first component of $\mathbf{u_0}$, and following what we have done for Rayleigh–Taylor, we get the following one-dimensional question.

Can we analyze in function of the parameters (k_1, k_2, g, γ, k) with $k^2 = k_1^2 + k_2^2$, if the operator

$$\mathcal{P}_{KH}(\gamma, k_1, k_2, g) :=$$
$$-\frac{d}{dx} \varrho_0 \frac{d}{dx} (\gamma + ik_1 u_0(x))^2 + k^2 \varrho_0(x)(\gamma + ik_1 u_0(x))^2$$
$$-ik_1(\gamma + ik_1 u_0(x)) \frac{d}{dx} \varrho_0 u_0'(x) + gk^2 \varrho_0'((x)$$

is approximately injective (say for large values of k).

Like in Rayleigh–Taylor which corresponds to $k_1 = 0$ (or actually to $u_0 = 0$), our semi-classical parameter will be $h = \frac{1}{k}$. The parameter $\gamma = \Gamma_0 + i\Gamma_1$ is not necessarily real but we are interested in approximate null solutions for which Γ_0 is as large as possible (or complementarily) to show that Γ_0 should necessarily remain bounded in the regime k large.

So we divide by k^2 in the equation above and meet the following semi-classical operator:

$$
\begin{aligned}
\mathcal{P}_k(x, hD_x) := \ & -h\tfrac{d}{dx}\varrho_0 h\tfrac{d}{dx}(\gamma + ik_1 u_0(x))^2 \\
& +\varrho_0(x)(\gamma + ik_1 u_0(x))^2 \\
& -ihk_1(\gamma + ik_1 u_0(x))h\tfrac{d}{dx}\varrho_0 u_0'(x) \\
& +g\varrho_0'(x) .
\end{aligned}
$$

So in this regime k_1 is fixed such that $|k_1| \leq k = \frac{1}{h}$. This last inequality will not be a restriction in the semi-classical regime.

Semi-classically, the principal symbol is given by

$$
p_0(x, \xi) := \varrho_0(1 + \xi^2)(\gamma + ik_1 u_0(x))^2 + g\varrho_0'(x) . \tag{52}
$$

This symbol is not real, hence the associated operator is clearly not symmetric and cannot be extended as a selfadjoint operator. Our aim is to describe a rather systematic strategy for constructing approximate null solutions or to decide that we can not construct such solutions. This question is naturally related to the notion of pseudo-spectra for families (depending in particular on h but also on other parameters) and adapted to the analysis of h-pseudodifferential operators. This is what we will explain now before to treat the various physical examples including this one.

3.2 Around the ϵ-Pseudo-Spectrum

Definition 3.1. If A is a closed operator with dense domain $D(A)$ in an Hilbert space $\mathcal{H}$, the ϵ-pseudospectrum $\sigma_\epsilon(A)$ of A is defined by

$$
\sigma_\epsilon(A) := \{z \in \mathbb{C} \mid \|(zI - A)^{-1}\| \geq \tfrac{1}{\epsilon}\}.
$$

We take the convention that $\|(zI - A)^{-1}\| = +\infty$ if $z \in \sigma(A)$, where $\sigma(A)$ denotes the spectrum of A, so it is clear that we always have:

$$
\sigma(A) \subset \sigma_\epsilon(A) .
$$

When A is selfadjoint (or more generally normal), $\sigma_\epsilon(A)$ satisfies, by the Spectral Theorem

$$
\sigma_\epsilon(A) = \{z \in \mathbb{C} \mid d(z, \sigma(A)) \leq \epsilon\} .
$$

So this is only in the case of non self-adjoint operators that this new concept (first appearing in numerical analysis, see Trefethen [Tref]) becomes interesting.

Although formulated in a rather abstract way, the following result by Roch–Silbermann [RoSi] explains rather well to what corresponds the pseudo-spectrum

$$\sigma_\epsilon(A) = \bigcup_{\{\delta A \in \mathcal{L}(\mathcal{H}) \text{ s. t. } ||\delta A||_{\mathcal{L}(\mathcal{H})} \leq \epsilon\}} \sigma(A + \delta A) \,.$$

In other words, z is in the ϵ-pseudo-spectrum of A if z is in the spectrum of some perturbation $A + \delta A$ of A with $||\delta A|| \leq \epsilon$. This is indeed a natural notion thinking of the fact that the models we are analyzing are only approximations of the real problem and of the fact that the numerical analysis of the model goes through the analysis of explicitly computable approximate problems.

3.3 Around the h-Family-Pseudospectrum

We are mainly interested in the semiclassical version of this concept attached to a family (indexed by $h \in]0, h_0]$) of operators A_h. Here we are inspired by various presentations of the subject including [Sj2], [DeSjZw] and [Pra3], without to necessary follow their terminology.

For a given $\mu \geq 0$, the h-family-pseudospectrum of index μ of a family A_h (indexed by $h \in]0, h_0]$) (of closed operators with a dense domain $D(A_h)$ in a fixed Hilbert Space $\mathcal{H}$) is defined by

$$\begin{aligned} &\Psi_\mu((A_h)) \\ &:= \{z \in \mathbb{C} \mid \forall C > 0, \forall h_0 > 0 \, s.t. \, \exists h \in]0, h_0], \\ &\qquad ||(A_h - z)^{-1}|| \geq C \, h^{-\mu}\} \,. \end{aligned} \qquad (53)$$

We can then define

$$\Psi_\infty((A_h)) = \bigcap_{\mu \geq 0} \Psi_\mu((A_h)) \,. \qquad (54)$$

May be it is easier to understand the quantifiers by observing that the h-family pseudoresolvent set corresponds to the z such that $\exists C > 0$ and $h_0 > 0$ such that $\forall h \in]0, h_0]$

$$||(A_h - z)^{-1}|| \leq C \, h^{-\mu} \,.$$

If one thinks of applications to Physics, these concepts are more stable by perturbation than the corresponding notion of spectrum and they are for this reason particularly relevant in the non self-adjoint case. Practically, one will exhibit the existence of this h-family-pseudo-spectrum by constructing quasimodes or approximate solutions. This leads to another natural definition.

For a given $\mu \geq 0$, the h-family-quasispectrum of index μ of the family A_h is defined by

$$
\begin{aligned}
\psi_\mu((A_h)) \\
:= \{z \in \mathbb{C} \mid \forall C > 0, \forall h_0 > 0 \, s.t. \, \exists h \in]0, h_0], \\
\exists u_h \in D(A_h) \setminus \{0\} \, s.t. \\
\|(A_h - z)u_h\| \leq C \, h^\mu \|u_h\|\} \, .
\end{aligned}
\tag{55}
$$

We can then define

$$
\psi_\infty((A_h)) = \bigcap_{\mu \geq 0} \psi_\mu((A_h)) \, .
\tag{56}
$$

The main point is then that

$$
\psi_\mu((A_h)) \subset \Psi_\mu((A_h)) \, .
$$

Note that the converse is not true (see the discussion in [Pra3]) in general.

We will be particularly interested in using these tools when A_h is actually an h-pseudodifferential operator.

The elliptic theory (with suitable conditions at ∞) for h-pseudodifferential operators says for example that

Proposition 3.1 (see the book of D. Robert).
If $z \notin \Sigma(p)$, where

$$
\Sigma(p) := \{\lambda \in \mathbb{C} \, , \mid \exists (x_n, \xi_n) \, s.t. \, \lambda = \lim_{n \to +\infty} p(x_n, \xi_n)\} \, ,
\tag{57}
$$

then $z \notin \Psi_\mu(\, Op_h(p))$.

This will actually also be true for any $A_h = \mathrm{Op}_h(p_h)$, for which the principal symbol of A_h is p.

The proof is very easy once an h-pseudodifferential calculus has been constructed. It is enough to use $\mathrm{Op}_h((p-z)^{-1})$ as first approximate inverse and then to use a Neumann series. The reader can look at the end of Sect. 2.4 for more details.

So the first natural thing to do when analyzing the h-pseudospectrum of the family is to analyze the numerical range $\Sigma(p)$ of its principal symbol.

3.4 The Davies Example by Hand

We present a variant of the proof of the generalization, by Pravda-Starov [Pra1], of the Davies result on the h-family pseudo-spectrum for the Schrödinger operator

$$
A_h := -h^2 \frac{d^2}{dx^2} + V(x) \, .
$$

This proof is inspired by similar proofs in [HelLaf2, Mar].

Remark 3.1. Davies treats a particular case by hand. Then Zworski observes that it can be interpreted as a semi-classical version of a result for operators of principal type (Hörmander [Ho1], [Ho2], Duistermaat–Sjöstrand [DuSj]). This was pushed further by Dencker–Sjöstrand–Zworski [DeSjZw], N. Lerner (together with collaborators) (see in [Le] and references therein), Pravda-Starov [Pra1].

One should of course compare with the selfadjoint result at the bottom of the well but here what is crucial is the non-selfadjointness!!

Theorem 3.1 (Davies–Pravda). *Let us assume that there exist x_0 and z such that*

$$z - V(x_0) \in \mathbb{R}^+ , \tag{58}$$

and such that, for an even $k \geq 0$,

$$\mathrm{Im}\, V^{(j)}(x_0) = 0 \,, \forall j \leq k \,, \tag{59}$$

and

$$\mathrm{Im}\, V^{(k+1)}(x_0) \neq 0 \,. \tag{60}$$

Then $z \in \psi_\infty((A_h))$.

Some Elementary Proof by a WKB Construction

The crucial point is that there exists $\xi_0 > 0$ such that

$$\xi_0^2 + V(x_0) = z \,.$$

In other words, there exists (x_0, ξ_0) such that $p(x_0, \xi_0) = z$. Hence, $z \in \Sigma(p)$ as defined in (57) and we are not at the boundary of $\Sigma(p)$.

Following the construction described in the first Lecture (see (37)–(40)), we look for a solution in the form

$$u(x, h) = a(x, h) \exp - \frac{\varphi(x)}{h} \tag{61}$$

near x_0 with

$$a(x, h) \sim \sum_{j \geq 0} h^j a_j(x) \,, \tag{62}$$

such that

$$\exp \frac{\varphi}{h}(A_h - z_0)u(\cdot; h) \sim 0 \,. \tag{63}$$

Let us emphasize that (conversely to what was done in the analysis of the Rayleigh–Taylor model) we keep z_0 fixed and did not look for an expansion $z(h) \sim \sum_{j \geq 0} z_j h^j$.

Expanding in powers of h and expressing the cancellation of each coefficient of h^ℓ, we first get an eikonal equation. The phase φ (appearing in (61)) should satisfy (we can after a change of notations assume that $z = 0$:

$$-\varphi'(x)^2 + V(x) = 0 \,, \tag{64}$$

where V satisfies by assumption $\operatorname{Re} V(x_0) < 0$, (59) and (60).
The existence of $\varphi(x)$, with $\varphi(x_0) = 0$ and $\varphi'(x_0) = i\xi_0$ is evident. So the important point, in order to have an approximate eigenfunction which is localized at x_0, is to verify that $\operatorname{Re}\varphi$ has actually a local minimum at x_0. Taking the real and imaginary parts in (64), we get

$$-\operatorname{Re}\varphi'(x)^2 + \operatorname{Im}\varphi'(x)^2 + \operatorname{Re} V(x) = 0 \,, \tag{65}$$

and

$$-2\operatorname{Re}\varphi'(x) \cdot \operatorname{Im}\varphi'(x) + \operatorname{Im} V(x) = 0 \,, \tag{66}$$

in a neighborhood of x_0.
In particular, this implies at x_0

$$\operatorname{Re}\varphi'(x_0) = 0, \ \xi_0^2 = \operatorname{Im}\varphi'(x_0)^2 = -\operatorname{Re} V(x_0) \,.$$

What we now need is to verify that the first non zero derivative of $\operatorname{Re}\varphi$ at x_0 is even and strictly positive.
We start from

$$\operatorname{Re}\varphi'(x) = \frac{\operatorname{Im} V(x)}{2\operatorname{Im}\varphi'(x)} \,.$$

But it is immediate from the assumptions that

$$\operatorname{Re}\varphi^{(j)}(x_0) = 0 \,, \quad \text{for } j \le k+1 \,,$$

and

$$\operatorname{Re}\varphi^{(k+2)}(x_0) = \frac{\operatorname{Im} V^{(k+1)}(x_0)}{2\operatorname{Im}\varphi'(x_0)} \,.$$

We can now choose the sign of ξ_0 in order to have

$$\operatorname{Re}\varphi^{(k+2)}(x_0) > 0 \,.$$

Due to the fact that $(\partial_\xi p)(x_0, \xi_0) = \xi_0 \ne 0$, the solution of the transport equations does not create problems like in the case of Rayleigh–Taylor and we can construct a solution $u_h = a(x, h) \exp -\frac{\varphi(x)}{h}$ in the neighborhood of x_0. Let us briefly show how to treat the cancellation of the coefficient of h which leads to the so-called first transport equation. This equation reads

$$2\varphi'(x)a_0'(x) + \varphi''(x)a_0(x) = 0 \,, \tag{67}$$

with as initial condition

$$a_0(x_0) = 1 \,.$$

But $\varphi'(x_0) = i\xi_0 \ne 0$, so it is immediate to find in a neighborhood of x_0 the main amplitude a_0 by

$$a_0(x) = \exp -\frac{1}{2}\left(\int_{x_0}^x \frac{\varphi''(\tau)}{\varphi'(\tau)}\,d\tau\right) .$$

The next equation has the same structure as in (67) except that there is a r.h.s. This equation reads

$$2\varphi'(x)a_1'(x) + \varphi''(x)a_1(x) = a_o''(x) , \qquad (68)$$

with as initial condition

$$a_1(x_0) = 0 ,$$

and has again a unique explicit solution. More generally all the successive equations read

$$2\varphi'(x)a_j'(x) + \varphi''(x)a_j(x) = a_{j-1}''(x) , \qquad (69)$$

with as initial condition

$$a_j(x_0) = 0 ,$$

and can be solve by recursion for $j \geq 2$.

Remark 3.2. K. Pravda-Starov constructs a solution in the form $\exp -\frac{\varphi(x,h)}{h}$ with $\varphi(x;h) \sim \sum_j h^j \varphi_j(x)$ but this is not really different when working with a groundstate which is supposed to have no zero.

Remark 3.3. Note that if $z \notin \overline{\Sigma(p)}$, then the elliptic theory says that it is impossible to construct an approximate solution, so it leaves open only the points at the boundary of $\Sigma(p)$.

3.5 Kelvin–Helmholtz II: Mathematical Analysis

We now come back to our motivating model and see if the ideas behind the treatment of Davies example are efficient.

Note also that our question is a little different and could be reformulated as: *For which values of the parameters is 0 in the h-family pseudospectrum of the family (with $h = \frac{1}{k}$)?*

So we have to analyze if 0 belongs to $\Sigma(p_0)$, where p_0 was defined in (52). We just do the local analysis (the analysis of the ellipticity at ∞ should be interesting to do). According to (52), we have:

$$\operatorname{Re} p_0(x,\xi) = \varrho_0(x)(\xi^2+1)(\Gamma_0^2 - (k_1 u_0(x) + \Gamma_1)^2) + g\varrho_0'(x) , \qquad (70)$$

$$\operatorname{Im} p_0(x,\xi) = 2\varrho_0(x)(\xi^2+1)\Gamma_0(k_1 u_0(x) + \Gamma_1) .$$

Assuming that

$$\Gamma_0 \neq 0 , \qquad (71)$$

and that

$$\varrho_0(x) > 0, \; \forall x \in \mathbb{R} \, , \tag{72}$$

we observe that

$$\operatorname{Im} p_0(x,\xi) = 0 \;\; \text{iff} \;\; k_1 u_0(x) + \Gamma_1 = 0 \, .$$

When this condition is satisfied, we get

$$\operatorname{Re} p_0(x,\xi) = \varrho_0(x)(\xi^2 + 1)\Gamma_0^2 + g\varrho_0'(x) \, .$$

If

$$\varrho_0' < 0, \;\; \text{on } \mathbb{R} \, , \tag{73}$$

then we see ($g > 0$), that, if

$$\Gamma_0^2 > g \max_x -\frac{\varrho_0'(x)}{\varrho_0(x)} \, ,$$

then the principal symbol is elliptic.
Hence no local approximate null solution can be constructed. 0 does not
belong to the h-family-pseudospectrum of the operator.

We also observe that this condition is the same as for Rayleigh–Taylor (see
for example (32), with in mind (24))!

Conversely, when

$$\Gamma_0^2 < g \max_x -\frac{\varrho_0'(x)}{\varrho_0(x)} \, ,$$

one can, for any x_0 such that

$$-g\frac{\varrho_0'(x_0)}{\varrho_0(x_0)} > \Gamma_0^2 \, ,$$

find some $\xi_0 \neq 0$ such that

$$\Gamma_0^2(1 + \xi_0^2) = -g\frac{\varrho_0'(x_0)}{\varrho_0(x_0)} \, .$$

We are now looking on the condition under which the operator A_h, which is
not elliptic at (x_0, ξ_0) which determines the parameter Γ_1 by,

$$\Gamma_1 = -k_1 u_0(x_0) \, ,$$

is not subelliptic at this point (we will explain later in the next lecture
(Theorem 4.1) what we can do in this case).

The computation of the bracket of $\operatorname{Re} p_0$ and $\operatorname{Im} p_0$ gives

$$\{\operatorname{Re} p_0, \operatorname{Im} p_0\}(x_0, \xi_0) = 4k_1\xi_0\varrho_0(x_0)^2 u_0'(x_0)\Gamma_0^3 \, . \tag{74}$$

So it is immediate by playing with the sign of k_1 (or of ξ_0) to get the condition (75) satisfied if $u'_0(x_0) \neq 0$.

A detailed analysis of what is going on for $\gamma = \Gamma_0 + i\Gamma_1$ with Γ_0 close to $\widetilde{\Gamma}_0$ with

$$\widetilde{\Gamma}_0^2 = g \max_x -\frac{\varrho'_0(x)}{\varrho_0(x)}$$

should surely be interesting. The techniques presented at the end of the last lecture will be helpful.

Here the simplest toy model should be

$$h^2 D_x^2 + ik_1 x \,,$$

the complex Airy operator, which is for $k_1 \neq 0$ a particular case of Davies example and can be also analyzed close to 0 by Dencker–Sjöstrand–Zworski result.

Let us explain more in detail how we guess this model. We do not try to be rigorous. For convenience we assume that ϱ' is strictly negative so the associated $K(h)$ (see (28)) appearing in the treatment of the Rayleigh–Taylor model is positive. At least locally near a maximum of $x \mapsto -\frac{\varrho'(x)}{\varrho(x)}$, one can (this is an interesting exercise in semi-classical analysis) modulo $\mathcal{O}(h^\infty)$ rewrite our problem of research of approximate null solutions in looking for which values of γ, the operator

$$\sqrt{K(h)} - ik_1 u_1(x) + hp_1(x, hD_x, h, k_1, \gamma)) - \gamma$$

has approximate null solutions.

There is a technique (functional calculus of Helffer–Robert ([Rob] and references therein) or direct approach for the square root) for recognizing $f(K(h))$ as an h-pseudodifferential operator if f is regular. In our case, one can use a C^∞-positive function coinciding with $\sqrt{t}$ on $[2\epsilon_0, +\infty[$ and equal to a strictly positive constant for $t \in] -\infty, \epsilon_0]$.

If we forget the dependence on γ in p_1, we are facing a very standard question of h-family-pseudospectrum.

The question becomes simply:

Is γ in the pseudospectrum of

$$\sqrt{K(h)} - ik_1 u_1(x) + hp_1(x, hD_x, h, k_1, \gamma))?$$

Taking the harmonic approximation of $\sqrt{K(h)}$ at a point where the principal symbol of $\sqrt{K(h)}$ (which is the square root of the principal symbol of $K(h)$) and the linear approximation of u_1 at x_0 leads (up to the constants) to the toy model.

3.6 Other Toy Models

Other toy models have been analyzed in detail. Let us mention

$$h^2 D_x^2 + ih D_x + x^2 \,,$$

whose symbol is $p(x, \xi) = \xi^2 + i\xi + x^2$ (See [DeSjZw], p. 3).

The spectrum is easy to determine as given by the sequence $\frac{1}{4} + (2n+1)h$ ($n \in \mathbb{N}$), the corresponding eigenfunctions being directly related with the Hermite functions. This permits to diagonalize the operator BUT in a non orthonormal basis.

The h-family pseudospectrum is given by the numerical range of the principal symbol of the operator:

$$\Sigma(p) = \{z \in \mathbb{C} \mid |\mathrm{Im}\, z|^2 \le \mathrm{Re}\, z\} \,.$$

More generally the h-family pseudospectrum of the Schrödinger operators $-h^2\Delta + V(x)$, with V quadratic has been analyzed in great detail in the PhD thesis of Pravda-Starov [Pra3].

Other models appear in connection with the analysis of the resolvent of the Fokker–Planck operator (see Risken (for the quadratic case), [Ris], Hérau–Nier [HerNi], Helffer–Nier [HelNi], Hérau–Sjöstrand–Stolk [HerSjSt]) or for other models (See Hager [Ha] and works in progress from Hager–Sjöstrand).

4 Lecture 3: On Semi-Classical Subellipticity

4.1 Introduction

The references for this lecture are papers by Davies [Da2], Zworski [Zw], Dencker–Sjöstrand–Zworski [DeSjZw], Lerner [Le] (and references therein).

We would like to show how the microlocal techniques (suitably adapted to the semi-classical context) permit to recover or complete the previous results. We will see in the last lecture how one can also analyze the transition between the elliptic region and the non elliptic one. We have already seen that many results of non-existence of approximate null solutions are just the consequence of "elliptic" semi-classical results. As a second step, we can look if, at non-elliptic points, some subellipticity condition is satisfied, starting by $\frac{1}{2}$-semi-classical subellipticity. This would again imply the same type of results.

Conversely, if the operator is not subelliptic, one can try to construct directly WKB solutions in the form $a(x, h) \exp{-\frac{\varphi(x)}{h}}$ with φ admitting a minimum at some point x_0 or to apply more general theorems in semi-classical analysis. We start in the next subsection by a typical result of the last alternative.

4.2 Non Subellipticity: Generic Result

The main relevant theorem in our context can be stated in the following way (see [DeSjZw]). One considers an h-pseudodifferential $A_h := a(x, hD_x)$ with principal symbol a_0 and one is looking for a simple criterion under which 0 belongs to the h family pseudospectrum of A_h.

Theorem 4.1. *Let us assume that at a point* (x_0, ξ_0)*, we have*

$$a_0(x_0, \xi_0) = 0 \, , \; \{Re \; a_0, Im \; a_0\}(x_0, \xi_0) < 0 \, . \tag{75}$$

Then there exists an L^2*-normalized solution* u_h*, whose* h*-wave front is* (x_0, ξ_0)*, and such that* (x_0, ξ_0) *is not in the* h*-wave front of* $A_h u_h$*.*

We recall that, for a bounded family of L^2 functions v_h, we say that a point (y, η) *is not* in the h-wave front set,[4] if there exists a C_0^∞ function χ equal to 1 in the neighborhood of y, such that $(\mathcal{F}_h \chi v_h)(\xi) := h^{-\frac{n}{2}} \widehat{\chi v_h}(\xi/h) = \mathcal{O}(h^\infty)$ in a neighborhood of η.

Another (equivalent) definition is to use the Fourier–Bros–Iagolnitzer (which will be familiar to the users of the Gabor transform) as intensively developed by J. Sjöstrand [DiSj].

We say that (x_0, ξ_0) is not in the h-Wave front set of a bounded family u_h in L^2 if the function

$$(x, \xi) \mapsto h^{-\frac{3n}{4}} \int \exp \frac{i}{h}(x - y) \cdot \xi \, \exp -\frac{(x - y)^2}{2h} \, u_h(y) \, dy \, ,$$

is $\mathcal{O}(h^\infty)$ in some (h-independent) neighborhood of (x_0, ξ_0).

Applications

Let us see what this theorem say for the two examples we have already met: the Davies example and the Kelvin–Helmholtz example.
In the first case, we have

$$Re \, a_0(x, \xi) = \xi^2 + Re \, V(x) - Re \, z_0 \, , \; Im \, a_0(x, \xi) = Im \, V(x) - Im \, z_0 \, . \tag{76}$$

The Poisson Bracket at (x_0, ξ_0) is

$$\{Re \, a_0, Im \, a_0\}(x_0, \xi_0) = 2\xi_0 Im \, V'(x_0) \, , \tag{77}$$

and we recall that $\xi_0 \neq 0$ with ξ_0^2 determined. So if $Im \, V'(x_0) \neq 0$, (which corresponds to $k = 0$ in Davies–Pravda theorem), the non-subelliptic theorem applies for the right choice of the sign x_0.
In the second case, we send back the reader to Formula (74).

---

[4] Another terminology used for example in [Rob] is to speak of frequency set.

4.3 Link with the Standard Non-Hypoellipticity Results for Operators of Principal Type

In the theory of Partial Differential Equations, Theorem 4.1 corresponds to a result of non-hypoellipticity. The basic simplest model is $D_x + ix D_t$, which is known to be non hypoelliptic microlocally at $(0,0)$ in the direction $(0,-1)$. Hence it is not hypoelliptic. But one should keep in mind that the link between the two problems is microlocal. As already explained in the lectures by N. Lerner [Le] (see also [Trev]), the link between the two theories is through the partial Fourier transform in the t-variable. For an operator in the form $D_x + ib(x) D_t$, we first get the family in τ, $D_x + ib(x)\tau$, that we have to analyze for $|\tau|$ large. With $h = \frac{1}{|\tau|}$, we get two semi-classical families of operators to analyze $hD_x \pm ib(x)$, each one corresponding to a microlocal analysis in the direction $(0,1)$ or $(0,-1)$.

4.4 Elementary Proof for the Non-Subelliptic Model

We give an elementary proof (cf. [Mar]) under the additional assumption that

$$a_0(x, i\xi) \in \mathbb{R} \ , \ \forall (x,\xi) \in \mathbb{R}^2 \ , \tag{78}$$

which appears to be satisfied for the two last physical models, which will be analyzed in the next section, but is not satisfied for the Davies example and the Kelvin–Helmholtz model.

In this case, we define the real symbol

$$q_0(x, \xi) = a_0(x, i\xi) \ , \ \forall (x,\xi) \in \mathbb{R}^2 \ ,$$

and we look for a point $(x_0, 0)$ such that

$$q_0(x_0, 0) = 0 \ ,$$

and for a non negative real phase φ defined in a neighborhood of x_0 such that $\varphi(x_0) = 0$ admitting at x_0 a local minimum and solution of

$$q_0(x, \varphi'(x)) = 0 \ . \tag{79}$$

Under the condition that $\partial_\xi q_0(x_0, 0)$ it is immediate to find φ by the implicit function theorem.

The first natural condition for having a minimum is then to see under which condition one has

$$\varphi''(x_0) > 0 \ .$$

Differentiating the eikonal equation (79), we obtain

$$(\partial_x q_0)(x, \varphi'(x)) + (\partial_\xi q_0)(x, \varphi'(x))\varphi''(x) = 0 ,$$

hence

$$\varphi''(x_0) = -\frac{\partial_x q_0(x_0, 0)}{\partial_\xi q_0(x_0, 0)} .$$

So we are done if the r.h.s. is strictly positive:

$$-\frac{\partial_x q_0(x_0, 0)}{\partial_\xi q_0(x_0, 0)} > 0 . \tag{80}$$

Let us now control that this condition can be recognized as the condition of the theorem.
From the relations

$$\partial_x q_0(x, \xi) = \partial_x a_0(x, i\xi) , \quad \partial_\xi q_0(x, \xi) = i\partial_\xi a_0(x, i\xi) ,$$

we get at any point $(x, 0)$:

$$\partial_x \operatorname{Im} a_0(x, 0) = 0 , \qquad \partial_\xi \operatorname{Re} a_0(x, 0) = 0$$
$$\partial_x \operatorname{Re} a_0(x, 0 = \partial_x q_0(x, 0) , \quad \partial_\xi \operatorname{Im} a_0(x, 0) = -\partial_\xi q_0(x, 0) .$$

So this gives the relation:

$$\{\operatorname{Re} a_0, \operatorname{Im} a_0\}(x, 0) = \partial_x q_0(x_0, 0)\partial_\xi q_0(x_0, 0) ,$$

and the result becomes clear.

The second step is to construct a quasimode in the form

$$u_h := b(x, h) \exp -\frac{\varphi(x)}{h} ,$$

with

$$b(x, h) \sim \sum_{j \geq 0} b_j(x)h^j .$$

The equation for b_0 reads

$$(\partial_\xi q_0)(x, \varphi'(x))b_0'(x) + \left(\frac{\varphi''(x)}{2}(\partial_\xi^2 q_0)(x, \varphi'(x)) + q_1(x, \varphi'(x)) \right) b_0(x) = 0 ,$$

where q_1 is the "subprincipal" symbol. One can always solve this equation with $b_0(x_0) = 1$ (see (67)).

Remark 4.1. When the first Poisson bracket of a_0 and $\overline{a_0}$ is 0 (which is equivalent to $\partial_x q_0(x, 0) = 0$), one can find a criterion involving higher order brackets. See [Pra3], [Mar] and the standard results on subelliptic operators obtained in the seventies.

We are in a particular case of the following more general situation. We look for solutions of $a(x, hD_x)u_h = \mathcal{O}(h^\infty)$ which are localized in a neighborhood of a point (x_0, ξ_0) such that

$$a_0(x_0, \xi_0) - z = 0 \;, \quad (\partial_\xi a_0)(x_0, \xi_0) \neq 0 \;.$$

In addition, we have

$$-i(\operatorname{ad} a_0)^k(\{a_0, \overline{a}_0\})(x_0, \xi_0) = 0 \;,$$

for $k < k_0$ and

$$-i(\operatorname{ad} a_0)^{k_0}(\{a_0, \overline{a}_0\})(x_0, \xi_0) > 0 \;,$$

where $\operatorname{ad} p$ is the operator of commutation

$$(\operatorname{ad} p)q = \{p, q\} \;.$$

This time we have to take a complex phase.

4.5 $\frac{1}{2}$ Semi-Classical Subellipticity

When the principal symbol is not elliptic, the best we can hope is a subelliptic result. The next theorem corresponds to the first (and the most generic) result of this type.

Theorem 4.2 ($\frac{1}{2}$-Subellipticity). *If $(u_h)_{h\in]0,h_0]}$ is an L^2 normalized solution in the domain of A_h such that $A_h u_h = \mathcal{O}(h^\infty)$, then if for some (x_0, ξ_0) we have*

$$a_0(x_0, \xi_0) = 0 \;, \quad \{Re \; a_0, Im \; a_0\}(x_0, \xi_0) > 0 \;,$$

then (x_0, ξ_0) does not belong to the h-wave front set of the family u_h.

Remark 4.2. In PDE theory this corresponds to the simplest result of microlocal hypoellipticity. The basic simplest model is $D_x + ixD_t$, which is known to be hypoelliptic (with loss of $\frac{1}{2}$ derivatives microlocally at $(0, 0)$ in the direction $(0, 1)$).

We will come back later in the last lecture to high order subellipticity.

Remark 4.3. Note that the elliptic theory simply says that if $z \notin \Sigma(p)$, then z is not in the pseudospectrum of $-h^2\Delta + V$. So what remains is simply a more precise analysis at $\partial\Sigma(p)$.

About the Proof
We refer to the lectures of N. Lerner [Le]. Let us just sketch the semi-classical proof. If we write

$$A_h = B_h + iC_h \,,$$

with B_h and C_h selfadjoint respectively of principal symbol $\operatorname{Re} a_0$ and $\operatorname{Im} a_0$, the basic point is that

$$A_h^* A_h = B_h^2 + C_h^2 + i[B_h, C_h] \,,$$

and to observe that $\frac{i}{h}[B_h, C_h]$ is positive elliptic at the points where A_h is not elliptic.

We can use rather weak forms of the Garding inequality. We refer to the lectures of N. Lerner ([Le]) for discussions around this point and the Fefferman–Phong inequality.

Remarks 4.3

- *Here we gave the impression that everything is done globally but let us now emphasize that one has to do very often the argument microlocally.*
- *Note that we do not really need this result. In the case of the symbol appearing in Kelvin–Helmholtz model the sign of the Poisson bracket at (x_0, ξ_0) is opposite to the sign at $(x_0, -\xi_0)$.*
  This will not be the case for the two next models for which we will have $\xi_0 = 0$ at the non-elliptic points.

5 Lecture 4: Other Non Self-Adjoint Models Coming from Hydrodynamics

5.1 Introduction

The two next models are deduced from the mass conservation and the momentum conservation equation of the Euler equation, and differ through the modelling of the energy equation. For simplicity the systems are written in $\mathbb{R}^2_{\tilde{x}, \tilde{y}} \times \mathbb{R}_t$ (instead of $\mathbb{R}^3_{\tilde{x}, \tilde{y}, \tilde{z}} \times \mathbb{R}_t$).

The density of the fluid satisfies, for some strictly positive constant $\rho_a > 0$,

$$\rho(\tilde{x}, \tilde{y}) \to \rho_a \quad \text{when } \tilde{x} \to +\infty \,,$$

and the velocity of the fluid satisfies, for some $V_a > 0$,

$$\mathbf{U} := (u, v) \to (-V_a, 0) \quad \text{when } \tilde{x} \to +\infty \,.$$

ρ_a is the density of the ablated fluid and V_a the modulus of the velocity of the ablated fluid.

The *Rayleigh model with convection* assumes that the perturbation of the velocity is incompressible. This means that there exists a function $\mathbf{U}_0(\tilde{x})$, called the convective velocity, such that

$$\text{div } (\mathbf{U} - \mathbf{U}_0) = 0 \, .$$

The system will be denoted by (RC) and writes

$$(RC) \quad \begin{cases} \partial_t \rho + \partial_{\tilde{x}}(\rho u) + \partial_{\tilde{y}}(\rho v) = 0 \, , \\ \partial_t(\rho u) + \partial_{\tilde{x}}(\rho u^2 + p) + \partial_{\tilde{y}}(\rho uv) = -\rho g \, , \\ \partial_t(\rho v) + \partial_{\tilde{x}}(\rho uv) + \partial_{\tilde{y}}(\rho v^2 + p) = 0 \, , \\ \text{div } (\mathbf{U} - \mathbf{U}_0) = 0 \, , \end{cases}$$

where the unknowns are the density ρ, the velocity (u, v) and the pressure p.

The *ablation front model* uses an energy equation with heat conduction. The enthalpy is defined by

$$h = C_p T \, , \tag{81}$$

With $T(t, \tilde{x}, \tilde{y})$ denoting the temperature of the fluid (at a point $\tilde{x}, \tilde{y}$ and a time t) and C_p being a constant characterizing the calorific capacity of the fluid, the enthalpy satisfies the equation:

$$\rho(\partial_t + \mathbf{U} \cdot \nabla)h - (\partial_t + \mathbf{U} \cdot \nabla)p = -\text{div } \mathbf{J}_q \tag{82}$$

Here $\mathbf{J}_q$ is the heat flux given by the Fourier conduction law

$$\mathbf{J}_q = -\lambda(T)\nabla T \, .$$

In this law, $\lambda(T)$ is proportional to a power of the temperature, that is satisfying, for some constants $\kappa > 0$ and $\nu > 0$,

$$\lambda(T) = \kappa T^\nu \, .$$

Note that these formulas assume that $T > 0$ and consequently, with p related with T as below in (83) to the condition $p > 0$. The parameter ν is called the conduction index.
We now write the perfect gas relation

$$p = \rho T(C_p - C_v) \, , \tag{83}$$

where C_v is the calorific capacity at constant volume. C_p/C_v is 5/3. Starting from (82) and then using (81), (83) and the first equation in (RC), we get:

$$C_p \rho T \text{ div } \mathbf{U} + C_v(\partial_t + \mathbf{U} \cdot \nabla)\rho T + \text{div } \mathbf{J}_q = 0 \, . \tag{84}$$

We shall not analyze this model, in particular because this model has no stationary solution. So the physicists use other models for which we can just explain (without being in any way rigorous) how they can be obtained.

5.2 Quasi-Isobaric Model (Kull and Anisimov)

The starting point consists in replacing the perfect gas relation by the relation:

$$\rho T = D_0 , \tag{85}$$

where D_0 is a constant.
Implementing (85) in (84) gives:

$$D_0 C_p \operatorname{div} \mathbf{U} + \operatorname{div} \mathbf{J}_q = 0 .$$

This constant is identified through the hypothesis that $T \to T_a$, $T_a > 0$, when $\tilde{x}$ goes to $+\infty$ (temperature of the ablated fluid).

Hence

$$D_0 = \rho_a T_a \quad \text{and} \quad T = \frac{\rho_a T_a}{\rho} .$$

For a derivation of this model, see [KullA], [Go, Mas, La3]. A similar model arises also in the Low Mach approximation (see [Li]).
The system of equations writes

$$(KA) \quad \begin{cases} \partial_t \rho + \partial_{\tilde{x}}(\rho u) + \partial_{\tilde{y}}(\rho v) = 0 , \\ \partial_t(\rho u) + \partial_{\tilde{x}}(\rho u^2 + p) + \partial_{\tilde{y}}(\rho uv) = -\rho g , \\ \partial_t(\rho v) + \partial_{\tilde{x}}(\rho uv) + \partial_{\tilde{y}}(\rho v^2 + p) = 0 , \\ \operatorname{div}\,(\mathbf{U} - \frac{\kappa}{C_p \rho_a} T_a^\nu (\frac{\rho_a}{\rho})^\nu \nabla \frac{\rho_a}{\rho}) = 0 , \end{cases}$$

where the unknowns are the functions $(t, \tilde{x}, \tilde{y}) \mapsto (\rho, u, v, p)$.
Of course we can recover T by the equation $\rho T = \rho_a T_a$, but in this approximation, we will no more impose that the perfect gas relation is satisfied when pursuing the analysis. So the solution of (KA) will not be satisfied with p constant as we could have thought by combining previous equations.

5.3 Stationary Laminar Solution

Both systems are studied around a stationary laminar (independent of $\tilde{y}$ and t) solution of the equations.
    For the system (RC), we are given an arbitrary convective velocity $\mathbf{U_0}$, and for the system (KA) it is deduced from the energy equation. In both cases a reference length L_0 plays an important role (for defining in which asymptotic regime we are).
    For the system of Rayleigh with convection,

$$\mathbf{U}_0(\tilde{x}) = (\tilde{u}_0(\tilde{x}), 0) ,$$

with

$$\tilde{u}_0(\tilde{x}) = u_0(\frac{\tilde{x}}{L_0}) \ .$$

For the ablation front model,

$$L_0 = \kappa \frac{T_a^{\nu+1}}{C_p \rho_a V_a} \ .$$

We use the rescaled variable

$$x := \frac{\tilde{x}}{L_0} \ .$$

The stationary laminar solution is given by

$$(\tilde{x}, \tilde{y}) \mapsto (\tilde{\rho}_0(\tilde{x}), \tilde{u}_0(\tilde{x}), 0, \tilde{p}_0(\tilde{x}))$$

with

$$\tilde{\rho}_0(\tilde{x}) = \rho_0(\tfrac{\tilde{x}}{L_0}) \ , \ \tilde{p}_0(\tilde{x}) = p_0(\tfrac{\tilde{x}}{L_0}) \ .$$

Here ρ_0, u_0, p_0 are functions on $\mathbb{R}$

$$\begin{cases} \rho_0(x) u_0(x) = -\rho_a V_a \ , \\ \frac{d}{dx}\left(\rho_0(x) u_0(x)^2 + p_0(x)\right) = -\rho_0(x) g L_0 \ . \end{cases}$$

Note that p_0 is determined modulo a constant C_0 by:

$$\rho_0(x) u_0(x)^2 + p_0(x) = -g L_0 \int_0^x \rho_0(t) dt + C_0 \ .$$

Finally, we introduce the adimensionalized density profile $\varrho(x)$ which is the function

$$\varrho(x) = \frac{\rho_0(x)}{\rho_a} \ .$$

5.4 From the Physical Parameters to the Relevant Mathematical Parameters

Following [CCLaRa], we can now associate with the physical parameters, g, L_0, V_a, k, the parameters

$$\alpha = \frac{\sqrt{gk}L_0}{V_a} \ , \ \beta = V_a\sqrt{\frac{k}{g}} \ ,$$

and the relevant constants of this study (the constant σ_c stands for the Rayleigh with convection model and the constant σ_a is characteristic of the ablation front model)

$$h = \frac{1}{kL_0} = \frac{1}{\alpha\beta} \ , \ \sigma_c = \frac{h^{\frac{1}{2}}}{\beta} \ , \ \sigma_a = \frac{h^2}{\beta^2} \ .$$

These constants are linked to the reduced wave number

$$\varepsilon = kL_0 \ ,$$

and the Froude number,

$$F_r = \frac{V_a^2}{gL_0} \ .$$

They are linked to α and β through

$$F_r = \frac{\beta}{\alpha} \ , \ \varepsilon = \alpha\beta \ .$$

From the growth rate $\bar{\gamma}$, we deduce two dimensionless growth rates

$$\gamma = \frac{\bar{\gamma}}{\sqrt{gk}} \ ,$$

and

$$\Gamma = \frac{\bar{\gamma}}{kV_a} = \frac{\gamma}{\beta} \ . \tag{86}$$

The growth rate γ is the growth rate generally used in the classical Rayleigh–Taylor analysis, and the growth rate Γ is the one relevant in the semiclassical regime, that we study here.

As a conclusion, *Semi-classical analysis can be applied when the Froude Number is small enough.*

5.5 The Convection Velocity Model

In our rescaled variable x, the linearized system writes (with $q_4 = r_4\varrho - q_1$):

$$(LRC) \begin{cases} \frac{dq_1}{dx} + \alpha\gamma(\varrho^2 r_4 - \varrho q_1) - \alpha\beta\varrho q_3 = 0 \ , \\ \frac{dq_2}{dx} + \alpha\gamma q_1 + \alpha\beta q_3 + \frac{\alpha}{\beta}(\varrho^2 r_4 - \varrho q_1) = 0 \ , \\ \frac{dq_3}{dx} - \alpha\beta(q_2 + \frac{2q_1 + q_4}{\varrho}) - \alpha\gamma\varrho q_3 = 0 \ , \\ \frac{dr_4}{dx} - \alpha\beta q_3 = 0 \ . \end{cases}$$

Here (q_1, q_2, q_3, q_4) correspond to infinitesimal variation of the new unknowns $(\rho u, \rho u^2 + p, \rho uv, u)$.

This system rewrites, with $d_h = h\frac{d}{dx}$,

$$d_h \begin{pmatrix} q_1 \\ q_2 \\ q_3 \\ r_4 \end{pmatrix} + \begin{pmatrix} -\Gamma\varrho & 0 & -\varrho & \Gamma\varrho^2 \\ \Gamma - \frac{\varrho}{\beta^2} & 0 & 1 & \frac{\varrho^2}{\beta^2} \\ -\frac{1}{\varrho} & -1 & -\Gamma\varrho & -1 \\ 0 & 0 & -1 & 0 \end{pmatrix} \begin{pmatrix} q_1 \\ q_2 \\ q_3 \\ r_4 \end{pmatrix} = 0 \, .$$

The main point is now that we can reduce the analysis of the system to the analysis of one equation.

Proposition 5.1. *The C^4-valued function (q_1, q_2, q_3, r_4) is a solution of the linearized system (LRC) if and only if $q_4 := r_4\varrho - q_1$ belongs to the kernel of the operator (ELRC),*

$$\begin{aligned} &\mathcal{P}_c(x, \tfrac{1}{i}h\tfrac{d}{dx}, h, \sigma_c, \Gamma) \\ &:= d_h[(d_h - \Gamma\varrho)(d_h(\tfrac{1}{\varrho'}(d_h - \Gamma\varrho))) - \tfrac{2}{\varrho'}(d_h - \Gamma\varrho) + \tfrac{h}{\varrho}] \\ &\quad + \sigma_c^2\varrho + d_h(\tfrac{1}{\varrho'}(d_h - \Gamma\varrho)) + \Gamma(\tfrac{\varrho}{\varrho'}(d_h - \Gamma\varrho) - h) \, . \end{aligned}$$

Here the interesting point is that we have only two effective parameters (h, σ_c) which will make the discussion about various asymptotic regimes easier. The semi-classical regime will correspond to fix $\sigma_c > 0$ and to analyze the question when $h \to 0$.

The semi-classical principal symbol is

$$(x, \xi) \mapsto \mathcal{P}_c^0(x, \xi) := -\frac{1}{\varrho'}(i\xi - \Gamma\varrho)^2(\xi^2 + 1) + \varrho\sigma_c^2 \, . \tag{87}$$

Assumption 1

The profile ϱ satisfies:

1. $\varrho \in C^\infty(\mathbb{R};]0, 1[)$
2. $\lim_{x \to -\infty} \varrho(x) = \varrho_- \geq 0$
3. $\lim_{x \to +\infty} \varrho(x) = \varrho_+ = 1$
4. $\varrho' > 0$
5. $\lim_{|x| \to +\infty} \frac{\varrho'(x)}{\varrho(x)} = 0$

Remark 5.1. The reader should be aware that, in comparison with the two first models, we have changed the convention in order to be coherent to the reference [HelLaf2] in which the reader can find additional details.

Assumption 2

The maximum of $\frac{\varrho'}{\varrho}$ is attained at a unique x_{max}:

$$0 < \frac{\varrho'}{\varrho}(x_{max}) := (\vartheta_c^{max})^2 \, ,$$

and the map $x \mapsto \frac{\varrho'(x)}{\varrho(x)}$ is strictly increasing over $] - \infty, x_{max}[$ and then strictly decreasing over $]x_{max}, +\infty[$.

Local Ellipticity Condition
The imaginary part of the symbol is

$$\operatorname{Im} \mathcal{P}_c^0(x, \xi) = \frac{2\xi}{\varrho'(x)} \Gamma \varrho(x)(\xi^2 + 1) .$$

It is non zero except for

$$\xi = 0 .$$

Looking at the real part restricted to $\xi = 0$, we obtain that

$$\operatorname{Re} \mathcal{P}_c^0(x, 0) = -\Gamma^2 \frac{\varrho^2(x)}{\varrho'(x)} + \varrho(x)\sigma_c^2 .$$

This leads us to the following local ellipticity condition:

$$\frac{\Gamma}{\sigma_c} > \vartheta_c^{max} .$$

5.6 The Model for the Ablation Regime

Similarly, the linearization of the system (KA) leads to the following system

$$(LKA) \quad d_h\mathbf{q} + M_0(\varrho(x))\mathbf{q} = 0 ,$$

where

$$\mathbf{q} = \begin{pmatrix} q_1 \\ q_2 \\ q_3 \\ p_4 \\ q_5 \end{pmatrix} .$$

and the matrix is

$$M_0(\varrho) = \begin{pmatrix} 0 & 0 & \varrho & h\Gamma\varrho^{\nu+2} & 0 \\ \Gamma & 0 & -1 & \frac{h}{\beta^2}\varrho^{\nu+2} & 0 \\ \frac{2}{\varrho} & 1 & -\Gamma\varrho & h\varrho^{\nu} & 0 \\ \frac{1}{\varrho} & 0 & 0 & h\varrho^{\nu} & -1 \\ 0 & 0 & 1 & -1 & 0 \end{pmatrix} .$$

Proposition 5.2. *The C^5-valued function $\mathbf{q}$ is a solution of (LKA) if and only if its fourth component p_4 is in the kernel of the operator (ELKA):*

$$\mathcal{P}_a(x, \tfrac{1}{i}d_h, h, \sigma_a, \Gamma) :=$$
$$\left[d_h(d_h - \Gamma \varrho)d_h - (d_h - \Gamma \varrho) \right] \times$$
$$\times \tfrac{\varrho}{\varrho'} \left[d_h(d_h + h\varrho^\nu) - 1 - h\Gamma \varrho^{\nu+1} \right]$$
$$+ h \big(d_h \, (d_h - \Gamma \varrho) \, (d_h(d_h + h\varrho^\nu) - 1) \big)$$
$$+ h(d_h^2 - 1) + \sigma_a \, \varrho^{\nu+2} \, .$$

The principal symbol (in the semi-classical sense) is

$$\mathcal{P}_a^0(x, \xi, \sigma_a, \Gamma) = \frac{\varrho(x)}{\varrho'(x)}(i\xi - \Gamma \varrho(x))(\xi^2 + 1)^2 + \sigma_a \varrho(x)^{\nu+2} \, . \qquad (88)$$

The analysis of the zeroes of the symbol is similar to the other model. We have:

$$\operatorname{Re} \mathcal{P}_a^0(x, \xi, \sigma_a, \Gamma) = \frac{\varrho(x)}{\varrho'(x)}(-\Gamma \varrho(x))(\xi^2 + 1)^2 + \sigma_a \varrho(x)^{\nu+2} \, ,$$

and

$$\operatorname{Im} \mathcal{P}_a^0(x, \xi, \sigma_a, \Gamma) = \frac{\varrho(x)}{\varrho'(x)}\xi(\xi^2 + 1)^2 \, .$$

The zero set of $\operatorname{Im} \mathcal{P}_a^0$ is in $\{\xi = 0\}$ and:

$$\operatorname{Re} \mathcal{P}_a^0(x, 0, \sigma_a, \Gamma) = \frac{\varrho(x)}{\varrho'(x)}(-\Gamma \varrho(x)) + \sigma_a \varrho(x)^{\nu+2} \, ,$$

which leads to the analysis of the solutions of:

$$\sigma_a \varrho(x)^\nu \varrho'(x) = \Gamma$$

or

$$\sigma_a \varrho(x)^{2\nu+1}(1 - \varrho(x)) = \Gamma \, .$$

Hence we have first to analyze the variation of the function:

$$[0, 1] \ni t \mapsto \theta(t) := (1 - t)t^{2\nu+1} \, . \qquad (89)$$

If $\nu > 0$, θ is an application from $]0, 1[$ onto $]0, \vartheta_a^{max}]$, with

$$\vartheta_a^{max} = \frac{(2\nu + 1)^{2\nu+1}}{(2\nu + 2)^{2\nu+2}} \, . \qquad (90)$$

$$0 < \vartheta_a^{max} < 1 \, ,$$

and the maximum in $]0, 1[$ is obtained at

$$t_a^{max} = \frac{2\nu + 1}{2\nu + 2} \, .$$

For $L \in]0, \vartheta_a^{max}[$, two solutions of $\theta(t) = L$, satisfying:

$$0 < t_-(L) < t_a^{max} < t_+(L) .$$

$x \mapsto \varrho(x)$ is a bijection of $\mathbb{R}$ onto $]0, 1[$.

For any $L \in]0, \vartheta_a^{max}[$, there exist two points $x_\pm(L)$ such that

$$\varrho(x_\pm(L)) = t_\pm(L) ,$$

and consequently

$$\theta(\varrho(x_\pm(L))) = L .$$

We note also that, when $\xi = 0$,

$$(\partial \mathcal{P}_a^0 / \partial \xi)(x, 0) = i \frac{\varrho(x)}{\varrho'(x)} \neq 0 ,$$

which shows that $\mathcal{P}_a^0$ is also of principal type.
Finally when $\frac{\Gamma}{\sigma_a} > \vartheta_a^{max}$, is satisfied, one gets the local ellipticity of the symbol $\mathcal{P}_a^0$.

5.7 Semi-Classical Regimes for the Ablation Models

Let us emphasize at this stage the analogies between the three last physical models. As in the case of the Kelvin–Helmholtz model, two different "effective" parameters have been exhibited corresponding to each situation of the convective velocity problem (parameter denoted by σ_c) and in the ablation front problem (parameter denoted by σ_a), together with h. Both problems lead to a h-differential equation on one of the unknowns, and consist in finding a function $u(x, h)$ such that

$$\mathcal{P}_p(x, \frac{1}{i}h\frac{d}{dx}, h, \sigma_p, \Gamma)u = 0 ,$$

where $\mathcal{P}_p$ is a fifth or fourth order h-differential operator. The main results will take the following form:

Under suitable relations on the reference density profile at $\tilde{x} \to \pm\infty$, then, if

$$\Gamma \in]0, \vartheta_p^{max} \sigma_p[,$$

then 0 belongs to the h-family-pseudospectrum of

$$\mathcal{P}_p(x, \frac{1}{i}h\frac{d}{dx}, h, \sigma_p, \Gamma) .$$

More precisely there exists $x_p(\Gamma, \sigma_p)$ such that there exists a WKB solution of

$$\mathcal{P}_p u = \mathcal{O}(h^\infty)$$

localized in the neighborhood of the point $x_p(\Gamma, \sigma_p)$.

Note that in the three models there is no quantization of Γ. The result is with this respect quite different from the solution of the problem linked with pure Rayleigh–Taylor instability.

The assumptions are essentially optimal in this semi-classical regime:
Under the same assumptions on the density profile, and, for $\Gamma > \vartheta_p^{max} \sigma_p$, no approximate (in the WKB sense) bounded solution can be constructed, if h is small enough.

This was a consequence of the ellipticity of the operator for this regime of operators. Let us now look at what is obtained by application of Theorem 4.1.

5.7.1 Application to the (ELRC) Model

We start from $a_0 = \mathcal{Q}_c^0$:

$$a_0(x, \xi) = (\xi + i\Gamma\varrho)^2(\xi^2 + 1) + \varrho\varrho'\sigma_c^2 \,.$$

We obtain

$$\operatorname{Re} a_0(x, \xi) = (\xi^2 - \Gamma^2\varrho^2)(\xi^2 + 1) + \varrho\varrho'\sigma_c^2 \,,$$

and

$$\operatorname{Im} a_0(x, \xi) = 2\Gamma\varrho\xi(\xi^2 + 1) \,.$$

Let us compute the Poisson bracket at $(x_c, 0)$

$$\begin{aligned}
\{\operatorname{Re} a_0, \operatorname{Im} a_0\}(x_c, 0) \\
= -2\Gamma\varrho(x_c)[-2\Gamma^2\varrho(x_c)\varrho'(x_c) + \sigma_c^2(\varrho\varrho')'(x_c)] \,,
\end{aligned}$$

which is effectively strictly negative and Theorem 4.1 can be applied.

5.7.2 Application to the (ELKA) Model

The principal symbol is here:

$$\mathcal{P}_a^0(x, \xi) = \frac{\varrho(x)}{\varrho'(x)}(i\xi - \Gamma\varrho(x))(\xi^2 + 1)^2 + \sigma_a\varrho^{\nu+2} \,. \tag{91}$$

Because we are interested in null solutions, it is equivalent to apply the criterion for

$$a_0(x, \xi) = (i\xi - \Gamma\varrho(x))(\xi^2 + 1)^2 + \sigma_a\varrho(x)^{2\nu+2}(1 - \varrho(x)) \,.$$

We get

$$\text{Re } a_0 = -\Gamma \varrho(x)(\xi^2 + 1)^2 + \sigma_a \varrho(x)^{2\nu+2}(1 - \varrho(x)) ,$$

and

$$\text{Im } a_0 = \xi(\xi^2 + 1)^2 .$$

A point in $a_0^{-1}(0)$ should satisfy $\xi = 0$, and for the real part:

$$-\Gamma \varrho(x_0) + \sigma_a \varrho(x_0)^{2\nu+1}(1 - \varrho(x_0)) = 0 .$$

Let us compute the Poisson bracket at $(x_0, 0)$:

$$\begin{aligned} \{\text{Re } a_0, \text{Im } a_0\}(x_0, \xi_0) = &\ \Gamma \varrho'(x_0) \\ &-\sigma_a(2\nu + 2)\varrho'(x_0)\varrho^{2\nu+1}(x_0) \\ &+\sigma_a(2\nu + 3)\varrho'(x_0)\varrho^{2\nu+2}(x_0) . \end{aligned}$$

Dividing by $\varrho'(x_0)$ (which is positive), we get that this bracket is negative if:

$$\begin{aligned} \frac{\Gamma}{\sigma_a} &< (2\nu + 2)\varrho^{2\nu+1}(x_0) - (2\nu + 3)\varrho^{2\nu+2}(x_0) \\ &= \varrho^{2\nu+1}(x_0)\left((2\nu + 2) - (2\nu + 3)\varrho(x_0)\right) . \end{aligned} \tag{92}$$

Hence Theorem 4.1 can be applied if this last condition is verified.

5.8 Subellipticity II: At the Boundary of $\Sigma(a_0)$

In the case of our example the neighborhood of the maximal Γ, for which one can construct quasimodes can be analyzed by analyzing the iterated brackets. One can then apply the results, which were recalled in [DeSjZw] which are related to the much older theory of the subelliptic operators (see [Ho3] and references therein). More recent work have been performed by N. Lerner (See his lectures in this conference) and by K. Pravda-Starov in his quite recent PhD [Pra2].

The theorem in [DeSjZw] reads:

Theorem 5.1. *We assume that a_0 is a C^∞ bounded function together with all its derivatives and that our operator is an h-pseudodifferential operator with principal symbol $(x, \xi) \mapsto a_0(x, \xi)$. Then if $z_0 \in \partial\Sigma(a_0)$ is of finite type for a_0 of order $k \geq 1$, then k is even and there exists $C > 0$ such that, for h small enough,*

$$\|(A(h) - z_0)^{-1}\| \leq C h^{-\frac{k}{k+1}} . \tag{93}$$

Here $\Sigma(a_0)$ is the closure of the numerical range of a_0.

The condition that a_0 is of finite type for the value z_0 is that a_0 is of principal type (i.e. $\nabla_{x,\xi} a_0(x, \xi) \neq 0$) at any point (x, ξ) such that $a_0(x, \xi) = z_0$ and that at these points there is at least one non zero (possibly iterated) bracket of $\text{Re } a_0$ and $\text{Im } a_0$.

Remarks 5.2

- *The authors in [DeSjZw] mention that one can reduce more general cases to this one by use of the functional calculus. This can be verified more directly in our case.*
- *In the case of (ELRC), it is enough to compose on the left by* $(I - h^2 \Delta)^{-2}$. *In the second case, the situation is a little more delicate. See [HelLaf2].*

Let us show how this theorem can be applied in this case, with $k = 2$.

Application to (ELRC) Model
Coming back to this model, we first observe that

$$\{\operatorname{Re} a_0, \operatorname{Im} a_0\}(x, \xi) = -2\Gamma \varrho[-2\Gamma^2 \varrho \varrho' + \sigma_c^2 (\varrho \varrho')'] + \mathcal{O}(\xi^2), \qquad (94)$$

When

$$\Gamma = \Gamma_c := \vartheta_c^{max} \sigma_c, \qquad (95)$$

we can verify that

$$a_0(x_c, 0) = 0, \quad \{\operatorname{Re} a_0, \operatorname{Im} a_0\}(x_c, 0) = 0,$$

and that, under the additional assumption that the point x_c is a non degenerate maximum of $\frac{\varrho'}{\varrho}$,

$$\{\operatorname{Im} a_0, \{\operatorname{Re} a_0, \operatorname{Im} a_0\}\}(x_0, 0) \neq 0. \qquad (96)$$

This implies that the operator is of type 2.

Application to the (ELKA) Model
We consider, after a small change, as principal symbol the function:

$$(x, \xi) \mapsto -\Gamma \varrho(x) + \sigma_a \varrho(x)^{2\nu+2}(1 - \varrho(x))(1 + \xi^2)^{-2} + i\xi. \qquad (97)$$

Here we choose $\Gamma/\sigma_a = \vartheta_a^{max}$, where ϑ_a^{max} is defined in (90). The Poisson bracket $\{\operatorname{Re} a_0, \operatorname{Im} a_0\}$ vanishes at $(x_0, 0)$, where x_0 is the point such as $\varrho(x_0) = \frac{2\nu+1}{2\nu+2}$. Now the computation of the first iterated bracket gives

$$\{\operatorname{Im} a_0, \{\operatorname{Im} a_0, \operatorname{Re} a_0\}\}(x_0, 0) = (2\nu + 1)\varrho'(x_0)^2 \varrho(x_0)^{2\nu} \neq 0. \qquad (98)$$

As in the case of the ellipticity zone, one can eliminate the problem at ∞.

Remark 5.2. The Dencker–Sjöstrand–Zworski Theorem shows that there exists $C > 0$ and h_0 such that, when Γ belongs to $]\Gamma_p - Ch^{\frac{2}{3}}, \Gamma_p]$ and $h \in]0, h_0]$, then no approximate solution in the kernel of $\mathcal{P}_p(x, \frac{1}{i} d_h, h, \sigma_p, \Gamma)$ exists.

Acknowledgements My first thanks are for O. Lafitte for introducing me to the subject and for fruitful collaboration [HelLaf1, HelLaf2]. Many preliminary versions of this course have been presented to various audiences and in different forms together with him (see for example [He2], [La2]). I also acknowledge partial support by the programme "Instabilités hydrodynamiques en fusion par confinement inertiel" supported by the CEA, the IRPHE and the CNRS.

References

[Ag]   S. Agmon. Lectures on exponential decay of solutions of second order elliptic equations. Bounds on eigenfunctions of N-body Schrödinger operators. Mathematical Notes of Princeton University.

[BeSh]   F.A. Berezin, and M.A. Shubin. The Schrödinger equation. Mathematics and its Applications. Kluwer Academic Publishers (1991).

[BrHe]   M. Brunaud, B. Helffer. Un problème de double puits provenant de la théorie statistico-mécanique des changements de phase, (ou relecture d'un cours de M. Kac). LMENS 1991.

[Bo]   L.S. Boulton. Non-selfadjoint harmonic oscillator semi-groups and pseudospectra. J. Operator Theory 47, p. 413–429 (2002).

[BudkoL]   A.B. Budko and M.A. Liberman. Stabilization of the Rayleigh–Taylor instability by convection in smooth density gradient: W.K.B. analysis. Phys. Fluids, p. 3499–3506 (1992).

[Cha]   S. Chandrasekhar. Hydrodynamic and Hydromagnetic stability. Dover publications, inc., New York (1981).

[ChLa]   C. Cherfils, and O. Lafitte. Analytic solutions of the Rayleigh equation for linear density profiles. Physical Review E 62 (2), p. 2967–2970 (2000).

[CCLaRa]   C. Cherfils-Clerouin, O. Lafitte, and P-A. Raviart. Asymptotics results for the linear stage of the Rayleigh–Taylor instability. In Advances in Mathematical Fluid Mechanics (Birkhäuser) (2001).

[CCLa]   J. Cahen, R. Chong-Techer, and O. Lafitte. Expression of the linear groth rate for a Kelvin–Helmholtz instability appearing in a moving mixing layer. To appear in M^2AN 2006.

[Col]   P. Collet. Leçons sur les systèmes étendus. Unpublished (2005).

[Da1]   E.B. Davies. Pseudo-spectra, the harmonic oscillator and complex resonances. Proc. R. Soc. Lond. A, p. 585–599 (1999).

[Da2]   E.B. Davies. Semi-classical states for non self-adjoint Schrödinger operators. Comm. Math. Phys. 200, p. 35–41 (1999).

[Da3]   E.B. Davies. Pseudo-spectra of differential operators. J. Operator theory 43 (2), p. 243–262 (2000).

[DeSjZw]   N. Dencker, J. Sjöstrand, and M. Zworski. Pseudo-spectra of semi-classical (Pseudo)differential operators. Comm. in Pure and Applied Mathematics 57(4), p. 384–415 (2004).

[DiSj]   M. Dimassi and J. Sjöstrand. Spectral asymptotics in the semi-classical limit. London Mathematical Society Lecture Note Series 269. Cambridge University Press, Cambridge (1999).

[DuSj]   J. Duistermaat and J. Sjöstrand. A global construction for pseudo-differential operators with non-involutive characteristics. Invent. Math. 20, p. 209–225 (1973)

[Eg]   Y.V. Egorov. Subelliptic pseudodifferential operators. Soviet Math. Dok. 10, p. 1056–1059 (1969).

[Go]   V.N. Goncharov. Selfconsistent stability analysis of ablation fronts in inertial confinement fusion. PHD of Rochester University (1998).

[GH]   Y. Guo and H.J. Hwang. On the dynamical Rayleigh–Taylor instability. Arch.
            Ration. Mech. Anal. 167, no. 3, p. 235–253 (2003).

[Ha]   M. Hager. Instabilité spectrale semi-classique d'opérateurs non-autoadjoints.
            PHD Ecole Polytechnique (2005).

[He1]   B. Helffer : Introduction to the semiclassical analysis for the Schrödinger op-
            erator and applications. Springer lecture Notes in Math., n^0 1336 (1988).

[He2]   B. Helffer. Analyse semi-classique et instabilité en hydrodynamique. Talk at
            "Journées de GrandMaison" Nov. 2003. http://www.math.u-psud.fr/∼ helffer.

[HelLaf1]   B. Helffer and O. Lafitte. Asymptotic growth rate for the linearized Rayleigh
            equation for the Rayleigh–Taylor instability.   Asymptot. Anal. 33 (3–4),
            p. 189–235 (2003).

[HelLaf2]   B. Helffer and O. Lafitte. Study of the semi-classical regime for ablation
            front models. Archive for Rational Mechanics and Applications. Vol 183 (3),
            p. 371–409 (2007).

[HelNi]   B. Helffer and F. Nier Hypoelliptic estimates and spectral theory for Fokker–
            Planck operators and Witten Laplacian. Lecture Notes in Mathematics 1862
            (2005).

[HePa]   B. Helffer and B. Parisse : Effet tunnel pour Klein-Gordon, Annales de l'IHP,
            Section Physique théorique, Vol. 60, n^{0}2, p. 147–187 (1994).

[HeRo1]   B. Helffer and D. Robert. Calcul fonctionnel par la transformée de Mellin et
            applications. Journal of functional Analysis, Vol. 53, n$^\circ$3, oct. 1983.

[HeRo2]   B. Helffer and D. Robert. Puits de potentiel généralisés et asymptotique semi-
            classique. Annales de l'IHP (section Physique théorique), Vol. 41, n$^\circ$3, p.
            291–331 (1984).

[HelSj1]   B. Helffer, J. Sjöstrand. Multiple wells in the semi-classical limit I. Comm. in
            PDE 9(4), p. 337–408, (1984).

[HelSj2]   B. Helffer, J. Sjöstrand. Analyse semi-classique pour l'équation de Harper (avec
            application à l'étude de l'équation de Schrödinger avec champ magnétique)
            Mémoire de la SMF, n^{0}34, Tome 116, Fasc. 4, (1988).

[HerNi]   F. Hérau, F. Nier. Isotropic hypoellipticity and trend to equilibrium for the
            Fokker–Planck equation with high degree potential. Arch. Rat. Mech. Anal.
            171(2), p. 151–218 (2004).

[HerSjSt]   F. Hérau, J. Sjöstrand and C.C. Stolk Semi-classical subelliptic estimates and
            the Kramers–Fokker–Planck equation. Comm. Partial Differential Equations
            30, no. 4–6, p. 689–760 (2005).

[Ho1]   L. Hörmander. Differential operators of principal type. Math. Ann. 140,
            p. 124–146 (1960).

[Ho2]   L. Hörmander. Differential operators without solutions. Math. Ann. 140, p.
            169–173 (1960).

[Ho3]   L. Hörmander. The analysis of Pseudo-differential operators. Grundlehren der
            mathematischen Wissenschaften 275, Springer, Berlin (1983–1985).

[KeSu]   M. Kelbert and I. Suzonov. Pulses and other wave processes in fluids. Kluwer.
            Acad. Pub. London Soc.

[KlSc90]   M. Klein and E. Schwarz. An elementary approach to formal WKB expansions
            in R^n. Rev. Math. Phys. 2 (4), p. 441–456 (1990).

[Kull]   H.J. Kull. Incompressible description of Rayleigh–Taylor instabilities in laser-
            ablated plasmas. Phys. Fluids B 1, p. 170–182 (1989).

[KullA]   H.J. Kull and S.I. Anisimov. Ablative stabilization in the incompressible
            Rayleigh–Taylor instability. Phys. Fluids 29 (7), p. 2067–2075 (1986).

[La1]   O. Lafitte. Sur la phase linéaire de l'instabilité de Rayleigh–Taylor. Séminaire
            à l'Ecole Polytechnique, Exp. No. XXI, Sémin. Equ. Dériv. Partielles, Ecole
            Polytech., Palaiseau (2001).

[La2]   O. Lafitte. Quelques rappels sur les instabilités linéaires. Talk at "Journées de
            GrandMaison" Nov. 2003.

[La3]   O. Lafitte. Linear ablation growth rate for the quasi-isobaric model of Euler equations with thermal conductivity. In preparation (2006).

[Le]   N. Lerner. Some facts about the Wick calculus. Cime Course in Cetraro (June 2006).

[Li]   P.-L. Lions. *Mathematical topics in fluid mechanics.* Volume 1 Incompressible models. Oxford Science Publications (1996).

[Mar]   J. Martinet. Personal communication and work in progress.

[Mas]   L. Masse. Etude linéaire de l'instabilité du front d'ablation en fusion par confinement inertiel. Thèse de doctorat de l'IRPHE (2001).

[Pra1]   K. Pravda-Starov. A general result about pseudo-spectrum for Schrödinger operators. Proc. R. Soc. Lond. A 460, p. 471–477 (2004).

[Pra2]   K. Pravda-Starov. A complete study of the pseudo-spectrum for the rotated harmonic oscillator. Journal of the London Math. Soc. (2) 73, p. 745–761 (2006).

[Pra3]   K. Pravda-Starov. Etude du pseudo-spectre d'opérateurs non auto-adjoints. PHD University of Rennes (June 2006).

[Ris]   H. Risken. The Fokker–Planck equation. Vol. 18. Springer-Verlag, Berlin (1989).

[Rob]   D. Robert. Autour de l'analyse semi-classique. Progress in Mathematics, Birkhäuser (1987).

[RoSi]   S. Roch and B. Silbermann. C^*-algebras techniques in numerical analysis. J. Oper. Theory 35, p. 241–280 (1996).

[Si1]   B. Simon. Functional Integration and Quantum Physics. Academic Press (1979).

[Si2]   B. Simon. Semi-classical analysis of low lying eigenvalues I. Non degenerate minima: Asymptotic expansions. Ann. Inst. Henri Poincaré 38, p. 295–307 (1983).

[Sj1]   J. Sjöstrand. Singularités analytiques microlocales. Astérisque 95, p. 1–166 (1982).

[Sj2]   J. Sjöstrand. Pseudospectrum for differential operators. Séminaire à l'Ecole Polytechnique, Exp. No. XVI, Sémin. Equ. Dériv. Partielles, Ecole Polytech., Palaiseau (2003).

[St]   J.W. Strutt (Lord Rayleigh). Investigation of the character of the equilibrium of an Incompressible Heavy Fluid of Variable Density. Proc. London Math. Society 14, p. 170–177 (1883).

[Tay]   G. Taylor. The instability of liquid surfaces when accelerated in a direction perpendicular to their planes. Proc. Roy. Soc. A 301, p. 192–196 (1950).

[Tref]   L.N. Trefethen. Pseudospectra of linear operators. Siam Review 39, p. 383–400 (1997).

[Trev]   F. Trèves. A new proof of subelliptic estimates. Comm. Pure Appl. Math. 24, p. 71–115 (1971).

[Zw]   M. Zworski. A remark on a paper of E.B. Davies. Proc. Amer. Math. Soc. 129 (10), p. 2955–2957 (2001).

An Introduction to Numerical Methods of Pseudodifferential Operators

M.P. Lamoureux and G.F. Margrave

Abstract Pseudodifferential operators were introduced in the mid 1900s as a powerful new tool in the development of the theory of partial differential equations. More recently, it has been observed that these operators may form the basis for novel numerical techniques used in the analysis and simulation of physical systems including wave propagation and medical imaging, as well as for advances in signal processing. This course will focus on the numerical implementations of pseudodifferential operators and practical applications. Of particular interest are: the variety of ways to implement these operators, including via fast transforms, decomposition into product-convolution operators, Gabor multipliers, and wavelet transform; speed of implementations; relation to asymptotic expansions; real experience with numerical implementations including in geophysical applications.

1 Signal Processing and Pseudodifferential Operators

1.1 Introduction to Seismic Imaging

These lectures are motivated by a specific physical problem, the imaging of the earth's subsurface using seismic waves. Mathematically, this problem can be approximated to be the study of the three-dimensional acoustic wave equation

Michael P. Lamoureux
University of Calgary, 2500 University Drive NW, Calgary, AB, Canada T2N 1N4
e-mail: mikel@ucalgary.ca

Gary Margrave
University of Calgary, 2500 University Drive NW, Calgary, AB, Canada T2N 1N4
e-mail: margrave@ucalgary.ca

L. Rodino, M.W. Wong (eds.) *Pseudo-Differential Operators.* Lecture Notes in Mathematics 1949.
© Springer-Verlag Berlin Heidelberg 2008

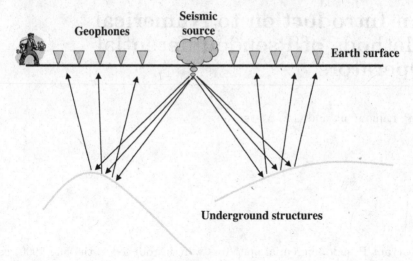

Fig. 1 Seismic wave experiment

$$\frac{\partial^2 \varphi}{\partial x^2} + \frac{\partial^2 \varphi}{\partial y^2} + \frac{\partial^2 \varphi}{\partial z^2} = \frac{1}{c^2}\frac{\partial^2 \varphi}{\partial t^2},$$

where $\varphi(x, y, z, t)$ is the wave function and $c = c(x, y, z)$ is the (non-constant) speed of propagation of the seismic wave. Numerical calculations based on this equation are key to recovering an image of the subsurface.

In practice, real seismic experiments are performed by exploding dynamite on the surface (or near surface) of the earth, and recording the vibrations produced by the explosive energy using sensitive geophones. The signals of interest are those acoustic waves that have propagated down to some interesting geological formation and been reflected back to the surface. Figure 1 shows a simplified seismic setup, with instruments placed on the surface of the earth, and seismic energy traveling along raypaths within the earth. The geophones are typically placed at the surface of the earth,[1] and are sensitive enough to record vibrations that have traveled from the dynamite source, down five kilometers or more through rock, and return the same distance back to the surface. Hundreds of geophones are monitored and the signal data collected from them are recorded in a computer; dozens of dynamite blasts are recorded, exploded at different locations, and independently at different times. This recorded data is then processed to create an image of the subsurface, and is often used in the search for hydrocarbons (oil and gas).

In marine seismic imaging, the experiments take place at sea rather than on land. Typically, thousands of hydrophones attached to floating cables are towed behind a ship traveling back and forth across a target area. A signal

---

[1] In Vertical Seismic Profiling, or VSP, the geophones may be placed deep within the earth, usually down the borehole of an oilwell.

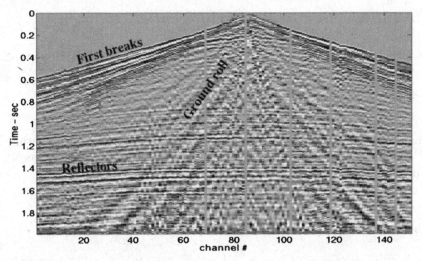

Fig. 2 Data collected from a seismic experiment. Reflectors are the images of interest

is initiated by setting off an air gun at the ocean surface, which starts an acoustic wave traveling down through the water and into the ocean floor, which then propagates through the rock in the form of a seismic wave. Energy that is reflected back travels through rock, then water, and then to the hydrophones where the data is recorded. It is also possible to use an Ocean Bottom Cable (OBC), where a string of geophones is actually placed on the ocean floor and data recorded directly from the ocean bottom. This is a much more expensive setup. In either case, a given marine survey may collect data over several days, covering dozens of square kilometers of territory. A huge amount of data can be collected in one ocean survey, often amounting to terabytes of computer files.

The raw data contains much information, and much noise. In Fig. 2, we have organized the time series data from a sequence of geophones placed in a line, so that a useful raw image appears. The first breaks and ground roll are due to surface propagation of seismic energy, and are considered noise. The hyperbolas represent reflections from interesting geological structures, and contain the information of a true image of the subsurface. Many of the developments in seismic data processing are techniques in signal processing and include steps to remove noise due to first breaks and ground roll (muting, f–k filtering), to straighten the hyperbolas into proper reflectors (migration), and to sharpen the image (deconvolution), among others.

Thus, this problem combines aspects of physical modeling via PDEs, and signal processing. Time–frequency analysis, and in particular pseudodifferential operators, are well-suited for combining these approaches.

1.2 Introduction to Pseudodifferential Operators

We begin with a function f of a single variable x on the real line $\mathbb{R}$, say time, and define its Fourier transform

$$\hat{f}(\xi) = \int_{\mathbb{R}} f(x)e^{-2\pi i x \cdot \xi}\, dx,$$

which represents the function in a dual variable ξ, which we usually think of as frequency.[2] The inverse Fourier transform returns f from its frequency representation, so

$$f(x) = \int_{\mathbb{R}} \hat{f}(\xi)e^{2\pi i x \cdot \xi}\, d\xi.$$

The convolution of two functions f, g is given by the integral

$$(f * g)(x) = \int_{\mathbb{R}} f(x - y)g(y)\, dy$$

and it will be useful to note that the Fourier transform converts a convolution of functions in time to a product of functions in frequency

$$\widehat{f * g} = \hat{f} \cdot \hat{g}$$

and conversely,

$$\widehat{f \cdot g} = \hat{f} * \hat{g}.$$

Two simple ways to modify a function is to multiply by a function in the time domain, $f(x) \mapsto m(x)f(x)$, or to multiply by a function in the frequency domain $\hat{f}(\xi) \mapsto n(\xi)\hat{f}(\xi)$, which by the convolution property is equivalent to convolving $f(x)$ with the corresponding function $g(x)$ whose Fourier transform is $\hat{g}(\xi) = n(\xi)$. Convolution operators can be characterized as the only linear shift-invariant operators. Multiplication operators, on the other hand, are precisely those that preserve support (in the x domain).

A pseudodifferential operator is a linear operator $K = K_\sigma$ on the space of functions $f(x)$ that combines both these notions of time and frequency modifications using a single function of two variables $\sigma(x, \xi)$. The operator K_σ is defined by inserting the function $\sigma(x, \xi)$ into the Fourier inversion formula, with

$$(K_\sigma f)(x) = \int_{\mathbb{R}} \sigma(x, \xi)\hat{f}(\xi)e^{2\pi i x \cdot \xi}\, d\xi.$$

---

[2] The choice to place the factor 2π in the integral determines a correspondence of physical units. If x is measured in seconds, then ξ is measured in Hertz, or cycles per second. If x is measured in meters, ξ is spatial frequency and is measured in wavenumber, the reciprocal of wavelength in meters. This choice also makes the Fourier transform a unitary operator. Not all authors use this convention.

Without worrying too much about what types of functions are used here, note that this formula makes sense for any Schwartz class function f and any reasonable choice of σ, including a tempered distribution. Intuitively, one can think of the symbol $\sigma(x, \xi)$ as a prescription of how to modify the function f simultaneously in the time domain x and the frequency domain ξ.

This particular representation of a pseudodifferential operator is called the Kohn–Nirenberg form, which will be the focus in these lectures. It is particularly well-suited for our study of the wave equation, and in general for non-constant coefficient linear PDEs. Let us mention a few other useful forms. First, let us expand the Kohn–Nirenberg form into a double integral, by writing out the Fourier transform, so

$$(K_\sigma f)(x) = \int_\mathbb{R} \int_\mathbb{R} \sigma(x, \xi) f(y) e^{2\pi i (x-y)\cdot \xi} \, d\xi.$$

The adjoint form, or right quantization, is given by the formula

$$(A_\sigma f)(x) = \int_\mathbb{R} \int_\mathbb{R} \sigma(y, \xi) f(y) e^{2\pi i (x-y)\cdot \xi} \, dy d\xi.$$

Notice here we start with the function $f(y)$ in the time domain, modify in that same domain by multiplying by $\sigma(y, \xi)$ and then moving to the frequency domain by taking the forward Fourier transform of the result. The second integration is the inverse Fourier transform. The Feynman quantization take the average of the two, so

$$(F_\sigma f)(x) = \int_\mathbb{R} \int_\mathbb{R} \frac{\sigma(x, \xi) + \sigma(y, \xi)}{2} f(y) e^{2\pi i (x-y)\cdot \xi} \, dy d\xi.$$

The Weyl form interpolates between the x, y variables, so

$$(W_\sigma f)(x) = \int_\mathbb{R} \int_\mathbb{R} \sigma(\frac{x+y}{2}, \xi) f(y) e^{2\pi i (x-y)\cdot \xi} \, dy \, d\xi.$$

This can be seen as an interpolation between the Kohn–Nirenberg and adjoint forms, as any convex combination of x, y in the integral will give a useful operator. We define a general Weyl form as

$$(W_{\sigma,t} f)(x) = \int_\mathbb{R} \int_\mathbb{R} \sigma((1-t)x + ty, \xi) f(y) e^{2\pi i (x-y)\cdot \xi} \, dy \, d\xi$$

for any real parameter t. Note that $W_{\sigma,0}$ is the Kohn–Nirenberg form, $W_{\sigma,1}$ is the adjoint form, and $W_{\sigma,1/2}$ is the usual Weyl operator. What is special about the $1/2$ is that the Weyl operator W_σ is self-adjoint when σ is real-valued; this is not the case for the K–N and adjoint forms. Perhaps for this reason, the Weyl form is often used in quantum mechanics.

We introduce here a new form, called the L-form or line-averaging representation. The idea is to simply average along the line connecting x to y, thus define

$$L_\sigma = \int_0^1 W_{\sigma,t}\, dt.$$

More generally, one could include a weighting function $g(t)$ and set

$$L_{\sigma,g} = \int_0^1 g(t) W_{\sigma,t}\, dt.$$

However, a more useful generalized L-form includes a weighting function of three variables, $g(x,y,t)$ where we define the operator $L_{\sigma,g}$ via a weighted sum in the Weyl double integral. That is, we set

$$(L_{\sigma,g}f)(x) = \int_\mathbb{R} \int_\mathbb{R} \int_0^1 g(x,y,t)\sigma(x + t(y - x), \xi) f(y) e^{2\pi i(x-y)\cdot\xi}\, dt\, dy\, d\xi.$$

In fact, it would be very useful to define the L-form using an average along a curved path connecting points x and y. Such a definition has physical significance. For instance, with seismic waves, the path connecting two points could be chosen as the path of least time (Fermat's principle). In numerical work, geophysicists already have good computer code for computing such paths, so it is reasonable to include them in our calculations.

We have the following:

Theorem 1.1. *Suppose the function $g(x,y,t)$ is non-negative, normalized ($\int_0^1 g(x,y,t)dt = 1$) and satisfies reciprocity ($g(y,x,1-t) = g(x,y,t)$). The following hold:*

- *If $\sigma(x,\xi) = m(x) + n(\xi) + cx \cdot \xi$ then $L_{\sigma,g} = W_\sigma = F_\sigma$.*
- *If σ is real-valued, then $L_{\sigma,g}$ is self-adjoint.*

This definition is motivated by our experience in wavefield extrapolation. With y representing the source position of a wavefield, and x the target position of the extrapolated wavefield, each of the above operators corresponds to propagating a wave using local information about the velocity, at different positions. So K_σ uses velocity at target x to propagate; A_σ uses the velocity at the source y. Neither is physically correct, but they are often close enough to be useful. The Feynman form uses the average of those two operators, while the Weyl form uses velocity information at the midpoint $(x+y)/2$. The L-form uses a (weighted) average of operators all along the line from x to y (Fig. 3).

None of these is exactly correct for wave propagation, but each step gives a better approximation. It is worth noting that the L-form is (numerically) not much more difficult to compute than the Weyl form. To see this, observe that there is a pseudodifferential operator P_τ defined for symbols of three variables, $\tau(x,y,\xi)$, as

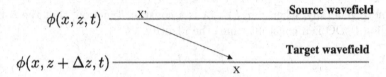

$$\phi(x, z, t) \quad \underline{\text{X'}} \qquad\qquad\qquad\quad \textbf{Source wavefield}$$

$$\phi(x, z + \Delta z, t) \quad\underline{\qquad\qquad\qquad\qquad}\qquad \textbf{Target wavefield}$$
$$\qquad\qquad\qquad\qquad\qquad\qquad \text{x}$$

Normal form takes velocity at x (target).

Adjoint form takes velocity at x' (source).

Feynman form takes average of the two forms.

Weyl form takes the velocity at midpoint (x+x')/2.

L-form takes the average velocity along the line connecting x, x'.

Fig. 3 Wavefield extrapolation and the physical meaning of various PsDOs

$$(P_\tau)(x) = \int_{\mathbb{R}} \int_{\mathbb{R}} \tau(x, y, \xi) f(y) e^{2\pi i (x-y)\cdot\xi} \, dy \, d\xi.$$

Given a symbol of two variables $\sigma(x, \xi)$, we can line-average to get a function of three variables,

$$\tilde{\sigma}(x, y, \xi) = \int_0^1 g(x, y, t)\sigma(x + t(y - x), \xi) \, dt.$$

It is simple to verify that

$$L_{\sigma,g} = P_{\tilde{\sigma}}.$$

It might seem that symbols of three variables gives the possibility of more types of operators. It doesn't. The following result follows from Schwartz's kernel theorem.

Theorem 1.2. *Let $B : \mathcal{S} \to \mathcal{S}'$ be a continuous linear operator. Then there exist symbols (tempered distributions) $\sigma_1, \sigma_2, \sigma_3, \tau$ such that*

$$B = K_{\sigma_1} = A_{\sigma_2} = W_{\sigma_3} = P_\tau.$$

It would seem these results are true for the Feynman and L-forms as well, although we have not explicitly checked this.

We return to the Kohn–Nirenberg form. From the Fourier inversion formula, it is easy to see that when the symbol $\sigma(x, \xi)$ is the constant function one, $\sigma(x, \xi) \equiv 1$, that the corresponding linear operator is the identity,

$$K_1 = I,$$

and more generally, when $\sigma(x,\xi) \equiv \lambda$ is a constant function, that the corresponding PsDO is a constant times the identity,

$$K_\lambda = \lambda I.$$

If $\sigma(x,\xi) = m(x)$ is a function of the time variable x only, the PsDO reduces to a simple multiplier function, since by the Fourier inversion formula, we have

$$(K_m f)(x) = \int_{\mathbb{R}} m(x)\hat{f}(\xi)e^{2\pi i x \cdot \xi}\, d\xi = m(x)\int_{\mathbb{R}} \hat{f}(\xi)e^{2\pi i x \cdot \xi}\, d\xi = m(x)f(x).$$

On the other hand, when $\sigma(x,\xi) = \hat{g}(\xi)$ is a function of the frequency variable ξ only, then we obtain an operator which is simply multiplication by the $\hat{g}(\xi)$ in the frequency domain, which reduces to a convolution by the function of time g,

$$(K_{\hat{g}} f)(x) = \int_{\mathbb{R}} \hat{g}(\xi)\hat{f}(\xi)e^{2\pi i x \cdot \xi}\, d\xi = \int_{\mathbb{R}} \widehat{g * f}(\xi)e^{2\pi i x \cdot \xi}\, d\xi = (g * f)(x).$$

An interesting case combining these two operations is obtained by setting $\sigma(x,\xi) = m(x)\hat{g}(\xi)$, in which case

$$K_\sigma(f)(x) = m(x)(f * g)(x) = (K_m K_{\hat{g}})f,$$

which is a product-convolution operator. It is somewhat encouraging that the symbol for the product $K_m K_{\hat{g}}$ is simply the product of the symbols m and $\hat{g}$; however, the order of the product of operators is important. In general, $K_m K_{\hat{g}} \neq K_{\hat{g}} K_m$, because the operations of multiplication and convolution do not commute. This also indicates that there will be some complications in creating a functional calculus for composing operators and symbols.

We are ignoring for the moment important issues such as what are the spaces on which these operators are defined, and what is the class of symbols $\sigma(x,\xi)$ for which this becomes a useful theory. In the above examples, it is easy enough to see that everything is well-defined on the Hilbert space of square integrable functions $L^2(\mathbb{R})$, and the exact norms on the operators K_m and $K_{\hat{g}}$ are given by the supremum norms $||m||_\infty$ and $||\hat{g}||_\infty \leq ||g||_{L^1}$, respectively. However, a simple product of these two simple cases, say $\sigma(x,\xi) = m(x)\hat{g}(\xi)$ gives a product-convolution operator, whose norm can be extremely difficult to compute. So it is clear there is some complicated properties contained in these simple symbols.

Pseudodifferential operators generalize the notion of differential operators, which we can see by differentiating $f(x)$ in the Fourier inverse formula,

$$\frac{d}{dx}f(x) = \frac{d}{dx}\int_{\mathbb{R}} \hat{f}(\xi)e^{2\pi i x \cdot \xi}\, d\xi = 2\pi i \int_{\mathbb{R}} \xi \hat{f}(\xi)e^{2\pi i x \cdot \xi}\, d\xi = 2\pi i (K_\xi f)(x).$$

That is, with $\sigma(x,\xi) = \xi$, the corresponding PsDO is the linear differential operator

$$K_\xi = \frac{1}{2\pi i}\frac{d}{dx}.$$

More generally, with $\sigma(x,\xi) = a_0(x) + a_1(x)\xi + a_2(x)\xi^2 + \cdots + a_m(x)\xi^m$ a polynomial in ξ with coefficients $a_\alpha(x)$, the corresponding Kohn–Nirenberg pseudodifferential operator is an ordinary m-th order differential operator with non-constant coefficients, given by

$$K_\sigma f(x) =$$

$$= a_0(x)f(x) + a_1(x)\frac{1}{2\pi i}f'(x) + a_2(x)\frac{1}{(2\pi i)^2}f''(x) + \cdots + a_m(x)\frac{1}{(2\pi i)^m}f^{(m)}(x).$$

Thus we have a prescription for translating nonconstant coefficient linear differential operators into pseudodifferential operators, creating a symbol which is a polynomial in ξ. Pseudodifferential operators are more general, since we are not restricted only to polynomials for the symbol. Note that the derivative operator is not bounded on $L^2(\mathbb{R})$, which is not surprising since the function $\sigma(x,\xi) = \xi$ is not bounded on its domain. However, the operator norm could be controlled by controlling the size of the first derivative. This is our first hint the Sobolev spaces may be important in controlling the norms of these operators,

1.3 A Jump in Dimension

Pseudodifferential operators are effective for the study of partial differential operators, and thus we must apply these ideas to functions of several variables. It is very convenient to set some notation so that the dimensionality of the problem remains rather hidden: our formulas with be correct, but the complications hidden by a really nifty trick.

The variable x will now represent a point in n-dimensional space $\mathbb{R}^n$, with $x = (x_1, x_2, \ldots, x_n)$. The Fourier dual variable ξ is of the same dimension, $\xi = (\xi_1, \xi_2, \ldots, \xi_n)$ and their dot product is given as

$$x \cdot \xi = x_1\xi_1 + x_2\xi_2 + \cdots x_n\xi_n.$$

The Fourier transform and inverse formula remain the same, except now we integrate over n-dimensional space, so

$$\hat{f}(\xi) = \int_{\mathbb{R}^n} f(x)e^{-2\pi i x \cdot \xi}\, dx,$$

where of course $dx = dx_1\, dx_2\, \ldots dx_n$.

A multiindex $\alpha = (\alpha_1, \alpha_2, \ldots, \alpha_n)$ is an n-tuple of non-negative integers, whose length is defined as the sum $|\alpha| = \alpha_1 + \cdots + \alpha_n$ and whose factorial is defined as the product $\alpha! = (\alpha_1!)(\alpha_2!) \cdots (\alpha_n!)$. The set of multiindices is partially ordered by the relation $\alpha \geq \beta$ if and only if $\alpha_k \geq \beta_k$ for each k. When $\alpha \geq \beta$ we can define the binomial coefficient as

$$\binom{\alpha}{\beta} = \frac{\alpha!}{\beta!(\alpha - \beta)!},$$

which is simply the product of the one-dimensional binomial coefficients $\binom{\alpha_k}{\beta_k}$; when it is not the case that $\alpha \geq \beta$, we define $\binom{\alpha}{\beta}$ to be zero.

With the multiindex notation, we define monomials in x as

$$x^\alpha = x_1^{\alpha_1} x_2^{\alpha_2} \cdots x_n^{\alpha_n}$$

and mixed partial derivatives as

$$\partial^\alpha = \partial_1^{\alpha_1} \partial_2^{\alpha_2} \cdots \partial_n^{\alpha_n} = \frac{\partial^{|\alpha|}}{\partial x_1^{\alpha_1} \partial x_2^{\alpha_2} \cdots \partial x_n^{\alpha_n}}.$$

The utility of this compact notation becomes clear when we obtain the following multivariable formulations of the Binomial, Leibniz, and Taylor formulas.

Theorem 1.3 (Binomial formula). *For multiindex α and points x, y in* $\mathbb{R}^n$,

$$(x + y)^\alpha = \sum_\beta \binom{\alpha}{\beta} x^\beta y^{\alpha - \beta}.$$

Theorem 1.4 (Leibniz's formula). *For multiindex α and C^∞ functions f and g,*

$$\partial^\alpha(fg) = \sum_\beta \binom{\alpha}{\beta} (\partial^\beta f)(\partial^{\alpha - \beta} g).$$

Theorem 1.5 (Taylor's formula). *For a C^∞ function f defined on R^n, any two points x, y in $\mathbb{R}^n$, and any positive integer k, we have*

$$f(x + y) = \sum_{|\alpha| < k} \frac{y^\alpha}{\alpha!} \partial^\alpha f(x) + \sum_{|\alpha| = k} k \frac{y^\alpha}{\alpha!} \int_0^1 (1 - t)^{k-1} \partial^\alpha f(x + ty) \, dt.$$

Again, the amazing utility of this multiindex formulation is that the above formulas, so familiar from calculus in one variable, appear in exactly the same form in multidimensions, with the notation of multiindices hiding the complications.

Given a function $\sigma(x, \xi)$ of n-dimensional variables $x, \xi \in \mathbb{R}^n$, we define the pseudodifferential operator K_σ on functions $f(x)$ by the formula

$$(K_\sigma f)(x) = \int_{\mathbb{R}^d} \sigma(x,\xi)\hat{f}(\xi)e^{2\pi i x \cdot \xi}\, d\xi.$$

Again, we are ignoring the question of when these integrals are defined, but for reasonably smooth functions σ and f, with the f vanishing sufficiently rapidly at infinity, there are no problems.

At some point we have to deal with the fact that our pseudodifferential operators carry along an extra factor of $2\pi i$ in the derivatives. It is convenient to define a differential operator D and its multiindex powers via the scaling

$$D^\alpha = \frac{1}{(2\pi i)^{|\alpha|}} \frac{\partial^{|\alpha|}}{\partial x_1^{\alpha_1} \partial x_2^{\alpha_2} \cdots \partial x_n^{\alpha_n}}.$$

We then see the monomial symbol $\sigma(x,\xi) = \xi^\alpha$ produces a pseudodifferential operator which is a scaled version of the mixed partials, so

$$K_{\xi^\alpha} = D^\alpha.$$

This is a particularly convenient scaling.

More generally, a linear, non-constant coefficient partial differential operator of order m can be expressed in the form

$$f \mapsto \sum_{|\alpha| \le m} a_\alpha(x) D^\alpha f(x)$$

which is represented exactly by the pseudodifferential operator K_σ with polynomial symbol

$$\sigma(x,\xi) = \sum_{|\alpha| \le m} a_\alpha(x)\xi^\alpha.$$

By hiding the factor of $2\pi i$ in the definition of D, we get a simpler correspondence between symbols and operators.

1.4 Boundedness of the Operators

If we have any hope of evaluating these operators numerically, we need to know that they are bounded operators. Informally, this means if the input function f is small in some norm, then the output function $K_\sigma f$ is also relatively small, perhaps in another norm. More precisely, if the error in the input is small (and numerically, there is always error), then the error in the output is small. With an unbounded operator, the output error could be arbitrarily large, and so the numerical computation becomes meaningless.

In many practical situations, one assumes the symbol $\sigma(x,\xi)$ is zero outside some compact set, and otherwise continuous, or at least measurable and reasonably bounded. One might assume so on the grounds that we never observe signals for arbitrarily large times, or for arbitrarily high frequencies, and thus it is meaningless to apply an operator that modifies the signal based on these unmeasurable characteristics. It is easy to see that such an operator is then a bounded operator on the Hilbert space of square integrable functions.

Theorem 1.6. *Suppose the symbol $\sigma(x, \xi)$ is square integrable on $\mathbb{R}^n \times \mathbb{R}^n$. Then the corresponding pseudodifferential operator K_σ is a bounded operator from $L^2(\mathbb{R}^n)$ into $L^2(\mathbb{R}^n)$, with operator norm bounded by the L^2 norm of σ.*

The proof is obtained by computing the inner product of $K_\sigma f$ with another L^2 function g, and observing this can be rewritten as an inner product in $L^2(\mathbb{R}^n \times \mathbb{R}^n)$ of the function $\sigma(x, \xi)e^{2\pi i x \cdot \xi}$ with the elementary tensor $(g \otimes \hat{f})(x, \xi) = g(x)\hat{f}(\xi)$. That is,

$$\langle K_\sigma f, g \rangle_{L^2(\mathbb{R}^n)} = \int \int \sigma(x, \xi)\hat{f}(\xi)e^{2\pi i x \cdot \xi}\overline{g}(x) \, dx \, d\xi =$$

$$= \langle \sigma(x, \xi)e^{2\pi i x \cdot \xi}, g(x)\overline{\hat{f}(\xi)} \rangle_{L^2(\mathbb{R}^n \times \mathbb{R}^n)},$$

and thus

$$|\langle K_\sigma f, g \rangle| \le ||\sigma||_2 ||f||_2 ||g||_2,$$

from which the operator norm bound follows.

It is worth noting that this L^2 norm is a gross overestimate of the operator norm; for instance, in the case $\sigma(x, \xi) \equiv 1$, the L^2 norm is unbounded, but the identity operator has norm one. In fact, it is not too hard to push the above proof to show the L^2 norm of the symbol is a bound on the Hilbert–Schmidt norm of the operator K_σ. Thus, in the case of finite L^2 norms, the PsDO is in fact a compact operator.[3]

A much deeper result is the following:

Theorem 1.7 (Calderon–Vaillancourt). *Suppose $\sigma(x, \xi)$ and all its derivatives are bounded. Then K_σ is a bounded operator from $L^2(\mathbb{R}^n)$ into $L^2(\mathbb{R}^n)$. The norm of K_σ can be computed from the sup of σ and a finite number of its derivatives.*

We refer the reader to [4] for the details of this theorem and its proof. Notice the theorem tells us a lot about boundedness. One might ask, however, why do the derivative of the symbol have to be bounded? After all, the previous result on Hilbert–Schmidt operators didn't look at derivatives. The reason is simply that the symbol must not oscillate too much, or some bad

---

[3] Without getting too technical, a compact operator is an operator that is well-approximated by finite dimensional linear operators, which of course are matrices. The Hilbert–Schmidt operators are compact operators whose singular values form a square-summable sequence, which says something important about how quickly those finite matrices can approximate.

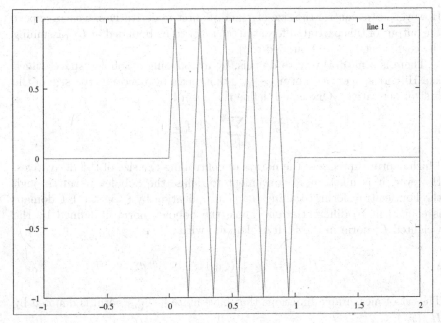

Fig. 4 A function with small L^2 norm, but large derivative

behaviour could occur. For instance, a symbol of the form $\sigma(x,\xi) = e^{-2\pi i x \cdot \xi}$ is nicely bounded, but its derivatives are not. And this symbol is especially chosen to cancel the complex exponential built into the definition of the pseudodifferential operator. So here, $K_\sigma = \delta_0$, the Dirac delta function at zero. That is, $(K_\sigma f)(x) = f(0)$, which is surely not continuous on L^2; neither is the output a square integrable function. So yes, the derivatives of the symbol really do need to be bounded, if we hope to get a bounded operator.

To go further on boundedness result, one must first face the fact that differential operators aren't usually bounded on a function space like $L^2(\mathbb{R})$. For instance, in Fig. 4, we see a sawtooth function which is relatively small in the L^2 norm (norm of about one), but as the slopes of the linear sections are high, the L^2 norm of the derivative will be large (norm of about 20). With more zigzags in the function, the higher the norm of the derivative will be. Thus, we expect a differential map of the form

$$f \mapsto \frac{df}{dx}$$

to be unbounded on $L^2(\mathbb{R})$, as would be a more general partial differential operator such as

$$f \mapsto \sum_{|\alpha| \leq m} a_\alpha(x) D^\alpha f(x).$$

However, if we control the size of f and its first m derivatives, we can expect the output of this partial differential operator to be bounded in L^2 (assuming the coefficients a_α are bounded too).

There is a natural way to do this, by introducing a Sobolev space, which is a Hilbert space with norm $||\cdot||_{(m)}$ that takes into account the size of the first m derivatives. One could define the norm by

$$||f||^2_{(m)} = \sum_{|\alpha| \leq m} ||D^\alpha f(x)||^2_{L^2}$$

which captures precisely the notion of controlling the size of the derivatives. However, it is much more convenient to define the Sobolev norm through the Fourier transform, recalling that multiplication by ξ^α in the FT domain is equivalent to differentiation. Thus, the Sobolev norm is defined by the weighted L^2 norm in the Fourier domain, with

$$||f||^2_{(m)} = \int_{\mathbb{R}^n} |\hat{f}(\xi)|^2 (1 + |\xi|^2)^m \, d\xi.$$

The set of measurable functions that have finite $||\cdot||_{(m)}$ norm is denoted by $H_m(\mathbb{R}^d)$.

This definition extends to any real number m, although for negative m this becomes a set of distributions. Note that $H_0(\mathbb{R}^n) = L^2(\mathbb{R}^n)$ and the space H_{-m} is the dual space of H_m. It is also worth noting the following.

Theorem 1.8. $f \in H_m(\mathbb{R}^n)$ iff $f, \partial_1 f, \partial_2 f, \ldots, \partial_n f \in H_{m-1}(\mathbb{R}_n)$

That is, a function is in a particular order of Sobolev space if and only if the function and its first partial derivatives are in the next lowest order space. In particular, f is in $H_m(\mathbb{R}^d)$ if and only if f and its m-th order partial derivatives are square-integrable.

With this in mind, it is easy to see that if a function f is in $H_m(\mathbb{R}^d)$, then a sum of the form

$$\sum_{|\alpha| \leq m} a_\alpha(x) D^\alpha f(x)$$

would be square integrable (again, assuming the a_α are bounded). That is, in the case of a linear partial differential operator of order m, with bounded coefficients, we obtain a bounded linear operator $K_\sigma : H_m(\mathbb{R}^d) \to H_0(\mathbb{R}^d)$, where of course σ is the polynomial symbol for the operator,

$$\sigma(x, \xi) = \sum_{|\alpha| \leq m} a_\alpha(x) \xi^\alpha.$$

Noting that the PsDO drops the order of the Sobolev spaces by m, which is the order of the operator, we might expect a more general result, that the operator

$$K_\sigma : H_s(\mathbb{R}^d) \to H_{s-m}(\mathbb{R}^d)$$

is continuous, for any Sobolev index s. A moment's reflection reveals, however, that not only do the coefficients a_α need to be bounded, but so should their derivatives of various orders.[4]

Thus we are led naturally to the notion of symbols spaces, that is, those functions $\sigma(x,\xi)$ which grow no faster than polynomially in ξ, as with all their derivatives. Roughly speaking, we want $\sigma(x,\xi)$ to behave like a polynomial of some order m in variable ξ, so we expect the order of growth to *decrease* as we differentiate in ξ. More precisely, we say that $\sigma(x,\xi)$ is a symbol of order m, and write $\sigma \in S^m$, if the function and all its derivatives are of a specific order

$$\partial_x^\alpha \partial_\xi^\beta \sigma(x,\xi) \sim O(|\xi|^{m-|\beta|}) \text{ as } \xi \to \infty.$$

With this definition in hand, a standard boundedness result is the following:

Theorem 1.9. *Suppose $\sigma(x,\xi)$ is a symbol of order m. Then for every real number s, the pseudodifferential operator*

$$K_\sigma : H_s(\mathbb{R}^d) \to H_{s-m}(\mathbb{R}^d)$$

is bounded.

A proof of this result is available in [4] or [20]. An important step in the proof shows how to reduce these operators to symbols of order zero. Skipping the adjoints for simplicity, one sets $b = \lambda^{s-m} \sharp \sigma \sharp \lambda^{-s}$, where $\lambda(x,\xi) = (1 + |\xi|^2)^{1/2}$. The corresponding pseudodifferential operator $K_b = \lambda^{s-m}(D)K_\sigma \lambda^{-s}(D)$ is order zero. The $\lambda^{-s}(D)$ will map H_0 onto H_s, then K_σ maps H_s to H_{s-m} and finally $\lambda^{s-m}(D)$ maps things back to H_0. This strongly suggests, then, that we can implement any pseudodifferential operator by moving up and down Sobolev spaces with powers of $\lambda(D)$, and really do the hard work with an order zero operator on H_0.

Thus, all our difficulties with derivatives can be dealt with using the elliptic, constant coefficient operator $\lambda(D)$. And we are left with implementing a bounded operator on Hilbert space, from the zero order pseudodifferential operator.

2 Manipulating Pseudodifferential Operators

2.1 Composition of Operators

We need formulas for composing pseudodifferential operators, since we always will be reducing to order zero using the $\lambda(D)$ operator. And there are other

---

[4] Since derivatives of the output will involve derivatives of the coefficients, by the product rule.

specific applications we have in mind, such as trying to find a square root of an operator for use in wave propagation models.

First, we should observe that there really is some difficulty here. Because of our choice to use the Kohn–Nirenberg form for the pseudodifferential operators, certain products of operator are easy. For instance, with one multiplier symbol $\sigma_1(x, \xi) = m(x)$ and one differential symbol $\sigma_2(x, \xi) = \xi^\alpha$, we have a simple operator product and simple symbol product

$$K_m K_{\xi^\alpha} = K_{m(x)\xi^\alpha}.$$

That is, the symbol for the product operator $K_\sigma = K_m K_{\xi^\alpha}$ is simply the pointwise product $\sigma(x, \xi) = m(x)\xi^\alpha$. However, if we reverse the order of the operator product, a more complicated symbol results, since by Leibniz's formula,

$$K_{\xi^\alpha} K_m f = D^\alpha(m \cdot f) = \sum_{\beta \le \alpha} \binom{\alpha}{\beta} (D^\beta m)(D^{\alpha - \beta} f),$$

and thus the corresponding symbol for the product $K_\sigma = K_{\xi^\alpha} K_m$, in this order, is the polynomial

$$\sigma(x, \xi) = \sum_{\beta \le \alpha} \binom{\alpha}{\beta} (D^\beta m)\xi^{\alpha - \beta} = \sum_{\beta \le \alpha} \frac{1}{\beta!} (D^\beta m)\partial_\xi^\beta \xi^\alpha.$$

For operator products with more general symbols, we can expect a more complicated result. Here it is, and the formula for the adjoint:

Theorem 2.1. *Let σ be a symbol of order m, and ρ a symbol of order l. Define the functions σ^* and $\sigma\sharp\rho$ by the integrals*

$$\sigma^*(x, \xi) = \int_{\mathbb{R}^n \times \mathbb{R}^n} e^{-2\pi i y \cdot \eta} \overline{\sigma}(x - y, \xi - \eta)\, dy\, d\eta$$

$$\sigma\sharp\rho(x, \xi) = \int_{\mathbb{R}^n \times \mathbb{R}^n} e^{-2\pi i y \cdot \eta} \sigma(x, \xi - \eta)\rho(x - y, \xi)\, dy\, d\eta.$$

Then σ^* is symbol of order m while $\sigma\sharp\rho$ is a symbol of order $m + l$. Furthermore, the corresponding pseudodifferential operators satisfy

$$K_\sigma^* = K_{\sigma^*} \quad \text{and} \quad K_\sigma K_\rho = K_{\sigma\sharp\rho}.$$

That is, we have an exact formula for the adjoint of a symbol, and for the composition of two symbols, so that the corresponding operators represent the corresponding compositions of operators. In the next section, we will see some approximate formulas, but these ones here really are exact. It is a simple exercise to verify that these formulas work for the simple cases $m(x)\sharp\xi^\alpha$ and

$\xi^\alpha \sharp m(x)$. Try it![5] It is also interesting to note that the adjoint of a Kohn–Nirenberg operator K_σ has a simple representation in the adjoint form (or right quantization), for we can check that

$$K_\sigma^* = A_{\overline{\sigma}},$$

where $\overline{\sigma}$ is simply the pointwise complex conjugate of the function σ. Hence the name for the adjoint form. The Weyl form is even easier, as we can also verify that

$$W_\sigma^* = W_{\overline{\sigma}}.$$

However, the composition formulas for these two forms also involve rather complex integrals.

2.2 Asymptotic Series

The exact composition formula for the simple case where $\sigma(x, \xi) = \xi^\alpha$ and $\rho(x, \xi) = m(x)$ gives a finite sum

$$\sigma \sharp \rho(x, \xi) = \sum_{\beta \leq \alpha} \frac{1}{\beta!} (\partial_\xi^\beta \xi^\alpha)(D^\beta m)(x).$$

This suggests we might have a general formula for the composition of symbols, involving only derivatives of the symbols, perhaps something such as

$$\sigma \sharp \rho \sim \sum_\beta \frac{1}{\beta!} (\partial_\xi^\beta \sigma)(D^\beta \rho).$$

In fact this formula is true, provided we take an infinite "sum" over all β. And provided we understand the notation $\sim$ does not mean equality, but equivalence via an asymptotic expansion, which we will define below. The adjoint symbol also has an asymptotic expansion, as

$$\sigma^* \sim \sum_\beta \frac{1}{\beta!} \partial_\xi^\beta D_x^\beta \overline{\sigma}.$$

So, what's an asymptotic expansion? The idea is to expand an order-m symbol $\sigma \in S^m$ into a sum of symbols $\sigma_0, \sigma_1, \sigma_2, \ldots$ of decreasing order, such that the partial sums $\sum_{j<k} \sigma_j$ capture the highest order parts of σ. The series picks off the polynomial parts of the symbol, starting with the highest order terms and working its way down.

---

[5] You may wish to read ahead on oscillatory integrals, since a delta function appears here which may seem troublesome. It's not.

More precisely, we say $\sigma \in S^m$ is asymptotic to a series $\sum_{j=0}^{\infty} \sigma_j$ if each $\sigma_j \in S^{m-j}$, supp $\sigma \subset \cup_j$supp σ_j, and

$$\sigma - \sum_{j=0}^{k-1} \sigma_j \in S^{m-k}.$$

Remarkably, given ANY selection of symbols $\sigma_j \in S^{m-j}$ we can always find some symbol $\sigma \in S^m$ with $\sigma \sim \sum \sigma_j$. In fact, σ is unique modulo $S^{-\infty} = \cap_m S^m$.

Are these things useful? Well, the idea is that the symbol σ differs from $\sum_{j<k} \sigma_j$ by only a low order symbol, and so the difference between the corresponding pseudodifferential operators will be a smoothing operator. Which is good, in some ways. But, it could be a really large (bounded) smoothing operator, which could present difficulties for numerical calculations. If one is simply tracking singularities, though, this is not such a bad thing.

Note that here, conditions on the symbol σ are required so that higher order derivatives have proper order behaviour like $|\xi|^{m-k}$. So to obtain a good asymptotic formula, one must choose the symbol class appropriately. If one is not concerned about asymptotic series, perhaps this careful choice of symbol class doesn't matter so much. For instance, in certain applications such as seismic imaging, mathematical models of the earth include nondifferentiable, discontinuous coefficients in the pseudodifferential operators, which do not fit well into the classical theory. But these are perfectly acceptable operators for numerical work.

2.3 Oscillatory Integrals

The integrals

$$\sigma^*(x, \xi) = \int_{\mathbb{R}^n \times \mathbb{R}^n} e^{-2\pi i y \cdot \eta} \overline{\sigma}(x - y, \xi - \eta) \, dy \, d\eta$$

$$\sigma \sharp \rho(x, \xi) = \int_{\mathbb{R}^n \times \mathbb{R}^n} e^{-2\pi i y \cdot \eta} \sigma(x, \xi - \eta) \rho(x - y, \xi) \, dy \, d\eta.$$

are special cases of oscillatory integrals. The integrals are, in general, not absolutely convergent, but the oscillatory factor $e^{-2\pi i y \cdot \eta}$ make these kinds of integrals particularly well-behaved. In fact, the usual "nice" integral properties (change of variable, integration by parts, etc.) will hold.

We abstract a little, and get a slightly more general result. Instead of the space $\mathbb{R}^n \times \mathbb{R}^n$, we use any Euclidean space $\mathbb{R}^d$. The function $(y, \eta) \mapsto y \cdot \eta$ is a special case of a nondegenerate quadratic form, so we replace it with any real, nondegenerate quadratic form $x \mapsto q(x)$ on $\mathbb{R}^d$. The symbols are replaced with a class of functions called amplitudes, where a function $a(x)$ is called an amplitude of order m, written $a \in A^m$, if the ratios

$$\frac{\partial^\alpha a(x)}{(1 + |x|^2)^{m/2}}$$

are bounded in $x \in \mathbb{R}^d$, for all multiindices α. This is similar to the symbol class S^m discussed in the previous section, except polynomial growth is required in all variables.

The corresponding oscillatory integral is defined via a limit, as

$$\int e^{iq(x)} a(x)\, dx = \lim_{\epsilon \to 0} \int e^{iq(x)} a(x)\varphi(\epsilon x)\, dx,$$

where $\varphi(x)$ is any smooth, Schwartz class function with $\varphi(0) = 1$. The first result of oscillatory integrals is that this definition is independent of the choice of φ and always gives a finite result. In the special case where the amplitude a is an integrable function, $a \in L^1(\mathbb{R}^d)$, the integrand $e^{iq(x)} a(x)$ is absolutely convergent and the limit agree with the usual Lebesgue integral.

Here are the results we want:

Theorem 2.2.

1. Change of variables: if A is a real, invertible matrix, then

$$\int e^{iq(Ay)} a(Ay)|\det A|\, dy = \int e^{iq(x)} a(x)\, dx.$$

2. Integration by parts:

$$\int e^{iq(x)} a(x)\partial^\alpha b(x)\, dx = \int b(x)(-\partial)^\alpha (e^{iq(x)} a(x))\, dx.$$

3. Interchange of differentiation, integral:

$$\partial_y^\alpha \int e^{iq(x)} a(x, y)\, dx = \int e^{iq(x)} \partial_y^\alpha a(x, y)\, dx.$$

4. Fubini's theorem:

$$\int e^{ir(y)} \left(\int e^{iq(x)} a(x, y)\, dx \right) dy = \int e^{iq(x)+ir(y)} a(x, y)\, dx\, dy.$$

Why is this interesting? Mainly because it is useful. Physicists use these results all the time. For instance, they are very happy to write the identity

$$\int_{\mathbb{R}^n} e^{2\pi i x \cdot \xi}\, d\xi = \delta_0(x),$$

which suggests a certain parameterized, non-convergent integral is equal to the Dirac delta function (which is actually a distribution, not a proper function). At first glance, this equality seems completely at odds with our usual understanding of Lebesgue integrals. In particular, the integral on the right does not converge in Lebesgue theory: the integrand is not absolutely convergent. However, let's see how oscillatory integral work for this. Let $u(x)$ be a Schwartz class function, and integrate it against the above, to obtain the double integral

$$\int \int e^{2\pi i x \cdot \xi} u(x) \, d\xi dx.$$

Since u is Schartz class, it is an amplitude (in A^0), and the function $(x, \xi) \mapsto 2\pi x \cdot \xi$ is a real quadratic form on $\mathbb{R}^{2n}$, so this double integral is in fact an oscillatory integral. We evaluate it by inserting a smooth function $\varphi(x, \xi) = \chi(\xi)\chi(x)$ say, with $\chi(0) = 1$, and we compute by definition that

$$\int \int e^{2\pi i x \cdot \xi} u(x) \, d\xi dx = \lim_{\varepsilon \to 0} \int \int e^{2\pi i x \cdot \xi} \chi(\varepsilon \xi) \chi(\varepsilon x) u(x) \, d\xi dx;$$

integrate over $d\xi$ for a Fourier transform

$$= \lim_{\varepsilon \to 0} \int \varepsilon^{-n} \hat{\chi}(-x/\varepsilon) \chi(\varepsilon x) u(x) \, dx$$

and now we change variables, $x/\varepsilon \mapsto x$

$$= \lim_{\varepsilon \to 0} \int \hat{\chi}(-x) \chi(\varepsilon^2 x) u(\varepsilon x) \, dx$$

then take the limit inside

$$= \chi(0) u(0) \int \hat{\chi}(-x) \, dx$$

compute the inverse Fourier transform

$$= \chi^2(0) u(0) = u(0).$$

In particular, integrating against $\int_{\mathbb{R}^n} e^{2\pi i x \cdot \xi} \, d\xi$ is the same as integrating against the Dirac delta function, hence we say this integral is equal, as a distribution, to the delta function. Notice this computation via oscillatory integrals gives the correct constants in the Dirac delta function, a computation which can be tricky when performed by other means.

Of course, one might prefer this computation:

$$\int \int e^{2\pi i x \cdot \xi} u(x) \, d\xi dx = \int \left(\int e^{2\pi i x \cdot \xi} u(x) \, dx \right) d\xi$$

$$= \int \hat{u}(-\xi) d\xi$$

$$= \int e^{i0 \cdot \xi} \hat{u}(-\xi) d\xi$$

$$= u(0).$$

And what's wrong with that? Well, the problem is at the first equation, we applied Fubini's theorem to change the order of integration, and we don't really know that it is applicable here. So this computation is suspect.

A very nice illustration of the utility of oscillatory integral is to demonstrate the Poisson summation formula, which can be written as a property about the Fourier transform of a comb of Dirac delta functions:

$$\sum_{n\in\mathbb{Z}} \hat{\delta}_n = \sum_{n\in\mathbb{Z}} \delta_n.$$

For this author (and many mathematicians), it seems like this would take some good analytical theory to prove. For others (including many physicists), it is a simple application of what they know in principle from oscillatory integrals.

There is a rich theory to oscillatory integrals, the idea is always the same: the integrals have a well-described oscillating part, multiplied by an amplitude that is not too badly behaved. For example, the function

$$E(t,x) = \int \frac{e^{2\pi i(ct|\xi|+x\cdot\xi)} - e^{2\pi i(-ct|\xi|+x\cdot\xi)}}{-4\pi c|\xi|} \, d\xi$$

is the difference of two (more general) oscillatory integrals, which satisfies the Cauchy problem for the wave equation

$$\frac{1}{c^2}\frac{\partial^2 E}{\partial t^2} - \Delta E = 0 \qquad \text{in } \mathbb{R}^{1+n}$$

with initial conditions

$$E(0,x) = 0, \qquad \frac{\partial E}{\partial t}(0,x) = \delta_0(x).$$

One can check this by differentiating under the integral sign, which is legal for such oscillatory integrals. This example is due to Hormander.

2.4 Other Pseudo-Topics

In most introductory texts on pseudodifferential operators, there is a lot of attention given to elliptic operators, in particular the question of how to invert them, or find approximate inverses. This follows analogous theories for elliptic partial differential equations. However, in our applications in seismic imaging we are almost always working with hyperbolic equations, such as the wave equation, and we rarely are inverting these operators. So there is not much to say on our practical approach to elliptic equations. It is worth noting, though, that much of the classical theory focuses on elliptic operators, even though many of the ideas of PsDOs work perfectly well with other operators.

Pseudodifferential operators are a subset of Fourier integral operators and Calderon–Zygmund operators. If there were time in these lectures, this would be an appropriate place to expand on these operators. We refer the interested readers to [4] and [18] for further reference.

3 Numerical Implementations

3.1 Sampling and Quantization Error in Signal Processing

If you have never tried to digitally processing a signal before, it seem audacious that anyone even thinks they can! There are a number of sources of significant errors involved in getting information about a signal $f(x) \in \mathbb{R}$ into a computer, besides even the simple measurement errors. We quickly review these issues here. A more complete discussion is available in any good text on digital signal processing, such as [19].

First, there is the aperture problem. Given a function $f(x)$, say of time, you can only observe it for a limited amount of time. You cannot observe it all the way out to infinity, because eventually you will get bored, or die. So in practice, we can really only observe a signal on some finite interval $[a, b]$.

Next is the sampling problem. We can only observe the signal at a finite number of times, as we always have limitation on how much data our computer can hold, as well as physical limitation to how often we can perform a measurement.[6] Usually the assumption is that the data is sampled on some uniform lattice or even a rectangular grid; in one dimension, we samples at linearly spaced points

$$a, a + \Delta_x, a + 2\Delta_x, a + 3\Delta_x, \ldots, a + N\Delta_x = b$$

making measurements of function values

$$f(a), f(a + \Delta_x), f(a + 2\Delta_x), f(a + 3\Delta_x), \ldots, f(a + N\Delta_x).$$

Obviously, there is a lot of missing data in those gaps between sample points. However, the Shannon sampling theorem tells us that if the signal f is bandlimited, then it can be reconstructed exactly from those data points, provided Δ_x is small enough. How small? Smaller than half the highest frequency in the support of the Fourier transform $\hat{f}$, so

$$\Delta_x \leq \frac{1}{2} \frac{1}{max|\xi|},$$

---

[6] Modern instruments can sample at the Gigahertz rate. But that still is a limit!

where $max|\xi|$ means simply the maximum frequency that appears in the support of $\hat{f}$.[7] We could also write this as

$$max|\xi| \leq \frac{1}{2}\frac{1}{\Delta_x},$$

where the quantity on the right is called the Nyquist rate or Nyquist frequency.

Third, there is the quantization problem: a particular sample $y_n = f(a + n\Delta x)$ is a real number, which must be stored to only finite precision in our finite computer. The range of the sample is also limited, both by the computer storage space but also the physical device that measures it.[8] So the actual number stored will not be the same as the number measured.

The technical details of how one samples, quantizes, and reconstructs are left up to the engineers who typical build real physical devices to do these measurements. Unfortunately, the devices never perform exactly, since (1) one cannot make a measurement instantaneously,[9] (2) one cannot make measurements exactly, (3) the measurements must be quantized before sending to the computer, and (4) the reconstruction formulas involve basis functions of infinite length (e.g. sinc functions) which cannot be reproduced exactly with physical devices.

There are also some annoying mathematical details, such as the fact that pointwise evaluation of an L^2 function is meaningless, and the observation that the only L^2 function which is both of finite aperture and bandlimited is the zero function.

However, the underlying methodology is that given a signal f, we can stored in the computer a finite string of quantized numbers $y_0, y_1, \ldots, y_N$ and then compute an approximation to f at any time, let's call it $\tilde{f}$, such that the error

$$f - \tilde{f},$$

is small. In what sense small? Typically, one assumes small in $L^2(\mathbb{R})$, but more realistically one should use a weighted L^2 space, even $L^2[a, b]$, or weighted Sobolev space, as appropriate. In a badly designed system, of course these errors can be very large, on the order of the size of the original signal! Hopefully we won't work with too many badly designed systems.[10]

In summary, the point is that many approximations are swept under the rug by the small error assumption. We will continue in this tradition using our pseudodifferential methods.

---

[7] Notice again here that our good choice of scaling in the Fourier transform eliminates any extraneous factors of 2π in the sampling theorem.

[8] Try imagining measuring the voltage of a lightening strike using your digital multimeter!

[9] Usually an approximation is made instead, of the form $\frac{1}{h}\int_{x_n}^{x_n+h} f(x)\,dx$ for some small h. A circuit of switches, capacitors, and samplers can compute this integral.

[10] In seismic, a typical seismic wave is often undersampled in the spatial domain, which could be a significant source of error. This can cause problems in imaging.

3.2 The Discrete Fourier Transform and Periodization Errors

A somewhat more subtle error is introduced by the common usage of the Discrete Fourier Transform (DFT) and its more efficient implement via the Fast Fourier Transform (FFT).[11] This error is called a periodization error, and results from replacing the function $f(x)$ with a periodic version of itself.

Recall the Discrete Fourier Transform of a sequence $\{y_0, y_1, y_2, \ldots y_{N-1}\}$ of N complex numbers is given by

$$\hat{y}_k = \frac{1}{\sqrt{N}} \sum_{j=0}^{N-1} y_k e^{-2\pi i jk/N}.$$

The $\sqrt{N}$ normalizes this to be a unitary transformation, and so the inverse is given by a similar formula, with the minus sign omitted in the complex exponential. Note the important use of the factor $2\pi i$ which agrees with our earlier usage.[12]

The periodization error appears when we attempt to convolve two func- tions $f(x)$ and $g(x)$ using the DFT applied to the sample sequences

$$y_k = f(0 + k\Delta_x), \qquad z_k = g(0 + k\Delta_x),$$

where we've assumed a zero offset for simplicity. By the convolution theorem, we know that $f * g$ is the inverse Fourier transform of the product of their Fourier transforms, so

$$f * g(x) = iFT(\hat{f} \cdot \hat{g}).$$

One hopes for the same thing with the sample sequences, but in fact we have a circular convolution,

$$(y \otimes z)_k = \frac{1}{\sqrt{N}} \sum_{j=0}^{N-1} y_j z_{mod(k-j)},$$

where $mod(k - j)$ is the difference computed modulo N. That is, the convo- lution is computed as if the function g (and equivalently, f) were repeated periodically with a period of $N\Delta_x$. Or another way to say it, if we try to implement a convolution in the frequency domain through the DFT, we get that

$$y \otimes z = iDFT(\hat{y}\hat{z}) \neq \text{ sampled version of } f * g.$$

---

[11] Popularized by Cooley and Tukey in 1965, but known to Gauss in 1805. Unfortunately, Gauss neglected to tell his colleagues in computer science about this great numerical time saver, and thus remained some obscure mathematician untainted by patent royalties.

[12] Some numerical packages, notably MATLAB, omit the factor of $1\sqrt{N}$ in the forward transform, and compensate by introducing a factor of $1/N$ in the inverse transform. In such a case, neither transform is unitary.

The way to avoid periodization errors is to recognize that a convolution will expand the support set of two functions. If f is supported on $[0, a]$ and g is supported on$[0, b]$ then the (non-periodic) convolution is supported on $[0, a + b]$. So, apply the DFT to sampled signals, where we let the samples cover the longer interval $[0, a + b]$. In this case,

$$y \otimes z = iDFT(\hat{y}\hat{z}) \approx \text{ sampled version of } f * g.$$

Remember that the sample spacing Δ_x still depends on the frequency content of the signals.

The dual problem occurs with multiplication in the time domain. A simple example explains the phenomena. Suppose $f(x) = e^{2\pi i x}$ is sampled at precisely Nyquist, producing samples

$$y_k = e^{k\pi i} = (-1)^k.$$

With $z_k = y_k$ a sampled version of the same signal, we obtain the pointwise product

$$w_k = y_k z_k \equiv 1,$$

the constant function. Yet we were expecting the high frequency product

$$f(x)f(x) = e^{4\pi i x}.$$

What happened was that this high frequency signal is aliased back to zero frequency (a constant), which is an error since it does not represent the true product. In the frequency domain, what has happened is that the support of the signal was in the interval $[-1/2, 1/2]$, but under the product this support interval gets expanded to $[-1, 1]$ which will be aliased. To avoid this aliasing, we must increase the sample rate.

3.3 Direct Numerical Implementation via the DFT

The direct numerical implementation of a pseudodifferential operator K_σ for a given symbol $\sigma(x, \xi)$ is to take the integral definition

$$(K_\sigma f)(\xi) = \int_{\mathbb{R}} \sigma(x, \xi) \hat{f}(\xi) e^{2\pi i x \cdot \xi} \, d\xi,$$

and replace all the functions with their apertured, bandlimited, sampled versions.

For simplicity, we use the same N for the number of samples in both f and $\hat{f}$ with sample samplings related through the Nyquist relationship. More specifically, suppose input function f and its output $K_\sigma f$ are apertured so we that are interested in its values in the time interval $[-A, A]$, and $\hat{f}$ is bandlimited to the frequency interval $[-B, B]$. The sampling in x is determined by the Nyquist frequency, so

$$\Delta_x = \frac{1}{2B} = \frac{2A}{N},$$

where the second equality comes from the fact that interval $[-A, A]$ is cut up into N subintervals. Similarly,

$$\Delta_\xi = \frac{1}{2A} = \frac{2B}{N},$$

and combining these two we see that

$$N = 4AB \text{ and } \Delta_x \Delta_\xi = \frac{1}{N}.$$

That is, the aperture and bandwidth determine the number of samples N and the sample spacing in both the time and frequency domains. The function $\sigma(x, \xi)$ is sampled on a rectangular grid of $N \times N$ points with spacing $\Delta_x \times \Delta_\xi$. The numerical implementation of K_σ is thus defined by the finite Riemann sum

$$(K_{\sigma,N} f)(-A + j\Delta_x) =$$

$$= \frac{2B}{N} \sum_{j=0}^{N-1} \sigma(-A + j\Delta_x - B + k\Delta_\xi) \hat{f}(-B + j\Delta_\xi) e^{2\pi i(-A+j\Delta_x)\cdot(-B+k\Delta_\xi)}.$$

Note the scaling constant $\frac{2B}{N}$ comes from the spacing in the partitions of the Riemann sum. In the exponential, the $j \cdot k$ term expands to

$$e^{2\pi i j \cdot k \Delta_x \Delta_y} = e^{2\pi i j k / N},$$

which is precisely the exponential factors that appear in the Discrete Fourier Transform. Thus this sum can be computed using the DFT at each x value.

The point is, this direct computation gives a good approximation for operators with symbols in class S^0.

Theorem 3.1. *Suppose σ is in symbol class S^0, and f is a Schwartz class function. Then*

$$K_{\sigma,N} f \to K_\sigma f \text{ as } N, A, B \to \infty.$$

More precisely, we mean the usual smooth construction of an L^2 function from the sample points in $K_{\sigma,N} f$ is a function close to $K_\sigma f$ in the L^2.

The proof proceeds by noting that K_σ is a continuous linear operator on L^2, so small L^2 errors in any approximations will remain small. Since f is Schwartz class, its Fourier transform $\hat{f}$ is smooth and rapidly decreasing. Thus, the difference between a piecewise constant approximation to $\hat{f}$ truncated to interval $[-B, B]$, and $\hat{f}$ itself, will be small in the L^2 norm, for B

and N large enough (and Δ_ξ small enough). Note the values of $\hat{f}$ outside this interval will be exponentially small, so the operator K_σ will be small on this part, so we can throw away that part of the integration. The Riemann sum on the remaining finite interval will be a good approximation for small sample size. And the reconstruction brings us close to the final output in the L^2 sense.

It would be useful to have some rule of thumb on how large to choose A, B and N. Usually, the signals of interest have some natural aperture $[-A, A]$ and effective bandwidth $[-B, B]$. The symbol $\sigma(x, \xi)$ should not operate on parts of the signal we don't see, so it is reasonable to assume the effective support of $\sigma(x, \xi)$ is in the rectangle $[-A, A] \times [-B, B]$, and if it is not, it should be truncated to this region. To avoid periodization errors, we should then double the aperture to $[-2A, 2A]$ (to avoid Fourier wrap around) and double the bandwidth by reducing the sample size by one half, $\Delta_x = 1/4B$ (to avoid the aliasing problem).

In many applications, these operators are iterated many times. It is not acceptable to keep doubling the aperture and bandwidth. So some other analysis is required. Although in many practical situations, one learns to live with the aliasing and wrap around – not because it is correct, or of small magnitude, but simply because practitioners are used to them.

What about symbols of higher order, $\sigma \in S^m$, for $m > 0$? Direct discretization will not work. This fact is already understood in signal processing applications, since the operation of differentiation is not continuous on L^2, and thus we expect problems with any pseudodifferential operator of positive order. The problem, of course, is that one cannot integrate out to infinity using the direct method. The usual fix, though, is to recognize that we will have to ignore these arbitrarily high frequencies, and must truncate or window the symbol $\sigma(x, \xi)$ to a zero-th order symbol. One slick trick is to replace σ with the symbol

$$\sigma_B(x, \xi) = B^m \frac{\sigma(x, \xi)}{(B^2 + |\xi|^2)^{m/2}},$$

which reduces an order m symbol to an order 0 symbol, and doesn't change the frequency response too much in the range $[-B, B]$. If this is too abstract, just imagine the one-dimensional derivative with symbol

$$\sigma(x, \xi) = 2\pi i \xi$$

which will convert to the zero-th order symbol

$$\sigma_B(x, \xi) = 2\pi i B \frac{\xi}{(B^2 + |\xi|^2)^{1/2}}$$

and K_{σ_B} will be a good approximation to the derivative for band limited signals. Other smooth windows could be used.

3.4 Operations Count

From the discretized formulas in the previous section, we see the vector samples of $K_{\sigma,N} f$ can be computed as follows:

1. A DFT produces samples of $\hat{f}$.
2. For each x_j, a pointwise product is formed of the row $\sigma(x_j, \xi_k)$ with $\hat{f}(\xi_k)$.
3. For each x_j, an inverse DFT is performed on the above pointwise product.
4. Repeat 2, 3 for all N values of j.

The DFT takes $N \log N$ operations. The pointwise product takes N operations, and repeated N times, this is N^2 operators. The inverse DFT takes $N \log N$ operations, and repeated N times this is $N^2 \log N$ operations.

   The total is thus dominated by the $N^2 \log N$ operations required to complete the many inverse DFTs. Which is a bit shocking, since a linear operator on N-dimensional space typically requires only N^2 operations using a simple matrix-vector multiply.[13] So why do we get so excited about this form, which appears to take many more operations to computer?

   This is an important question. One significant reason is that to convert the K–N form into an $N \times N$ matrix in the standard basis requires the multiplication of two matrices, the $\sigma(x_j, \xi_k)$ with the discrete Fourier matrix. In general, this matrix operation is order N^3, although we can reduce it to order $N^2 \log N$ since there is a Fourier matrix involved. So converting to the K–N operator to standard form takes about as much work as evaluating the K–N form directly. In many applications, including our seismic applications, the operator is created once and then thrown away, so there is no cost savings in converting the K–N operator to standard form (and trying to save operations using fast matrix-vector multiplies).

   Another reason is that the behaviour of the operator in the K–N form is quite nice. The DFT is a unitary operator, the coefficients coming from the symbols $\sigma(x, \xi)$ a smooth and well behaved, so the combination is a "good" operator. (Where good means, roughly, the numerical behaviour is satisfactory.) In contrast, when these K–N operators are converted to standard form, there are often singularities (both actual singularities, and numerical anomalies) along the diagonal, which will give problems in the numerics.

   Finally, the K–N form is preferred because it means something physical, as a time variant linear filter or approximation to a physical differential equation. So plugging in physically realistic symbols gives a physically interesting operator, and the direct implementation gives a means to compute it directly from the symbol. That has great appeal when one is following some physical intuition.

   There are other forms of pseudodifferential operators. The adjoint form also take $O(N^2 \log N)$ operations, because of the DFT in the implementation, just as with the K–N form. The Feynman quantization is the sum of the two,

---

[13] In fact, $O(N^{1+\epsilon})$ operations are possible using rapid matrix multiply algorithms.

and thus is also $O(N^2 \log N)$, although it is worth noting it takes twice as long to compute. The improvements are often worth it.

On the other hand, the Weyl implementation requires order N^3 operations, since each of the N output points requires an order N^2 double integral. Unfortunately, it does not appear that the DFT can be inserted to speed up this computation. The triple-symbol form is the same order of complexity as the Weyl, and so is the L-quantization, since it can be evaluated as a triple symbol.

3.5 Numerical Implementation via Product-Convolution Operators

It was noted earlier that a convolution operator can be implemented through the Fourier transform, thus for the finite sampled signals in a numerical implementation, a convolution can be done in order $N \log N$ operations. A component-wise vector product can be done in only $O(N)$ operations, and thus a single product-convolution operator with symbol $\sigma(x,\xi) = m(x)\hat{g}(\xi)$ can be implemented in order $N \log N$ steps. If a more general symbol $\sigma(x,\xi)$ can be expressed as a sum of M elementary products $m(x)\hat{g}(\xi)$, or is well-approximated by such a sum, then the corresponding pseudodifferential operator can be computed in order $MN \log N$ steps. In particular, if M is much smaller than our sample size, this can lead to very efficient computations.

Sometimes the problem is in a form where the approximation by product-convolution operators is obvious, or the symbol $\sigma(x,\xi)$ decomposes easily into a finite sum of elementary products. However, even when this is not the case, we can use a finite partition of unity to create an approximate decompositions.

For instance, let $p_1(x), p_2(x), \ldots p_J(x)$ be a sequence of non-negative functions that form a partition of unity in the time domain,

$$\sum_j p_j(x) \equiv 1,$$

and $q_1(\xi), q_2(\xi), \ldots q_K(\xi)$ a partition of unity in the frequency domain. Assume for each pair of indices j, k we can find a point (x_j, ξ_k) so that

$$\sigma(x,\xi) - \sigma(x_j, \xi_k)$$

is small for all (x,ξ) in the support of the function $p_j(x)q_k(\xi)$.[14] The finite sum of elementary products

$$\sigma_e(x,\xi) = \sum_{j,k} \sigma(x_j, \xi_k)p_j(x)q_k(\xi)$$

---

[14] Usually we start with σ, assume it is slowly varying, take a representative lattice of points (x_j, ξ_k) and build the partition of unity from that grid.

is close to σ by the partition of unity property, and leads to a sum of JK product-convolution operators with symbols

$$\sigma_{j,k}(x,\xi) = p_j(x)q_k(\xi).$$

Thus the approximation K_{σ_e} can be implemented in order $JKN\log N$ operations.

Note that having σ_e close in magnitude to σ is sufficient for the operators to be close if the symbols have finite support (by the Hilbert–Schmidt property). For more general operators of order zero, we need to know that σ_e is a symbol of order zero, and that the approximation is close both for function values, and some of the first few derivatives (by Calderon–Vailliancourt). This will require certain smoothness assumptions on the partitions of unity.

These ideas have been implemented by one of our graduate students in the case of wavefield extrapolation [11].

3.6 Almost Diagonalization via Wavelet and Gabor Bases

Diagonalizing a matrix is a great way to speed up the calculation of a linear operator. If K_{jk} is an $N \times N$ matrix, and x_k a vector of length N, the computation

$$y_j = \sum K_{jk}x_k, \text{ for } j = 1,\ldots,N$$

required order N^2 operations. If the matrix K_{jk} is diagonal, the operation count drops to only order N since the sum disappears, and we have

$$y_j = K_{jj}x_j, \text{ for } j = 1,\ldots,N.$$

The matrix is *almost* diagonal if the off-diagonal terms are small, in which case we can approximate the matrix-vector product by summing over a limited range of indices, such as

$$y_j \approx \sum_{|j-k|<R} K_{jk}x_k.$$

In this case, the order of operations is only RN, which will be much smaller than N^2 if the band range R is small.

Given a finite dimensional linear operator $K_{\sigma,N}$, we can hope to find an orthonormal basis in which the matrix is diagonal, or almost diagonal. That is, we look for basis functions f_j such that the matrix coefficients in this basis, given by

$$K_{jk} = \langle K_{\sigma,N}f_j, f_k \rangle,$$

are diagonal, or almost diagonal. Once this is done, it is a simple matter to convert any input vector f into this basis representation, apply the almost diagonal operator, and then convert back.

So it is a useful speed up, if you have one operator K_σ and many vectors f to push through it. However, in many of our applications, we tend to use the operator K_σ just once, so it is hardly worth the trouble of diagonalizing it. Even if we did, we would compute the diagonal values by somehow evaluating the linear operator against the particular basis vectors f_j so we are back at the same problem. The symbols we use are generated numerically, so there is no analytic expressions that give us the diagonalization directly. So, unfortunately, we do not have any direct experience in computing these almost diagonal forms.

Nevertheless, it is an interesting idea to pursue. It is well-known in the literature that Calderon–Zygmund operators are almost diagonalized by wavelets; an excellent reference is by Meyer and Coifman [18]. Pseudodifferential operators are a subset of the Calderon–Zygmund operators, and thus are also almost diagonalized by wavelets. Tachizawa has shown that the Wilson basis almost diagonalizes a pseudodifferential operator. Rochberg and Tachizawa show almost diagonalization of the Weyl operators in a local trigonometric basis, which we quote here in a simplified form.

Theorem 3.2 (Rochberg and Tachizawa). *Let σ be a symbol of order zero on $\mathbb{R}^d \times \mathbb{R}^d$. Then there is a local trigonometric basis $\{g_{mn}\}$ indexed by pairs $m, n \in \mathbb{Z}^d$ for which the Weyl operator W_σ is almost diagonal. In fact, the matrix entries in this basis satisfy*

$$|K_{mn,m'n'}| = |\langle W_\sigma g_{mn}, g_{m'n'} \rangle| \leq \frac{C}{(1 + |m - m'|)^{d+2}(1 + |n - n'|)^{d+2}}.$$

The version of the theorem given in [21] is quite a bit more general, but the above statement gives the basic idea. Notice as one moves away from the diagonal, the matrix entries rapidly, given by the polynomial term with power $d + 2$. The local trigonometric basis is given by the typical Gabor atoms,

$$g_{mn}(x) = g(x - m)e^{in \cdot x}$$

where the basic Gabor atom is a d-fold product,

$$g(x) = u(x_1) \times u(x_2) \times \cdots \times u(x_d,)$$

where the function u is non-negative, symmetric, supported in $[-1, 1]$ and its translates form a partition of unity

$$\sum_{k \in \mathbb{Z}} u(x - k)^2 = 1, \text{ for all } x \in \mathbb{R}.$$

One would imagine that similar results would hold for other Gabor-type basis elements.

4 Gabor Multipliers

For this introductory lecture, we will focus on one-dimensional problems, although all the technique apply immediately to higher dimensions. Indeed the theory of Gabor transforms and Gabor multipliers extends nicely to transforms on locally compact abelian groups; see for instance [8] and [9]. For more details on representation of linear operators by Gabor multipliers, see [7]. For details on Gabor transforms on non-uniform lattices, see [6].

4.1 Short Time Fourier Transforms and Their Multipliers

The short time Fourier transform (STFT) of a signal $f(x)$ is obtained by windowing the signal with translates of a basic window function $g(x)$ and performing the usual Fourier transform on the product. It is defined as a function $V_g f$ of two variables (x, ξ), which represents local time–frequency information about the signal. The definition of the transform is given by

$$(V_g f)(x, \xi) = \int_{\mathbb{R}} f(y)\overline{g}(y - x)e^{-2\pi i y \cdot \xi} \, dy.$$

For good signal processing behaviour, g is usually chosen as a smooth function of compact support, or nearly compact support, non-negative, and symmetric about zero. Something like a Gaussian

$$g(x) = e^{-x^2}$$

is a typical choice for a window. For reasonable choices of windows g, the operator V_g is a linear map from the Hilbert space $L^2(\mathbb{R})$ into $L^2(\mathbb{R}^2)$. The adjoint thus maps functions of two variables to functions of one variable, so with $u = u(x, \xi)$ we have the adjoint formula

$$(V_g^* u)(x) = \int \int u(y, \xi)g(x - y)e^{2\pi i y \cdot \xi} \, dy \, d\xi.$$

For two (possibly different) L^2 windows g, γ we have Moyal's formula

$$\langle V_g f_1, V_\gamma f_1 \rangle = \langle f_1, f_2 \rangle \langle \gamma, g \rangle,$$

from which it follows immediately that

$$V_\gamma^* V_g f = \langle \gamma, g \rangle f$$

as well as that the operator norm $\|V_g\| = \|g\|_2$, the L^2 norm of the window function. Thus we have an obvious inversion formula for the STFT provided that the inner product $\langle \gamma, g \rangle$ is non-zero. The function g is called the analysis window, and γ is the synthesis window. It can be advantageous to choose g, γ separately, to getting different behaviour the analysis and synthesis process. Within certain limits, the windows can be chosen as distributions, and the STFT then is applied to Schwartz class functions.

Once a signal has been transformed into the time–frequency domain, it is only natural to attempt to modify it using the physical intuition that the values $(V_g f)(x, \xi)$ at a particular point (x, ξ) represents a certain amplitude of energy localized around that particular point. We can modify the signal by changing the amplitude at that point, say by a factor of $\sigma(x, \xi)$, where σ is some symbol function. The map

$$(V_g f)(x, \xi) \mapsto \sigma(x, \xi)(V_g f)(x, \xi)$$

represents a modification of the signal in the time–frequency domain, as prescribed by the symbol σ. Mapping back to the time domain gives back a signal in time. The composition of the operators

$$S_\sigma = \frac{1}{\langle g, \gamma \rangle} V_\gamma^* M_\sigma V_g$$

defined the STFT multiplier S_σ as the product of the STFT transform V_g, a pointwise multiplier M_σ in the t-f domain, and the adjoint transform V_γ^*. Note that the definition of S_σ depends on the choice of g and γ.

Theorem 4.1. *Suppose $\gamma = g \neq 0$. Then considering S_σ as an operator on $L^2(\mathbb{R})$:*

- σ *bounded* $\Rightarrow \|S_\sigma\| \leq \|M_\sigma\| = \|\sigma\|_{L^\infty}$.
- $S_\sigma^* = S_{\bar\sigma}$ *where σ is the pointwise complex conjugate of σ.*
- σ *real-valued* $\Rightarrow S_\sigma$ *self-adjoint.*
- $\sigma \geq 0 \Rightarrow S_\sigma \geq 0$.

For unit vector g, the proof proceeds from the identity $S_\sigma = V_g^* M_\sigma V_g$, where the V_g is an isometry. For instance, the fourth statement follow by writing $S_\sigma = (M_{\sigma^{1/2}} V_g)^* (M_{\sigma^{1/2}} V_g)$, which is a positive operator. This is similar to results for Wick operators, as discussed in Lerner's lectures elsewhere in this text.

It is interesting to note that we obtain a functional calculus for these STFT multipliers:

Theorem 4.2. *Suppose $\gamma = g \neq 0$. Then considering S_σ as an operator on $L^2(\mathbb{R})$:*

- $S_0 = 0$, *the zero operator, and $S_1 = I$, the identity operator.*
- $S_\sigma + S_\tau = S_{\sigma+\tau}$.
- $S_\sigma^* = S_{\bar\sigma}$.
- *For Gaussian windows, $S_\sigma S_\tau = S_\lambda$, where symbol λ satisfies $\hat{\lambda} = \hat{\sigma} *^{1/2} \hat{\tau}$.*

The first three statements are proved directly from the definition of S_σ. The fourth statement is essentially contained in a paper by Du and Wong [2] which considers the special case of localization operators where the windows g, γ are Gaussians. The $1/2$ convolution is defined by

$$(f *^{1/2} g)(z) = \int_{\mathbb{C}} f(z - w)g(w)e^{(z \cdot \overline{w} - |w|^2)/2} \, dw,$$

where w, z are complex numbers representing the sum of the translation and modulation. We refer the reader to [2] for details. It is certainly hoped that a similar convolution result will hold for more general windows.

Given the work in our earlier lectures, it is natural to ask the question: when is a STFT multiplier S_σ equal to a pseudodifferential operator? And what is the connection between the symbols. The answer is remarkably straightforward.

Theorem 4.3. *Suppose σ, τ are smooth symbols and K_σ, S_τ the corresponding Kohn–Nirenberg pseudodifferential operator and STFT multiplier, respectively.*
Then $K_\sigma = S_\tau$ if and only if

$$\hat{\sigma}^s = \frac{1}{\langle g, \gamma \rangle}(V_g \gamma)\hat{\tau}^s,$$

where $\hat{\sigma}^s, \hat{\tau}^s$ are the symplectic Fourier transforms of the symbols.[15]

This result appears at various papers in the literature, including [4], [7], and [23]. What it says essentially is that given a symbol τ that defines a STFT multiplier, the corresponding pseudodifferential operator K_σ has a symbol that is a smoothed version of τ, since multiplication by the function $V_g \gamma$ in the Fourier domain is a smoothing operation.

In the case where g, γ are Gaussians, the factor $(V_g \gamma)(x, \xi)$ is a product of two Gaussians, one in x, one in ξ. This is an extreme smoother. Essentially this says the only classical pseudodifferential operators that can be implemented by STFT multipliers have symbols that are extremely smooth: in the Fourier domain, they decay like a Gaussian (times at most a polynomial, say). This is something like an analyticity assumption on the symbol. Note the symbols whose (symplectic) Fourier transform have compact support will fall under this class.

For other windows, it is not usually the case that the STFT factor $V_g \gamma$ factors as a product of two functions $h(x)$ and $k(\xi)$, although the smoothness

---

[15] The symplectic Fourier transform puts one plus, and one minus, in the exponential, so

$$\hat{\sigma}^s(y, \eta) = \int \int \sigma(x, \xi)e^{2\pi i(x \cdot \eta - y \cdot \xi)} \, dx \, d\xi.$$

result still holds. Another interesting example is when g, γ are extreme value windows, that is, functions of the form

$$g(x) = e^{t - e^{\alpha t}}, \text{ for some } \alpha > 0 .$$

This window has exponential decay in one direction, double exponential decay in the other.

This does suggest an interesting question, as to when the factor $V_g \gamma$ decomposes into the product of two functions, $h(x)$ and $k(\xi)$. The answer is interesting:

Theorem 4.4 (In dimension one). *Suppose the factor $V_g \gamma$ factors as a product of two functions $h(x)$ and $k(\xi)$. Then one of the following is true:*

1. *One of g, γ is a Dirac delta function and the other is a constant function.*
2. *Both g, γ are Gaussians.*
3. *Both g, γ are extreme value functions.*
4. *g, γ are translations, rescalings, and/or modulations of the above.*

The proof involved a reduction to the solution of a particular differential equation, which is done in detail in [7].

4.2 Gabor Transforms and Gabor Multipliers

In numerical work, we can never compute the STFT at all points in the time–frequency plane, so one is quickly led to the notion of sampling in the variables (x, ξ). The Gabor transform is a sampled version of the STFT, and the Gabor multiplier is the corresponding multiplier defined on these discrete atoms.[16]

Fixing a lattice $\Lambda \subset \mathbb{R}^{2d}$ and a window function $g = g(x)$, we define the Gabor transform T_g as the map from functions on R^d to functions on the lattice Λ by

$$(T_g f)(\lambda) = (V_g f)(\lambda), \text{ for all } \lambda = (x, \xi) \in \Lambda.$$

We say the window g has upper frame bound $B \geq 0$ if the operator $(T_g)^* T_g$ is bounded above by B, as an operator from $L^2(\mathbb{R}^d)$ to $l^2(\Lambda)$. It has lower frame bound A if that same positive operator is bounded below by A.

Another window γ is called a dual window to g (relative to the given lattice) if we have

$$T_\gamma^* T_g = I$$

---

[16] Thus we wish to make a careful distinction between the STFT with its multipliers, which has continuous variables (x, ξ), and the Gabor transform with its multipliers, which is the sampled version.

which happens if and only if the symmetric result holds: $T_g^* T_\gamma = I$. Much of Gabor theory relates to finding suitable dual windows g and γ for various lattices.

A Gabor multiplier is defined in the analogous ways as STFT multipliers, so

$$(G_\sigma f) = \frac{1}{\langle g, \gamma \rangle} T_\gamma^* M_\sigma T_g f.$$

Note that the operator $G_\sigma \equiv G_{\sigma,g,\gamma}$ depends on the choice of windows and the choice of lattice, as well as the symbol σ.

We have a result analogous to the STFT multiplier case.

Theorem 4.5. *Suppose windows g, γ form a dual frame for lattice Γ, and σ, τ are bounded symbols on Γ. Then the Gabor multiplier $G_\sigma = G_{\sigma,g,\gamma}$ is a bounded operator on $L^2(\mathbb{R}^d)$, where:*

- $G_0 = 0$, *the zero operator, and $G_1 = I$, the identity operator.*
- $G_\sigma + G_\tau = G_{\sigma+\tau}$.
- $G_{\sigma,g,\gamma}^* = G_{\overline{\sigma},\gamma,g}$.
- *For $g = \gamma$, we have: σ real-valued $\Rightarrow G_\sigma$ self-adjoint; $\sigma \geq 0 \Rightarrow G_\sigma$ positive.*

One might expect to have a product formula, $G_\sigma G_\tau = G_{\sigma \sharp \tau}$ for some suitable combination of σ and τ. But a moment's thought and one realizes this is impossible in general. For instance, if the windows are chosen with compact support, a Gabor multiplier can modify a signal only in a local area, which can be measured as a finite distance. The product of two multipliers can push out to twice that limit, which cannot be recaptured by a single multiplier (cf. the result of Du and Wong for localization operators, [2]).

However, one can obtain some useful approximations. We will say a pair of dual windows are centered if the function $C(\lambda, \mu) = \langle \pi(\mu)\gamma, \pi(\lambda)g \rangle$ defined on the lattice $\Lambda \times \Lambda$ is concentrated along the diagonal $\lambda = \mu$. It is not hard to construct such dual windows; for instance, g, γ could be Gaussians of different widths, but the same center.

Theorem 4.6. *Suppose the dual windows g, γ for lattice Λ are centered, and σ, τ are symbols with finite support on the lattice. Then the Gabor multipliers are almost diagonal in the t-f domain, so*

$$T_g G_\sigma f = M_\sigma T_g + \text{ small terms }.$$

And the product of two Gabor multipliers is almost the Gabor multiplier of their symbols,

$$G_\sigma G_\tau = G_{\sigma\tau} + \text{ small terms}$$

Admittedly, this is an imprecisely stated theorem, perhaps one might call it more of a philosophy. But the idea is the following. On the time–frequency lattice Λ, we can think of the function $T_g f$ as a discrete vector indexed by the lattice. The operator M_σ acts as a diagonal matrix in this representation, so using the notation $(-)_\lambda$ to index our vector, $(-)_{\lambda\mu}$ to index a matrix, we see

$$(M_\sigma T_g f)_\lambda = \sigma(\lambda)(T_g f)_\lambda.$$

and

$$(T_g T_\gamma^*)_{\lambda\mu} = \langle \pi(\mu)\gamma, \pi(\lambda)g \rangle = C(\lambda, \mu),$$

where C is the concentration function.

Under the Gabor transform the Gabor multiplier can be factored as

$$T_g G_\sigma f = (T_g T_\gamma^*) M_\sigma (T_g f),$$

which is the product of two discrete matrices $(T_g T_\gamma^*)$ and M_σ times the vector $(T_g f)$. This first matrix $(T_g T_\gamma^*)$ is the concentration matrix noted above. So it is almost diagonal, with diagonal entries $\langle \gamma, g \rangle$. The diagonal of this multiplies with the diagonal matrix M_σ, and all the off diagonal elements give something relatively small, because of the concentration.

An alternative formula for expressing the function $T_g G_\sigma f \in c_{00}(\Lambda)$ can be obtained, showing this as a twisted convolution of the functions $T_g \gamma$ and $M_\sigma T_g f$. The twisted convolution on the lattice is defined by

$$F *' G(\lambda) = \sum_{\mu \in \Lambda} e^{-2\pi i [\lambda - \mu, \mu]} F(\lambda - \mu) G(\mu)$$

where $[\lambda - \mu, \mu] = [(x - y, \xi - \eta), (y, \eta)] = y(\xi - \eta)$. We calculate the concentration function as

$$C(\lambda, \mu) = e^{-2\pi i [\lambda - \mu, \mu]} (V_g \gamma)(\lambda - \mu),$$

which is a twisted version of the function $T_g \gamma$. Then it is routine to verify that the following three lattice functions satisfy

$$(T_g G_\sigma f) = (T_g \gamma) *' (M_\sigma T_g f).$$

Both the matrix view and the twisted convolution are useful.

The product of multiplier formula follows from the equality

$$G_\sigma G_\tau = T_\gamma^* M_\sigma (T_g T_\gamma^*) M_\tau T_g,$$

where we see the concentration matrix $(T_g T_\gamma^*)$ appears between two diagonal matrices. So we can multiply diagonal, and have the off diagonal terms only contribute a small amount.

In our applications, we seem to implicitly believe the following:

Theorem 4.7. *In the limit, as lattice Γ gets denser in the time-frequency space,*

$$G_\sigma = G_{\sigma, g, \gamma} \to S_\sigma$$

for a suitable choice of windows g, γ that change with the lattice.

That is, we seem to believe that is we sample the symbol σ densely enough, we can choose a lattice and dual window pairs so that the Gabor multiplier G_σ is close to the STFT multiplier S_σ. Making this precise depends on the application. There are precise results of this form in the literature, for instance Feichtinger and Nowak [3] give conditions where G_σ converges to S_σ in the trace class operator norm.

5 Gabor Transforms in Practice

5.1 Sampled Space

Recall the Gabor transform is a sampled version of the STFT, which is a transform on functions on the continuous space $\mathbb{R}^d$. In typical applications, one is never able to measure a true STFT, as the data functions usually are presented as sampled data series, not functions on the real line. Thus, we cannot compute STFT or Gabor transforms, but something else. We will call this the Gabor transform on a discrete (sampled) space, or the discrete Gabor transform for simplicity.

The idea is essentially the same. In the one-dimensional case, data is represented as functions in $l^2(\mathbb{Z})$. A window function g defined on $\mathbb{Z}$ is set, and Gabor coefficients are computed as

$$(T_g f)(n, \xi) = \sum_{k=-\infty}^{\infty} f(k)\overline{g(k-n)}e^{2\pi i k \xi}.$$

Note the time–frequency space now is $\mathbb{Z} \times [0, 1]$, and we can sample along some discrete lattice within this space. Figure 5 shows a typical implementation, notice the window spacing is much greater than one sample size.

In practice, there are a few problems with this regular structure. First of all, it might be very useful to use a window that is not only spaced along by an integer number of sample; for instance, a Gaussian shifted by some fractional amount could be used. Once some freedom in windowing is allowed, we next notice that in certain areas the signal may be changing rapidly, while in other areas it may have a more uniform character.[17] It would be useful to use short windows in the first case, to accurately track those changes, and use larger windows in the second case. The third problem is that when applying a time-varying filter, typically we get edge effects near the sides of any window, and regularly spaced windows will give regularly spaced artifacts. It is important to point out, of course, that artifacts never appear if one only computes

---

[17] In the seismic problem, this could correspond to areas in the earth where the geology is complex and changing rapidly, versus other areas where the earth is of a more uniform character.

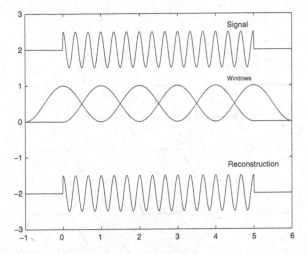

Fig. 5 Regular window translated along signal space

Gabor coefficients, and then inverts, as the numerical inversion is essentially exact. However, when modifying the Gabor coefficients before inverting (as when implementing a time-varying filter, or applying a Gabor multiplier), there will be changes in the output signal; unwanted changes we loosely describe as artifacts. Any artifacts, particularly regularly-spaced artifacts, can be confusing or annoying to the person viewing the Gabor coefficients and can obfuscate the processed data. For instance, in image processing, one often observes artifacts that lie on a rectangular boundary (the jpeg effect, see [22]). Perhaps a more optimal tiling of the image space could be useful; for instance, the hexagonal tiling shown in Fig. 6 corresponds to an optimal packing of discs in the plane. Even a non-periodic tiling might prove to be useful. In our seismic experiments, it is numerically troublesome to have artifacts that accumulate along the direction of wave propagation, so again, something other than regular, rectangular division of the image space is useful.

Thus, we propose generating a Gabor transform using a collection of windows $\{g_n\}$, and duals $\{\gamma_n\}$ which appropriately cover signal space, as in Fig. 7. Although the functions g_n are not related to each other by translation, we may still define the modulation appropriately, as

$$g_{mn} = M_m g_n$$

and form Gabor coefficients of signal f as

$$V_{mn}(f) = \langle f, g_{mn} \rangle,$$

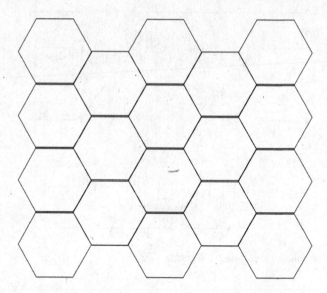

Fig. 6 A hexagonal tiling of the plane

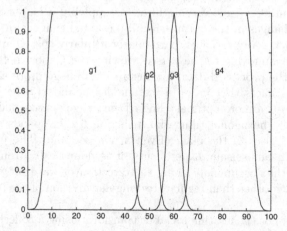

Fig. 7 A collection of four non-uniform windows covering signal space

and similarly for the reconstruction operator, using the general collection of dual windows $\{\gamma_n\}$ rather than a collection of translates.

The next section shows how we avoid rectangular lattices in the frequency domain, and Sect. 5.3 pulls this all together to create a general Gabor transform.

5.2 Sampling in the Frequency Domain

In the Gabor transform, the window functions are modified by modulation functions which, in one dimension, are simply complex exponential functions of the form

$$P_\alpha(j) = e^{2\pi i \alpha j} \qquad \text{for all } j \in \mathbb{Z},$$

for some fixed parameter α. In applications, it is common to choose only periodic modulation functions, obtained by setting parameter α to be a rational number, say $\alpha = m/M$. In higher dimensions, one may take a product of several one-dimensional modulation functions, and obtain functions

$$P_m(j) = e^{2\pi i m_1 j_1/M_1} e^{2\pi i m_2 j_2/M_2} \dots e^{2\pi i m_d j_d/M_d}$$

for all $j = (j_1, j_2, \dots j_d)$ in $\mathbb{Z}^d$, where $m = (m_1, m_2, \dots m_d)$ is an index in $\mathbb{Z}^d$ and $M_1 \dots M_d$ is some fixed choice of integer denominators. It is convenient to fix a diagonal matrix B with integer entries

$$B = \begin{bmatrix} M_1 & 0 & \dots & 0 \\ 0 & M_2 & \dots & 0 \\ \vdots & \vdots & \ddots & \vdots \\ 0 & 0 & \dots & M_d \end{bmatrix},$$

and then express the above modulation function via the standard inner product on $\mathbb{Z}^d$ as

$$P_m(j) = e^{2\pi i (B^{-1} m) \cdot j} \qquad \text{for all } j \in \mathbb{Z}^d.$$

In this form, it is clear that each modulation function is periodic in each component of j, and the "vector of frequencies" represented by function P_m is the vector $B^{-1}m$. Less obvious, but also true, is that there are only finitely many different functions P_m (as two different indices m, m' can give the functions $P_m = P_{m'}$), and a sum of the different P_m gives a delta function.

However, this is too restrictive a class of modulation functions, as essentially it restricts us to sampling in the frequency domain with a simple rectangular lattice, in this case the set of points $\frac{1}{M_1}\mathbb{Z} \times \frac{1}{M_2}\mathbb{Z} \times \dots \times \frac{1}{M_d}\mathbb{Z}$. Just as in the time domain, we wished to moved away from rectangular sample lattices, so here too we may do so in the frequency domain.[18] The key is using a more general form for matrix B and considering the lattices these matrices generate.

A lattice is, roughly speaking, a regularly spaced collection of points in Euclidean space $\mathbb{R}^d$. More precisely, it is a discrete subgroup of $\mathbb{R}^d$ under the operation of vector addition. The standard example of a discrete lattice is the subset $\mathbb{Z}^d$ of points in Euclidean space with integer coordinates. For the Gabor

---

[18] Our approach to lattices is quite a bit different than the symplectic case as considered by Gröchenig [9].

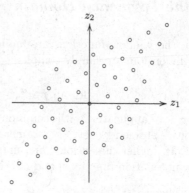

Fig. 8 A non-rectangular lattice in the plane

transform, we are interested in more general lattices that "fill out space," which is to say, are not confined to some hyperplane in $\mathbb{R}^d$, yet the points don't "bunch up" into a dense set; such lattices may always be represented as the image of the standard lattice $\mathbb{Z}^d$ under a linear transformation. That is, there is an invertible $d \times d$ matrix $A \in M_d(\mathbb{R})$ with the lattice given as the set

$$A\mathbb{Z}^d = \{Az : z \in \mathbb{Z}^d\}.$$

Figure 8 shows a portion of a typical lattice in the plane.

In the discussion on modulation functions, the diagonal matrix B defines a lattice of frequencies, namely the set

$$B^{-1}\mathbb{Z}^d = \{B^{-1}z : z \in \mathbb{Z}^d\}.$$

The matrix B also defines a lattice of periodicity for the modulation functions,

$$B\mathbb{Z}^d = \{Bz : z \in \mathbb{Z}^d\},$$

which (since B is an integer matrix), is a subset of the standard lattice $\mathbb{Z}^d$ in $\mathbb{R}^d$. By lattice of periodicity, we simply mean that if indices j and j' differ by some integer multiple of B (i.e. $j' - j \in B\mathbb{Z}^d$), then P_m takes the same values there, $P_m(j') = P_m(j)$.

To generalize the frequency sampling to non-rectangular lattices, fix B to be any invertible $d \times d$ matrix with integer entries, and for any index $m \in \mathbb{Z}^d$, define modulation function P_m by

$$P_m(j) = e^{2\pi i(B^{-1}m)\cdot j}, \qquad \text{for all } j \in \mathbb{Z}^d.$$

We also define two equivalence relations on the lattice $\mathbb{Z}^d$ by

$$m \sim m' \qquad \text{iff} \qquad m - m' \in B\mathbb{Z}^d,$$

and

$$m \sim^* m' \qquad \text{iff} \qquad m - m' \in B^* \mathbb{Z}^d,$$

where B^* is the transpose of matrix B.

The following proposition is easily proved in the case where B diagonal; the proof for general B is given in [6].

Theorem 5.1. *Let B be an invertible $d \times d$ matrix with integer entries, defining equivalence relations $\sim$ and $\sim^*$, and modulation functions $\{P_m : m \in \mathbb{Z}^d\}$ as above. Then:*

1. *Each function P_m is periodic, with $j \sim^* j' \Rightarrow P_m(j) = P_m(j')$.*
2. *$P_m = P_{m'}$ if and only if $m \sim m'$.*
3. *There are exactly $|\det B|$ different modulation functions P_m, which are uniquely indexed by some finite cube*

$$M = [0, \ldots, M_1 - 1] \times [0, \ldots, M_2 - 1] \times \ldots \times [0, \ldots, M_d - 1] \subset \mathbb{Z}^d.$$

4. *The finite sum of distinct modulation functions*

$$P = \sum_{m \in M} P_m$$

is a delta function, with

$$P(j) = \begin{cases} |\det(B)| & \text{if } j \in B^* \mathbb{Z}^d \\ 0 & \text{otherwise.} \end{cases}$$

In the following sections, there may be only one fixed modulation matrix B, but in general, the matrix may vary with the choice of window function.

5.3 Partitions of Unity and Frequency Subsampling

A *partition of unity* is a collection of functions which sum to the constant function one; that is, a collection of functions $\{w_n\}$ which sum as

$$\sum_n w_n(j) = 1 \qquad \text{for all } j \in \mathbb{Z}^d.$$

For typical signal processing applications, it will be useful to choose the w_n with particular characteristics; say smooth, non-negative, rapidly decreasing, and/or with a well-behaved Fourier transform. In [6] some typical smooth window based on splines are given. In the following, however, the only restriction we might make on each w_n is that it have compact (i.e. finite) support. Even this restriction may be dropped: see the remark at the end of the section.

Given a partition of unity $\{w_n\}$ on the set $\mathbb{Z}^d$, with each function w_n of compact support, choose for each index n an invertible integer matrix B_n so that the lattice $B_m^* \mathbb{Z}^d$ intersects the difference set[19]

$$\operatorname{supp}(w_n) - \operatorname{supp}(w_n) \equiv \{j - k : j, k \in \operatorname{supp}(w_n)\}$$

only at the origin. That is, choose B_n with entries large enough so that the non-zero entries in the lattice $B_m^* \mathbb{Z}^d$ are far enough from the (finite) set of differences of pairs of elements in $\operatorname{supp}(w_n)$.

Then, choose functions g_n and γ_n to form what we call a weighted factorization of the partition of unity; that is, g_n and γ_n are chosen with:

1. $\operatorname{supp}(g_n) = \operatorname{supp}(\gamma_n) = \operatorname{supp}(w_n)$
2. $\gamma_n \overline{g_n} |\det B_n| = w_n$ on the set $\mathbb{Z}^d$

For instance, with a non-negative partition of unity, one may choose

$$g_n = \left(\frac{w_n}{|\det B_n|} \right)^p$$

$$\gamma_n = \left(\frac{w_n}{|\det B_n|} \right)^{1-p}$$

for some real parameter $0 \le p \le 1$. (In applications, the choice of p is important in controlling the rate of roll-off for a window.)

For each window index n, and for each integer vector $m \in \mathbb{Z}^d$, the modulated version of the window functions are denoted by g_{mn} and γ_{mn} and are defined as

$$g_{mn}(j) = g_n(j) e^{2\pi i (B_n^{-1} m) \cdot j}$$

$$\gamma_{mn}(j) = \gamma_n(j) e^{2\pi i (B_n^{-1} m) \cdot j}$$

for all vectors $j \in \mathbb{Z}^d$. The Gabor transform of function $f \in l^2(\mathbb{Z}^d)$ is defined as

$$V_{mn}(f) = \langle f, g_{mn} \rangle$$

and the inverse transform given by

$$\tilde{f} = \sum_{m,n} V_{mn}(f) \gamma_{mn}.$$

By this careful choice of window functions, we have the following result.

Theorem 5.2. *With this choice of modulated window functions g_{mn} and γ_{mn} construction from a partition of unity, we have the exact reconstruction formula*

---

[19] The Minkowski difference.

$$f = \sum_{m,n} \langle f, g_{mn} \rangle \gamma_{mn}$$

for all functions $f \in l^2(\mathbb{Z}^d)$. That is, the Gabor transform with windows g_n has an inverse transform with windows γ_n.

Proof. We compute

$$\sum_{m,n} \langle f, g_{mn} \rangle \gamma_{mn}(j) = \sum_{m,n,k} f(k)\overline{g_n}(k)e^{-2\pi i(B_n^{-1}m)\cdot k}\gamma_n(j)e^{2\pi i(B_n^{-1}m)\cdot j}$$

$$= \sum_{n,k} f(k)\overline{g_n}(k)\gamma_n(j)\sum_{m} e^{2\pi i(B_n^{-1}m)\cdot(j-k)}$$

where the inner sum (over m) we recognize as the sum of modulation functions, which is non-zero only when $j - k$ is in the lattice $B^*\mathbb{Z}^d$. The term $\overline{g_n}(k)\gamma_n(j)$ is non-zero only when $j - k$ is in the finite difference set

$$\text{supp}(\gamma_n) - \text{supp}(g_n)$$

which, by construction, is the same set as $\text{supp}(w_n) - \text{supp}(w_n)$. But the intersection

$$B^*\mathbb{Z}^d \cap (\text{supp}(w_n) - \text{supp}(w_n))$$

is just the origin, and so the above sum collapses to the non-zero term with $j = k$, and becomes

$$= \sum_{n} f(j)\overline{g_n}(j)\gamma_n(j)|\det B_n|$$

$$= f(j)\sum_{n} w_n(j)$$

$$= f(j),$$

since the w_n sum to one. $\qquad\square$

Having chosen the analysis windows g_n with small support relative to matrix B_n, the frame operator is also in a particularly simply form, as indicated in the following.

Theorem 5.3. *With window functions g_m chosen as above, the frame operator*

$$Sf = \sum_{m,n} \langle f, g_{mn} \rangle g_{mn}$$

is a multiplication operator on $l^2(\mathbb{Z}^d)$, by values

$$s(j) = \sum_{n} |g_n(j)|^2 |\det B_n|.$$

Proof. We compute the pointwise values

$$(Sf)(j) = \sum_{m,n} \langle f, g_{mn} \rangle g_{mn}(j)$$

$$= \sum_{n,k} f(k)\overline{g_n}(k)g_n(j) \sum_m e^{2\pi i (B_n^{-1}m)\cdot(j-k)}$$

which collapses, as in the previous proof, to the sum

$$= \sum_n f(j)\overline{g_n}(j)g_n(j)|\det B_n|$$

$$= f(j)s(j)$$

with $s(j) = \sum_n |g_n(j)|^2|\det B_n|$. $\qquad\square$

There is a well-developed theory for the expansion of functions by non-orthogonal families of "basis" functions, namely frame theory, which is beyond the scope of this article (see for instance [1]). However, it is useful to insert the following observation, which is relevant to frame theory.

Corollary 5.1. *A tight frame is obtained if and only if the functions w_n' defined by*

$$w_n'(j) = |g_n(j)|^2|\det B_n|$$

form a partition of unity on $\mathbb{Z}^d$, times a fixed constant.

Proof. This is a consequence of the observation that the frame is tight if and only if the frame operator S is a scalar multiple of the identity; equivalently, that the multiplication operator S with values $s(j)$ is a constant function. Since $s(j) = \sum_n w_n'(j)$, tightness is equivalent to the w_n' forming a partition of unity, up to scaling by some constant. $\qquad\square$

It is worth observing that if the original partition of unity $\{w_n\}$ is non-negative, the choice of analysis window

$$g_n = \left(\frac{w_n}{|\det B_n|}\right)^{1/2}$$

will give a tight frame. It is also interesting to observe that once an analysis window g_n is chosen with suitably small support relative to $B_n^*\mathbb{Z}^d$, then *any* choice of dual window with small support (including the canonical dual $\gamma_n^c = S^{-1}g_n$) gives rise to a partition of unity, as noted in the following proposition.

Theorem 5.4. *Suppose analysis windows g_n and matrix B_n are chosen such that $supp(g_n)$ is finite, and the intersection*

$$B_n^*\mathbb{Z}^d \cap (supp(g_n) - supp(g_n))$$

contains only the origin. If γ'_n is any choice of dual windows, with support the same as the corresponding g_n, that satisfies the reconstruction

$$f = \sum_{m,n} \langle f, g_{mn} \rangle \gamma'_{mn} \quad \text{for all } f \in l^2(\mathbb{Z}^d),$$

then the functions

$$w'_n = \gamma' \overline{g_n} |\det B_n|$$

form a partition of unity for the set $\mathbb{Z}^d$.

Proof. As in the calculation above, we have

$$
\begin{aligned}
f(j) &= \sum_{m,n} \langle f, g_{mn} \rangle \gamma'_{mn}(j) \\
&= f(j) \sum_n \gamma'_n(j) \overline{g_n}(j) |\det B_n| \\
&= f(j) \sum_n w'_n(j).
\end{aligned}
$$

The only way this equality can hold for all f is if the w'_n sum to one. $\qquad\square$

For the canonical dual $\gamma^c_n = S^{-1} g_n$, since operator S is simply an invertible multiplication operator, the support of γ^c_n is the same as the support of g_n, and thus the canonical dual times $\overline{g_n} |\det B_n|$ forms a partition of unity $\{w'_n\}$. It is important to notice this is usually not the well-designed partition of unity $\{w_n\}$ that we started out with, and the canonical dual does not necessarily have the nice properties, such as smoothness, that we designed in the factorization. Again, Fig. 9 is a typical example of a poor canonical dual.

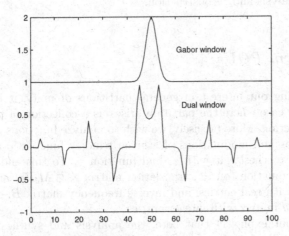

Fig. 9 A window and its spiky dual window

Thus, in standard Gabor theory, there often are partitions of unity lurking around in the background. Our approach in this lecture can be summarized as saying that we begin with a well-designed partition of unity, and create well-behaved windows from this partition.

Remark: In certain applications, it is convenient to chose some windows with non-finite support; for instance, in some filtering applications, it is useful to set $g_n = w_n$ and $\gamma_n \equiv 1$. Technically, this doesn't quite work, as one recovers (in the reconstruction) a periodization of the original signal. However, all is not lost: by a good choice of the frequency lattice, one can ensure the reconstruction is exact on the support of signal f, and just ignore the periodization that occurs outside the support. More precisely, in many applications, there is some reasonable finite set $F \subset \mathbb{Z}^d$ such that every signal f of interest has support in F; this set F may be used to truncate the window γ_n to finite support. Given a partition of unity $\{w_n\}$, we wish to chose a factorization with

$$w_n = \gamma_n \overline{g_n} |\det B_n| \qquad \text{on set } F,$$

where for the moment, matrix B_n is unspecified. One can choose supports for the windows to lie in the finite set F, and also

$$\operatorname{supp}(\gamma_n) \cap \operatorname{supp}(g_n) = \operatorname{supp}(w_n) \cap F.$$

Now choose matrix B_n so that

$$B_n^* \mathbb{Z}^d \cap (\operatorname{supp}(\gamma_n) - \operatorname{supp}(g_n)) = \{0\}.$$

All the result above apply. Thus larger support are possible for γ_n or g_n than specified by the partition of unity. However, the price paid is that matrix B_n may have a large determinant, which means many modulation functions are required in analysis and reconstruction.

5.4 Uniform POUs

Notwithstanding our interest in general partitions of unity, it is useful to restrict to the case where the partition arises as a collection of translations of a single function. More precisely, we wish to analyze functions f in $l^2(\mathbb{Z}^d)$, via translations and modulations of a given window function g along some lattice, and resynthesize it with a dual function γ. To this end, fix g and γ as bounded functions on $\mathbb{R}^d$, translation matrix $A \in M_d(\mathbb{R})$ an invertible $d \times d$ matrix with real entries, and inverse frequency matrix $B \in M_d(\mathbb{Z})$ an invertible $d \times d$ matrix with integer entries.

We point out explicitly that here, the analysis and synthesis windows are more generally functions on Euclidean space (for instance, Gaussians), and one is permitted to translate them by arbitrary real vectors – typically, these windows and their translates are computed on the fly, and there is no need to restrict the translations to the standard lattice. Indeed, in many applications, it is advantageous to use these general forms for the window functions. In contrast, the function f is defined on the regular lattice $\mathbb{Z}^d$, as it usually comes from sampled data in real applications. One must require some moderate conditions on the decay of g, γ to ensure the sum of their translates converge; piecewise continuous with compact support, or integrable (in $\mathbb{R}^d$) is sufficient for our purposes here. Also note the inverse frequency matrix B need not be diagonal, as again there may be advantages to sampling on a non-standard frequency lattice. Integer entries for B, however, are required for the reconstruction theorem.

The set of points $\{An : n \in \mathbb{Z}^d\}$ forms a discrete lattice in $\mathbb{R}^d$, while the set $\{Bm : m \in \mathbb{Z}^d\}$ forms a sublattice of the discrete lattice $\mathbb{Z}^d$. A translate of function g along the lattice $A\mathbb{Z}^d$ is defined by

$$g_n(x) = g(x - An), \text{ for all } x \in \mathbb{R}^d,$$

while the modulation of g_n by frequencies $B^{-1}m$ is defined as

$$g_{mn}(x) = g_n(x)e^{2\pi i(B^{-1}m)\cdot x}, \text{ for all } x \in \mathbb{R}^d.$$

The translations and modulation of the dual window γ are defined similarly. (The order of operation of translation and modulation is important. We have fixed it here with translations first, but observe that in some applications, it is relevant whether one is measuring phase relative to the signal, or relative to the translated window.)

The discrete Gabor transform (for given A, B, g and γ) is defined as a map of functions $f \in l^2(\mathbb{Z}^d)$ to $V(f) \in l^\infty(\mathbb{Z}^d \times \mathbb{Z}^d)$ using the Gabor coefficients

$$V_{mn}(f) = \langle f, g_{mn} \rangle = \sum_k f(k)\bar{g}(k - An)e^{-2\pi i(B^{-1}m)\cdot k},$$

where $\langle , \rangle$ denotes the usual inner product on $l^2(\mathbb{Z}^d)$. Since the inverse of matrix B occurs in the modulation, it is immediately clear that $V_{mn}(f)$ is periodic in m, with period B; that is $V_{mn}(f) = V_{m'n}(f)$ when $m \sim m'$ (modulo $B\mathbb{Z}^d$). Thus the Gabor transform is completely determined on the finite quotient set $\mathbb{Z}^d/B\mathbb{Z}^d$, and the reconstruction of f may be defined by the sum

$$\tilde{f} = \sum_{\substack{m \in \mathbb{Z}^d/B\mathbb{Z}^d, \\ n \in \mathbb{Z}^d}} V_{mn}(f)\gamma_{mn}.$$

That is, the reconstruction of f is obtained by taking a linear combination of translations and modulation γ_{mn} of the dual window function γ, using weights $V_{mn}(f)$, which come from the Gabor transform of f. The remarkable result is that this reconstruction depends only sparsely on the original function f; that is, the matrix representing this linear transformation from f to $\tilde{f}$ has many zeroes – we only need sum over equivalence classes in the sublattice $B^*\mathbb{Z}^d$.

Theorem 5.5. *The reconstruction $\tilde{f}$ satisfies*

$$\tilde{f}(j) = |\det(B)| \sum_{j'\sim^*j} c(j,j')f(j')$$

for all $j \in \mathbb{Z}^d$, where the sum is over all indices j' equivalent to j modulo $B^*\mathbb{Z}^d$, and c is the correlation function between the two functions γ, g over the lattice generated by A, given as

$$c(j,j') = \sum_n \gamma(j - An)\bar{g}(j' - An).$$

Proof. We compute:

$$\tilde{f}(j) = \sum_{m,n} V_{mn}(f)\gamma_{mn}(j)$$

$$= \sum_{m,n} V_{mn}(f)\gamma(j - An)e^{2\pi i(B^{-1}m)\cdot j}$$

$$= \sum_{m,n} (\sum_{j'} f(j')\bar{g}(j' - An)e^{-2\pi i(B^{-1}n)\cdot j'})\gamma(j - An)e^{2\pi i(B^{-1}m)\cdot j}$$

$$= \sum_{n,j'} f(j')\bar{g}(j' - An)\gamma(j - An)\left[\sum_m e^{2\pi i(B^{-1}m)\cdot(j-j')}\right]$$

$$= \sum_{n,j'} f(j')\bar{g}(j' - An)\gamma(j - An)P(j - j')$$

where we recognize the last sum over $m \in M$ as a sum of modulation functions as in Proposition 5.1, and thus this sum is equal to $|\det B|$ when $j-j' \in B^*\mathbb{Z}^d$, and zero otherwise. Thus, the sum over j' collapses to a sum over those j' equivalent to j modulo $B^*\mathbb{Z}^d$, and we write

$$\tilde{f}(j) = |\det B| \sum_{n,j'\sim^*j} f(j')\bar{g}(j' - An)\gamma(j - An)$$

$$= |\det B| \sum_{j'\sim^*j} \left[\sum_n \bar{g}(j' - An)\gamma(j - An)\right] f(j')$$

$$= |\det B| \sum_{j'\sim^*j} c(j,j')f(j')$$

where $c(j,j') = \sum_n \gamma(j - An)\bar{g}(j' - An)$ is the correlation function. $\qquad\square$

Combining with the results in Sect. 5.3, we obtain the following.

Corollary 5.2. *If invertible integer matrix B is chosen so that*

$$(supp(\gamma) - supp(g)) \cap B^*\mathbb{Z}^d = \{0\}$$

and the functions $w_n = \gamma_n \bar{g}_n$ form a partition of unity on $\mathbb{Z}^d$, then the reconstruction above satisfies

$$\tilde{f} = |\det B| f \quad \text{for all } f \in l^2(\mathbb{Z}^d).$$

Proof. As in the previous section, the correlation function

$$c(j, j') = \sum_n \gamma(j - An)\bar{g}(j' - An)$$

is non-zero only when $j - j'$ is in the set $\mathrm{supp}(\gamma) - \mathrm{supp}(g)$. In the sum for the reconstruction formula, we only get non-zero terms when $j - j'$ is in the lattice $B^*\mathbb{Z}^d$; combined with the observation about where c is non-zero, we conclude the sum collapses, and thus

$$\tilde{f}(j) = |\det B| c(j, j) f(j).$$

With $\gamma_n \bar{g}_n$ a partition of unity, these diagonal entries $c(j, j)$ are exactly one, and thus

$$\tilde{f}(j) = |\det B| f(j).$$

$\square$

One easy way to obtain a partition of unity over lattice translations is simply to fix a non-negative function $v : \mathbb{R}^d \to \mathbb{R}$ and symmetrize over the lattice $A\mathbb{Z}^d$. That is, we define

$$w(x) = \frac{v(x)}{\sum_{n \in \mathbb{Z}^d} v(x - An)} \quad \text{for all } x \in \mathbb{R}^d.$$

Provided the denominator is never zero, this gives rise to a partition of unity by translating along $A\mathbb{Z}^d$. A factorization of the partition of unity may be obtained by setting

$$g(x) = w(x)^p$$
$$\gamma = w(x)^{1-p}$$

for any real parameter $0 \le p \le 1$.

Further details on uniform and nonuniform, adaptive partitions of unity can be found in [5], [10], and [11].

6 Seismic Imaging

6.1 Wavefield Extrapolation

As mentioned earlier, much of our work on numerical PsDOs is motivated by the seismic imaging problem. One key process is to model the propagation of a seismic wave through the earth, using a numerical simulation. The main idea is to assume the wavefield is known at some particular plane in the earth, say $z = 0$ and then predict what the wavefield should be at a deeper level $z > 0$ (Fig. 10). (The positive z direction points downwards.)

We begin with the acoustic wave equation:

$$\frac{\partial^2 \varphi}{\partial x^2} + \frac{\partial^2 \varphi}{\partial y^2} + \frac{\partial^2 \varphi}{\partial z^2} = \frac{1}{c^2} \frac{\partial^2 \varphi}{\partial t^2},$$

where $\varphi(x, y, z, t)$ is the wave function and $c = c(x, y, z)$ is the (non-constant) speed of propagation of the seismic wave. We make a simplifying assumption that $c = c(x, y)$ depends only on the x and y variables, which is a reasonable assumption if we only propagate the wave through a thin layer. The wave equation can then be rewritten as

$$\frac{\partial^2 \varphi}{\partial z^2} = \frac{1}{c^2} \frac{\partial^2 \varphi}{\partial t^2} - \frac{\partial^2 \varphi}{\partial x^2} - \frac{\partial^2 \varphi}{\partial y^2}$$
$$= K_\sigma \varphi,$$

where K_σ is a pseudodifferential operator with symbol

$$\sigma(x, y, \xi_x, \xi_y, \omega) = \frac{1}{c^2(x, y)} (2\pi i)^2 \omega^2 - (2\pi i)^2 \xi_x^2 - (2\pi i)^2 \xi_y^2$$
$$= 4\pi^2 \left[(\xi_x^2 + \xi_y^2) - \frac{\omega^2}{c^2(x, y)} \right].$$

Note here that ξ_x is the Fourier dual variable to spatial coordinate x, ξ_y is the Fourier dual variable to coordinate y, and ω is the Fourier dual variable to time t. Thus ξ_x, ξ_y are measured in wavenumber, and ω is a frequency. Also note that the symbol σ does not depend on z nor its dual Fourier variable.

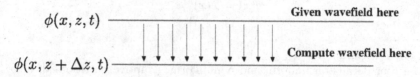

Fig. 10 The setup for wavefield extrapolation

Now, assume we can find a square root for the operator K_σ, which is also a pseudodifferential operator. This is a big assumption, but let's go with it.[20] Thus

$$K_\sigma = (K_\rho)^2 \approx K_{\rho^2},$$

where we have used the pseudodifferential calculus to approximate the square of an operator with the operator corresponding to the squared symbol ρ^2. We thus expect $\sigma \approx \rho^2$, and we can approximately solve for ρ as

$$\rho(x, y, \xi_x, \xi_y, \omega) = -2\pi\sqrt{\xi_x^2 + \xi_y^2 - \frac{\omega^2}{c^2(x, y)}},$$

where we choose the square root so that the symbol ρ is always negative when real, or negative when imaginary.[21] If there is a concern about the cusp at the zero of the square root, one can smooth it out there with a smooth approximation to the square root. For simplicity of discussion, we ignore that issue.

Of course, using asymptotic series, one can hope to get a better approximation for the symbol ρ. Or, using the oscillatory integral formula for combining symbols, one can hope to solve for ρ exactly, using the formula for

$$\sigma = \rho \sharp \rho.$$

For now, we go with the above approximation.

An exponential symbol $\tau = \tau_z$ is now defined, parameterized by variable z, as

$$\tau_z(x, y, \xi_x, \xi_y, \omega) = \exp(z \cdot \rho(x, y, z, \xi_x, \xi_y, \omega)).$$

We use z as a parameter, to make clear that the operator that we are creating actually acts on a space of functions of three variables x, y, t, and their duals, ξ_x, ξ_y, ω, and not including the z, ξ_z variables. By the choice of signs on the square root, we guarantee that τ_z is either oscillatory, or exponentially decaying, for each $z > 0$. We now define an extrapolated version of φ at level $z > 0$, which depends only on the values of φ at $z = 0$, by applying the operator K_{τ_z} to the function $\varphi(x, y, 0, t)$. That is, we set

$$\varphi(x, y, z, t) = K_{\tau_z}\varphi(x, y, 0, t).$$

To verify that φ is almost a solution to the wave equation, we differentiate, and find

---

[20] In fact, this is a Fourier integral operators.

[21] Or one can choose it to be negative when real, and positive when imaginary. It depends whether upgoing or downgoing waves are desired. But there is always a unique choice.

$$\frac{\partial^2 \varphi}{\partial z^2}(x,y,z,t) = K_{\rho^2 \tau_z} \varphi(x,y,0,t)$$
$$= K_{\sigma \tau_z} \varphi(x,y,0,t)$$
$$\approx K_\sigma K_{\tau_z} \varphi(x,y,0,t)$$
$$= K_\sigma \varphi(x,y,z,t)$$
$$= \frac{1}{c^2} \frac{\partial^2 \varphi}{\partial t^2} - \frac{\partial^2 \varphi}{\partial x^2} - \frac{\partial^2 \varphi}{\partial y^2}$$

as desired.

To summarize, we have an extrapolation formula for $\varphi(x,y,z,t)$ via a parameterized operator K_{τ_z}, given by

$$\varphi(x,y,z,t) =$$

$$= \int \exp\left(-2\pi z \sqrt{\xi_x^2 + \xi_y^2 - \frac{\omega^2}{c^2(x,y)}}\right) \hat{\varphi}(\xi_x, \xi_y, 0, \omega) e^{2\pi i (x\xi_x + y\xi_y + t\omega)} \, d\xi_x \, d\xi_y \, d\omega,$$

where by $\hat{\varphi}$ we mean the Fourier transform in only the three variables x, y, t.[22]

Practical applications of this technique for seismic imaging has been reported by our research group in [14]. Closely related to these ideas is the technique of wavefield extrapolation using Gabor multipliers (see [11]) as well as Gabor deconvolution (see [12], [16], [15], [13], [17]) which is a non-stationary generalization of Wiener deconvolution. Computer code based on these ideas have been implemented and are used competitively in the seismic industry.

Acknowledgements The authors gratefully acknowledge the support of the C.I.M.E. foundation, the Killam Trusts Foundation, MITACS, NSERC, and NuHAG for their assistance with this project.

References

1. I. DAUBECHIES, *Ten lectures on wavelets*, CBMS-NSF Regional conference series in applied mathematics, Academic Press, 1992.
2. J. DU AND M. W. WONG, *A product formula for localization operators*, Bulletin of the Korean Mathematical Society, 37 (2000), pp. 77–84.
3. H. G. FEICHTINGER AND K. NOWAK, *A first survey of Gabor multipliers*, Advances in Gabor analysis, Applied Numerical Harmonic Analysis, Birkhauser, Boston, 2003, pp. 99–128.

---

[22] Some authors prefer to choose the sign in the exponential of the Fourier transform to be the opposite for the time variable, than for space variables. Such a choice makes a difference between up and down going waves, which is a source of some confusion in the above formula.

4. G. B. FOLLAND, *Harmonic Analysis in Phase Space*, vol. 122 of Annals of Mathematical Studies, Princeton University Press, Princeton, New Jersey, 1989.

5. P. GIBSON AND M. LAMOUREUX, *Maximally symmetric, minimally redundant partitions of unity in the plane*, Comptes rendus mathematiques, 26 (2004), pp. 65–72.

6. P. C. GIBSON, J. GROSSMAN, M. P. LAMOUREUX, AND G. F. MARGRAVE, *A fast, discrete Gabor transform by partition of unity*. Preprint, 2002.

7. P. C. GIBSON, M. P. LAMOUREUX, AND G. F. MARGRAVE, *Representation of linear operators by Gabor multipliers*. Preprint, 2003.

8. K. GRÖCHENIG, *Aspects of Gabor analysis on locally compact abelian groups*, Gabor Analysis and Algorithms, H. Feichtinger and T. Strohmer, editors, Birkhauser, Boston, 1998, pp. 211–232.

9. ———, *Foundations of time–frequency analysis*, Birkhauser, Boston, 2001.

10. J. P. GROSSMAN, G. F. MARGRAVE, AND M. P. LAMOUREUX, *Constructing adaptive nonuniform Gabor frames from partitions of unity*, tech. rep., CREWES, University of Calgary, 2002.

11. ———, *Fast wavefield extrapolation by phase-shift in the nonuniform Gabor domain*, tech. rep., CREWES, University of Calgary, 2002.

12. J. P. GROSSMAN, G. F. MARGRAVE, M. P. LAMOUREUX, AND R. AGGARWALA, *Constant-Q wavelet estimation via a nonstationary Gabor spectral model*, tech. rep., CREWES, University of Calgary, 2001.

13. G. F. MARGRAVE, L. DONG, P. C. GIBSON, J. P. GROSSMAN, D. C. HENLEY, V. ILIESCU, AND M. P. LAMOUREUX, *Gabor deconvolution: extending Wiener's method to nonstationarity*, The CSEG Recorder, 28 (2003).

14. G. F. MARGRAVE, P. C. GIBSON, J. P. GROSSMAN, D. C. HENLEY, V. ILIESCU, AND M. P. LAMOUREUX, *The Gabor transform, pseudodifferential operators, and seismic deconvolution*, Integrated Computer-Aided Engineering, 9 (2004), pp. 1–13.

15. G. F. MARGRAVE, D. C. HENLEY, M. P. LAMOUREUX, V. ILIESCU, AND J. P. GROSSMAN, *An update on Gabor deconvolution*, tech. rep., CREWES, University of Calgary, 2002.

16. G. F. MARGRAVE AND M. P. LAMOUREUX, *Gabor deconvolution*, tech. rep., CREWES, University of Calgary, 2001.

17. ———, *Gabor deconvolution*, The CSEG Recorder, 2006 Special Issue (2006), pp. 30–37.

18. Y. MEYER AND R. COIFMAN, *Wavelets: Calderon–Zygmund and Multilinear Operators*, vol. 48 of Cambridge studies in advanced mathematics, Cambridge University Press, Cambridge, 1997.

19. A. V. OPPEHEIM AND R. W. SCHAFER, *Discrete-time Signal Processing*, Prentice Hall, New Jersey, 1998.

20. X. S. RAYMOND, *Elementary Introduction to the Theory of Pseudodifferential Operators*, CRC Press, Florida, 1991.

21. R. ROCHBERG AND K. TACHIZAWA, *Pseudodifferential operators, Gabor frames, and local trigonometric bases*, Gabor Analysis and Algorithms, H. Feichtinger and T. Strohmer, editors, Birkhauser, Boston, 1998, pp. 171–192.

22. G. STRANG AND T. NGUYEN, *Wavelets and Filter Banks*, Wellesley-Cambridge Press, 1997.

23. M. W. WONG, *Weyl transforms*, Springer-Verlag, 1998.

Some Facts About the Wick Calculus

N. Lerner

Abstract This is a slightly expanded version of a five-hour lecture series given at Cetraro during the CIME session of June 2006 dedicated to the topics of Pseudodifferential operators, Quantization and Signal.

1 Elementary Fourier Analysis via Wave Packets

1.1 The Fourier Transform of Gaussian Functions

Let u be a function in the Schwartz class of rapidly decreasing functions $\mathcal{S}(\mathbb{R}^n)$: it means that u is a C^∞ function on $\mathbb{R}^n$ such that for all multi-indices[1] α, β

$$\sup_{x \in \mathbb{R}^n} |x^\alpha \partial_x^\beta u(x)| = C_{\alpha\beta} < \infty.$$

A simple example of such a function is $e^{-|x|^2}$, ($|x|$ is the Euclidean norm of x) and more generally if A is a symmetric positive definite $n \times n$ matrix the function

$$v_A(x) = e^{-\pi \langle Ax, x \rangle} \tag{1}$$

belongs to the Schwartz class. The Fourier transform of u is defined as

$$\hat{u}(\xi) = \int_{\mathbb{R}^n} e^{-2i\pi x \cdot \xi} u(x) dx. \tag{2}$$

Nicolas Lerner
Institut de Mathématiques de Jussieu, Université Paris 6, 175 rue du Chevaleret
75013 Paris, France
e-mail: lerner@math.jussieu.fr

---

[1] $\alpha = (\alpha_1, \ldots, \alpha_n) \in \mathbb{N}^n, x^\alpha = x_1^{\alpha_1} \ldots x_n^{\alpha_n}, \ \beta \in \mathbb{N}^n, \partial_x^\beta = \partial_{x_1}^{\beta_1} \ldots \partial_{x_n}^{\beta_n}.$

L. Rodino, M.W. Wong (eds.) *Pseudo-Differential Operators*. Lecture Notes in Mathematics 1949.
© Springer-Verlag Berlin Heidelberg 2008

It is an easy matter to check that the Fourier transform sends $\mathcal{S}(\mathbb{R}^n)$ into itself.[2] Moreover, for A as above, we have

$$\widehat{v_A}(\xi) = (\det A)^{-1/2} e^{-\pi \langle A^{-1}\xi, \xi \rangle}. \tag{3}$$

In fact, diagonalizing the symmetric matrix A, it is enough to prove the one-dimensional version of (3), i.e. to check

$$\int e^{-2i\pi x\xi} e^{-\pi x^2} dx = \int e^{-\pi(x+i\xi)^2} dx\, e^{-\pi\xi^2} = e^{-\pi\xi^2},$$

where the second equality can be obtained by taking the ξ-derivative of $\int e^{-\pi(x+i\xi)^2} dx$. Using (3) we calculate for $u \in \mathcal{S}(\mathbb{R}^n)$ and $\epsilon > 0$, dealing with absolutely converging integrals,

$$\begin{aligned}
u_\epsilon(x) &= \int e^{2i\pi x\xi} \hat{u}(\xi) e^{-\pi\epsilon^2 |\xi|^2} d\xi \\
&= \iint e^{2i\pi x\xi} e^{-\pi\epsilon^2 |\xi|^2} u(y) e^{-2i\pi y\xi} dy d\xi \\
&= \int u(y) e^{-\pi\epsilon^{-2}|x-y|^2} \epsilon^{-n} dy \\
&= \int \underbrace{(u(x+\epsilon y) - u(x))}_{\text{with absolute value} \le \epsilon |y| \|u'\|_{L^\infty}} e^{-\pi|y|^2} dy + u(x).
\end{aligned}$$

Taking the limit when ϵ goes to zero, we get the Fourier inversion formula

$$u(x) = \int e^{2i\pi x\xi} \hat{u}(\xi) d\xi. \tag{4}$$

So far we have just proved that the Fourier transform is an isomorphism of the Schwartz class and provided an explicit inversion formula. This was devised to refresh our memory on this topic and we want now to move forward with the definition of our wave packets.

1.2 Wave Packets and the Poisson Summation Formula

We define for $x \in \mathbb{R}^n$, $(y, \eta) \in \mathbb{R}^n \times \mathbb{R}^n$

$$\varphi_{y,\eta}(x) = 2^{n/4} e^{-\pi(x-y)^2} e^{2i\pi(x-y)\cdot\eta} = 2^{n/4} e^{-\pi(x-y-i\eta)^2} e^{-\pi\eta^2} \tag{5}$$

---

[2] Just notice that

$$\xi^\alpha \partial_\xi^\beta \hat{u}(\xi) = \int e^{-2i\pi x\xi} \partial_x^\alpha (x^\beta u)(x) dx (2i\pi)^{|\beta|-|\alpha|} (-1)^{|\beta|}.$$

where for $\zeta = (\zeta_1, \ldots, \zeta_n) \in \mathbb{C}^n$, we set

$$\zeta^2 = \sum_{1 \le j \le n} \zeta_j^2. \tag{6}$$

We note that the function $\varphi_{y,\eta}$ is in $\mathcal{S}(\mathbb{R}^n)$ and with L^2 norm 1. In fact, $\varphi_{y,\eta}$ appears as a *phase translation* of a normalized Gaussian. The following lemma introduces the wave packets transform as a Gabor wavelet.

Lemma 1.1. *Let u be a function in the Schwartz class $S(\mathbb{R}^n)$. We define*

$$Wu(y,\eta) = \langle u, \varphi_{y,\eta} \rangle_{L^2(\mathbb{R}^n)} = 2^{n/4} \int u(x) e^{-\pi(x-y)^2} e^{-2i\pi(x-y)\cdot \eta} dx$$

$$= 2^{n/4} \int u(x) e^{-\pi(y-i\eta-x)^2} dx e^{-\pi\eta^2}. \tag{7}$$

For $u \in L^2(\mathbb{R}^n)$, the function Tu defined by

$$(Tu)(y+i\eta) = e^{\pi\eta^2} Wu(y,-\eta) = 2^{n/4} \int u(x) e^{-\pi(y+i\eta-x)^2} dx$$

is an entire function. The mapping $u \mapsto Wu$ is continuous from $S(\mathbb{R}^n)$ to $S(\mathbb{R}^{2n})$ and isometric from $L^2(\mathbb{R}^n)$ to $L^2(\mathbb{R}^{2n})$. Moreover, we have the reconstruction formula

$$u(x) = \iint_{\mathbb{R}^n \times \mathbb{R}^n} Wu(y,\eta) \varphi_{y,\eta}(x) dy d\eta. \tag{8}$$

Proof. For u in $S(\mathbb{R}^n)$, we have

$$Wu(y,\eta) = e^{2i\pi y\eta} \widehat{\Omega}^1(\eta, y)$$

where $\widehat{\Omega}^1$ is the Fourier transform with respect to the first variable of the $S(\mathbb{R}^{2n})$ function $\Omega(x,y) = u(x) e^{-\pi(x-y)^2} 2^{n/4}$. Thus the function Wu belongs to $S(\mathbb{R}^{2n})$. It makes sense to compute

$$2^{-n/2} \langle Wu, Wu \rangle_{L^2(\mathbb{R}^{2n})} =$$

$$= \lim_{\epsilon \to 0+} \int u(x_1) \overline{u}(x_2) e^{-\pi[(x_1-y)^2 + (x_2-y)^2 + 2i(x_1-x_2)\eta + \epsilon^2 \eta^2]} dy d\eta dx_1 dx_2.$$

Now the last integral on $\mathbb{R}^{4n}$ converges absolutely and we can use the Fubini theorem. Integrating with respect to η involves the Fourier transform of a Gaussian function and we get $\epsilon^{-n} e^{-\pi\epsilon^{-2}(x_1-x_2)^2}$. Since

$$2(x_1-y)^2 + 2(x_2-y)^2 = (x_1+x_2-2y)^2 + (x_1-x_2)^2,$$

integrating with respect to y yields a factor $2^{-n/2}$. We are left with

$$\langle Wu, Wu \rangle_{L^2(\mathbb{R}^{2n})} =$$

$$= \lim_{\epsilon \to 0_+} \int u(x_1)\, \overline{u}(x_2) e^{-\pi(x_1-x_2)^2/2} \epsilon^{-n} e^{-\pi\epsilon^{-2}(x_1-x_2)^2}\, dx_1 dx_2.$$

Changing the variables, the integral is

$$\lim_{\epsilon \to 0_+} \int u(s + \epsilon t/2)\, \overline{u}(s - \epsilon t/2) e^{-\pi\epsilon^2 t^2/2} e^{-\pi t^2}\, dt ds = \|u\|^2_{L^2(\mathbb{R}^n)}$$

by Lebesgue's dominated convergence theorem: the triangle inequality and the estimate $|u(x)| \leq C(1 + |x|)^{-n-1}$ imply, with $v = u/C$,

$$|v(s + \epsilon t/2)\, \overline{v}(s - \epsilon t/2)| \leq (1 + |s + \epsilon t/2|)^{-n-1}(1 + |s + \epsilon t/2|)^{-n-1}$$
$$\leq (1 + |s + \epsilon t/2| + |s - \epsilon t/2|)^{-n-1}$$
$$\leq (1 + 2|s|)^{-n-1}.$$

Eventually, this proves that

$$\|Wu\|^2_{L^2(\mathbb{R}^{2n})} = \|u\|^2_{L^2(\mathbb{R}^n)}, \tag{9}$$

i.e.

$$W : L^2(\mathbb{R}^n) \to L^2(\mathbb{R}^{2n}) \quad \text{with} \quad W^*W = \mathrm{id}_{L^2(\mathbb{R}^n)}.$$

Noticing first that $\iint Wu(y, \eta)\varphi_{y,\eta} dy d\eta$ belongs to $L^2(\mathbb{R}^n)$ (with a norm smaller than $\|Wu\|_{L^1(\mathbb{R}^{2n})}$) and applying Fubini's theorem, we get from the polarization of (9) for $u, v \in \mathcal{S}(\mathbb{R}^n)$,

$$\langle u, v \rangle_{L^2(\mathbb{R}^n)} = \langle Wu, Wv \rangle_{L^2(\mathbb{R}^{2n})}$$
$$= \iint Wu(y, \eta)\langle \varphi_{y,\eta}, v \rangle_{L^2(\mathbb{R}^n)} dy d\eta$$
$$= \left\langle \iint Wu(y, \eta)\varphi_{y,\eta} dy d\eta, v \right\rangle_{L^2(\mathbb{R}^n)}$$

yielding the result of the lemma $u = \iint Wu(y, \eta)\varphi_{y,\eta} dy d\eta$.   □

The following lemma is in fact the Poisson summation formula for Gaussian functions in one dimension.

Lemma 1.2. *For all complex numbers z, the following series are absolutely converging and*

$$\sum_{m \in \mathbb{Z}} e^{-\pi(z+m)^2} = \sum_{m \in \mathbb{Z}} e^{-\pi m^2} e^{2i\pi mz}. \tag{10}$$

Proof. We set $\omega(z) = \sum_{m \in \mathbb{Z}} e^{-\pi(z+m)^2}$. The function ω is entire and 1-periodic since for all $m \in \mathbb{Z}$, $z \mapsto e^{-\pi(z+m)^2}$ is entire and for $R > 0$

$$\sup_{|z| \le R} |e^{-\pi(z+m)^2}| \le \sup_{|z| \le R} |e^{-\pi z^2}| e^{-\pi m^2} e^{2\pi |m| R} \in l^1(\mathbb{Z}).$$

Consequently, for $z \in \mathbb{R}$, we obtain, expanding ω in Fourier series,[3]

$$\omega(z) = \sum_{k \in \mathbb{Z}} e^{2i\pi kz} \int_0^1 \omega(x) e^{-2i\pi kx} dx.$$

We also check, using Fubini's theorem on $L^1(0,1) \times l^1(\mathbb{Z})$

$$\int_0^1 \omega(x) e^{-2i\pi kx} dx = \sum_{m \in \mathbb{Z}} \int_0^1 e^{-\pi(x+m)^2} e^{-2i\pi kx} dx$$

$$= \sum_{m \in \mathbb{Z}} \int_m^{m+1} e^{-\pi t^2} e^{-2i\pi kt} dt$$

$$= \int_{\mathbb{R}} e^{-\pi t^2} e^{-2i\pi kt} = e^{-\pi k^2}.$$

So (10) is proved for real z and since both sides are entire functions, we conclude by analytic continuation. □

It is now straightforward to get the n-th dimensional version of Lemma 1.2: for all $z \in \mathbb{C}^n$, using the notation (6), we have

$$\sum_{m \in \mathbb{Z}^n} e^{-\pi(z+m)^2} = \sum_{m \in \mathbb{Z}^n} e^{-\pi m^2} e^{2i\pi m \cdot z}. \tag{11}$$

Theorem 1.1 (The Poisson summation formula). *Let n be a positive integer and u be a function in $S(\mathbb{R}^n)$. Then*

$$\sum_{k \in \mathbb{Z}^n} u(k) = \sum_{k \in \mathbb{Z}^n} \hat{u}(k), \tag{12}$$

where $\hat{u}$ stands for the Fourier transform (2).

Proof. We write, according to (8) and to Fubini's theorem

---

[3] Note that we use this expansion only for a C^∞ 1-periodic function. The proof is simple and requires only to compute $1 + 2 \operatorname{Re} \sum_{1 \le k \le N} e^{2i\pi kx} = \frac{\sin \pi(2N+1)x}{\sin \pi x}$. Then one has to show that for a smooth 1-periodic function ω such that $\omega(0) = 0$,

$$\lim_{\lambda \to +\infty} \int_0^1 \frac{\sin \lambda x}{\sin \pi x} \omega(x) dx = 0,$$

which is obvious since for a smooth ν (here we take $\nu(x) = \omega(x)/\sin \pi x$), $|\int_0^1 \nu(x) \sin \lambda x \, dx| = O(\lambda^{-1})$ by integration by parts.

$$\sum_{k \in \mathbb{Z}^n} u(k) = \sum_{k \in \mathbb{Z}^n} \iint Wu(y,\eta)\varphi_{y,\eta}(k)dyd\eta$$

$$= \iint Wu(y,\eta) \sum_{k \in \mathbb{Z}^n} \varphi_{y,\eta}(k)dyd\eta. \tag{13}$$

Now, (11), (5) and (3) give

$$\sum_{k \in \mathbb{Z}^n} \varphi_{y,\eta}(k) = \sum_{k \in \mathbb{Z}^n} \widehat{\varphi}_{y,\eta}(k),$$

so that (13), (8) and Fubini's theorem imply (12).   □

It is a simple matter to introduce at this point the dual space of the Fréchet $\mathcal{S}(\mathbb{R}^n)$, that is the space $\mathcal{S}'(\mathbb{R}^n)$ of tempered distributions (the continuous linear forms on $\mathcal{S}(\mathbb{R}^n)$). We can define the Fourier transform on $\mathcal{S}'(\mathbb{R}^n)$ by duality[4]:

$$\langle \widehat{T}, \varphi \rangle_{\mathcal{S}'(\mathbb{R}^n),\mathcal{S}(\mathbb{R}^n)} = \langle T, \widehat{\varphi} \rangle_{\mathcal{S}'(\mathbb{R}^n),\mathcal{S}(\mathbb{R}^n)}, \tag{14}$$

so that the inversion formula (4) still holds for $T \in \mathcal{S}'(\mathbb{R}^n)$ and reads

$$T = \widehat{\widecheck{T}}, \quad \text{with} \quad \langle \widecheck{T}, \varphi \rangle = \langle T, \widecheck{\varphi} \rangle, \quad \widecheck{\varphi}(x) = \varphi(-x).$$

Using duality, it is a matter of routine left to the reader to give a version of Lemma 1.1 for tempered distributions. Now Theorem 1.1 can be given a more compact version saying that the tempered distribution $D_0 = \sum_{k \in \mathbb{Z}^n} \delta_k$ is such that $\widehat{D_0} = D_0$.

We shall need as well a parametric version of wave packets, and we state here a lemma analogous to Lemma 1.1, whose proof is left to the reader.

1.3 Toeplitz Operators

We define for $x \in \mathbb{R}^n$, $(\lambda, y, \eta) \in \mathbb{R}_+^* \times \mathbb{R}^n \times \mathbb{R}^n$,

$$\varphi_{y,\eta}^\lambda(x) = (2\lambda)^{n/4}e^{-\pi\lambda(x-y)^2}e^{2i\pi(x-y)\cdot\eta} = (2\lambda)^{n/4}e^{-\pi\lambda(x-y-i\lambda^{-1}\eta)^2}e^{-\pi\lambda^{-1}\eta^2}. \tag{15}$$

We note that the function $\varphi_{y,\eta}^\lambda$ is in $\mathcal{S}(\mathbb{R}^n)$ and with L^2 norm 1.

Lemma 1.3. *Let u be a function in the Schwartz class $\mathcal{S}(\mathbb{R}^n)$. We define, for $(\lambda, y, \eta) \in \mathbb{R}_+^* \times \mathbb{R}^n \times \mathbb{R}^n$,*

---

[4] In the formula below, we deal with real duality, so that, if T, φ are in $L^2(\mathbb{R}^n)$, $\langle T, \overline{\varphi} \rangle_{\mathcal{S}'(\mathbb{R}^n),\mathcal{S}(\mathbb{R}^n)} = \langle T, \varphi \rangle_{L^2(\mathbb{R}^n)}.$

$$W_\lambda u(y,\eta) = \langle u, \varphi_{y,\eta}^\lambda \rangle_{L^2(\mathbb{R}^n)} = (2\lambda)^{n/4} \int u(x) e^{-\lambda\pi(x-y)^2} e^{-2i\pi(x-y)\cdot\eta} dx$$

$$= (2\lambda)^{n/4} \int u(x) e^{-\pi\lambda(y-i\lambda^{-1}\eta-x)^2} dx e^{-\pi\lambda^{-1}\eta^2}.$$

$$(16)$$

For $u \in L^2(\mathbb{R}^n)$, the function $T_\lambda u$ defined by

$$(T_\lambda u)(y+i\eta) = \lambda^{-n/4} e^{\pi\lambda\eta^2} W_\lambda u(y,-\lambda\eta) = 2^{n/4} \int u(x) e^{-\pi\lambda(y+i\eta-x)^2} dx$$

$$(17)$$

is an entire function. The mapping $u \mapsto W_\lambda u$ is continuous from $\mathcal{S}(\mathbb{R}^n)$ to $\mathcal{S}(\mathbb{R}^{2n})$ and isometric from $L^2(\mathbb{R}^n)$ to $L^2(\mathbb{R}^{2n})$. Moreover, we have the reconstruction formula for each positive λ,

$$u(x) = \iint_{\mathbb{R}^n \times \mathbb{R}^n} W_\lambda u(y,\eta) \varphi_{y,\eta}^\lambda(x) dy d\eta. \qquad (18)$$

We shall see in the sequel that the actual rôle of the Gaussian functions is in fact quite limited, except for the very explicit inversion formulas, essentially due to (3).

2 On the Weyl Calculus of Pseudodifferential Operators

2.1 A Few Classical Facts

Let $a(x,\xi)$ be a classical Hamiltonian defined on $\mathbb{R}^n \times \mathbb{R}^n$. The Weyl quantization rule associates to this function the operator a^w defined on functions $u(x)$ as

$$(a^w u)(x) = \iint e^{2i\pi(x-y)\cdot\xi} a\left(\frac{x+y}{2},\xi\right) u(y) dy d\xi. \qquad (19)$$

For instance we have $(x\cdot\xi)^w = (x\cdot D_x + D_x\cdot x)/2$, with $D_x = \frac{1}{2i\pi}\frac{\partial}{\partial x}$ whereas the *ordinary* quantization rule would map the Hamiltonian $x\cdot\xi$ to the operator $x\cdot D_x$. A nice feature of the Weyl quantization rule, introduced in 1928 by Hermann Weyl in [Wy], is the fact that real Hamiltonians get quantized by (formally) self-adjoint operators. Let us recall that the classical quantization of the Hamiltonian $a(x,\xi)$ is given by the operator Op(a) acting on functions $u(x)$ by

$$(\text{Op}(a)u)(x) = \int e^{2i\pi x\cdot\xi} a(x,\xi) \widehat{u}(\xi) d\xi. \qquad (20)$$

In fact, introducing the following one-parameter group $J^t = \exp 2i\pi t D_x \cdot D_\xi$, given by the integral formula

$$(J^t a)(x, \xi) = |t|^{-n} \iint e^{-2i\pi t^{-1} y \cdot \eta} a(x + y, \xi + \eta) dy d\eta,$$

we see that

$$(\mathrm{Op}(J^t a)u)(x) = \iint e^{2i\pi(x-y)\cdot\xi} \, a((1-t)x + ty, \xi) u(y) dy d\xi.$$

In particular one gets $a^w = \mathrm{Op}(J^{1/2}a)$. Moreover since $(\mathrm{Op}(a))^* = \mathrm{Op}(J\bar{a})$ we obtain

$$(a^w)^* = \mathrm{Op}(J(\overline{J^{1/2}a})) = \mathrm{Op}(J^{1/2}\bar{a}) = (\bar{a})^w,$$

yielding formal self-adjointness for real a.

Remark 2.1. Many other formulas of quantization yielding formal selfadjointness for real Hamiltonians have been used, e.g. the Feynman quantization $a \mapsto a^F$ defined by

$$(a^F u)(x) = \iint \frac{1}{2}\big(a(x, \xi) + a(y, \xi)\big) e^{2i\pi(x-y)\cdot\xi} u(y) dy d\xi.$$

Using the previous notations we see that $2a^F = \mathrm{Op}(a) + \mathrm{Op}(Ja)$, so that

$$(2a^F)^* = \mathrm{Op}(J\bar{a}) + \mathrm{Op}(J\,\overline{Ja}) = \mathrm{Op}(J\bar{a}) + \mathrm{Op}(\bar{a}) = 2\bar{a}^F$$

and for a real-valued, $(a^F)^* = a^F$. However, we shall see in Sect. 2.2 that the important property of symplectic invariance is true for the Weyl quantization and fails for the Feynman and the ordinary quantizations. Since it turns out that this symplectic invariance is actually a very important property, we shall stick with the Weyl quantization as our quantization of reference.

Formula (19) can be written as

$$(a^w u, v) = \iint a(x, \xi)\mathcal{H}(u, v)(x, \xi) dx d\xi, \tag{21}$$

where the *Wigner function* $\mathcal{H}$ is defined as

$$\mathcal{H}(u, v)(x, \xi) = \int u\big(x + \frac{y}{2}\big)\bar{v}\big(x - \frac{y}{2}\big) e^{-2i\pi y \cdot \xi} dy. \tag{22}$$

The mapping $(u, v) \mapsto \mathcal{H}(u, v)$ is sesquilinear continuous from $\mathcal{S}(\mathbb{R}^n) \times \mathcal{S}(\mathbb{R}^n)$ to $\mathcal{S}(\mathbb{R}^{2n})$ so that a^w makes sense for $a \in \mathcal{S}'(\mathbb{R}^{2n})$ (here $u, v \in \mathcal{S}(\mathbb{R}^n)$ and $\mathcal{S}^*$ stands for the antidual):

$$\langle a^w u, v\rangle_{\mathcal{S}^*(\mathbb{R}^n), \mathcal{S}(\mathbb{R}^n)} = \langle a, \mathcal{H}(u, v)\rangle_{\mathcal{S}'(\mathbb{R}^{2n}), \mathcal{S}(\mathbb{R}^{2n})}.$$

The Wigner function also satisfies, since $\mathcal{H}(u, v)$ is the partial Fourier transform of the function $(x, y) \mapsto u(x + y/2)\bar{v}(x - y/2)$,

$$\|\mathcal{H}(u,v)\|_{L^2(\mathbb{R}^{2n})} = \|u\|_{L^2(\mathbb{R}^n)} \|v\|_{L^2(\mathbb{R}^n)},$$
$$\mathcal{H}(u,v)(x,\xi) = 2^n \langle \sigma_{x,\xi} u, v \rangle_{L^2(\mathbb{R}^n)}, \tag{23}$$
$$\text{with} \quad (\sigma_{x,\xi} u)(y) = u(2x-y) \exp{-4i\pi(x-y)\cdot\xi}.$$

and the phase symmetries σ_X are unitary and selfadjoint operators on $L^2(\mathbb{R}^n)$. We have also ([U], [Wy]),

$$a^w = \int_{\mathbb{R}^{2n}} a(X) 2^n \sigma_X dX = \int_{\mathbb{R}^{2n}} \widehat{a}(\Xi) \exp(2i\pi\Xi \cdot M) d\Xi, \tag{24}$$

where $\Xi \cdot M = \hat{x}\cdot x + \hat{\xi}\cdot D_x$ (here $\Xi = (\hat{x},\hat{\xi})$). These formulas give in particular

$$\|a^w\|_{\mathcal{L}(L^2)} \le \min(2^n \|a\|_{L^1(\mathbb{R}^{2n})}, \|\widehat{a}\|_{L^1(\mathbb{R}^{2n})}), \tag{25}$$

where $\mathcal{L}(L^2)$ stands for the space of bounded linear maps from $L^2(\mathbb{R}^n)$ into itself.

2.2 Symplectic Invariance

As shown below, the symplectic invariance of the Weyl quantization is actually its most important property. Let us consider a finite dimensional real vector space E (the configuration space $\mathbb{R}^n_x$) and its dual space E^* (the momentum space $\mathbb{R}^n_\xi$). The phase space is defined as $\Phi = E \oplus E^*$; its running point will be denoted in general by a capital letter ($X = (x,\xi), Y = (y,\eta)$). The symplectic form on Φ is given by

$$[(x,\xi),(y,\eta)] = \langle \xi, y \rangle_{E^*,E} - \langle \eta, x \rangle_{E^*,E}, \tag{26}$$

where $\langle \cdot, \cdot \rangle_{E^*,E}$ stands for the bracket of duality. The symplectic group is the subgroup of the linear group of Φ preserving (26). With

$$\sigma = \begin{pmatrix} 0 & \mathrm{Id}(E^*) \\ -\mathrm{Id}(E) & 0 \end{pmatrix},$$

we have for $X, Y \in \Phi$, $[X,Y] = \langle \sigma X, Y \rangle_{\Phi^*,\Phi}$, so that the equation of the symplectic group is $A^* \sigma A = \sigma$. One can describe a set of generators for the symplectic group $Sp(n)$, identifying Φ with $\mathbb{R}^n_x \times \mathbb{R}^n_\xi$: the mappings

$$(x,\xi) \mapsto (Tx, {}^t T^{-1}\xi), \text{ where } T \text{ is an automorphism of } E, \tag{i}$$

$$(x_k,\xi_k) \mapsto (\xi_k, -x_k), \text{ and the other coordinates fixed}, \tag{ii}$$

$$(x,\xi) \mapsto (x, \xi + Sx), \text{ where } S \text{ is symmetric from } E \text{ to } E^*. \tag{iii}$$

We then describe the metaplectic group, introduced by André Weil [Wi]. The metaplectic group $Mp(n)$ is the subgroup of the group of unitary transformations of $L^2(\mathbb{R}^n)$ generated by

$$(M_T u)(x) = |\det T|^{-1/2} u(T^{-1}x), \text{ where } T \text{ is an automorphism of } E, \quad \text{(j)}$$

Partial Fourier transformation, with respect to x_k for $k = 1, \ldots, n$,   (jj)

Multiplication by $\exp(i\pi\langle Sx, x\rangle)$, where S is symmetric from E to E^*.   (jjj)

There exists a two-fold covering (the π_1 of both $Mp(n)$ and $Sp(n)$ is $\mathbb{Z}$)

$$\pi : Mp(n) \to Sp(n)$$

such that, if $\chi = \pi(M)$ and u, v are in $L^2(\mathbb{R}^n)$, $\mathcal{H}(u, v)$ is their Wigner function,

$$\mathcal{H}(Mu, Mv) = \mathcal{H}(u, v) \circ \chi^{-1}.$$

This is Segal formula [S] which could be rephrased as follows. Let $a \in \mathcal{S}'(\mathbb{R}^{2n})$ and $\chi \in Sp(n)$. There exists M in the fiber of χ such that

$$(a \circ \chi)^w = M^* a^w M. \tag{33}$$

In particular, the images by π of the transformations (j), (jj), (jjj) are respectively (i), (ii), (iii). Moreover, if χ is the phase translation, $\chi(x, \xi) = (x + x_0, \xi + \xi_0)$, (33) is fulfilled with $M = \tau_{x_0, \xi_0}$, the phase translation given by

$$(\tau_{x_0, \xi_0} u)(y) = u(y - x_0)\, e^{2i\pi\langle y - \frac{x_0}{2}, \xi_0\rangle}.$$

If χ is the symmetry with respect to (x_0, ξ_0), M in (33) is, up to a unit factor, the phase symmetry σ_{x_0, ξ_0} defined above.

Remark 2.2. Going back to the Remark 2.1 on the Feynman quantization, let us prove that this quantization is *not* invariant by the symplectic group: we assume that $n = 1$ and consider the symplectic mapping $\chi(x, \xi) = (x, \xi + Sx)$ where S is a non-zero real number. We shall prove now that one can find some $a \in \mathcal{S}(\mathbb{R}^{2n})$ such that

$$(a \circ \chi)^F \neq M^* a^F M,$$

where M is the unitary transformation of $L^2(\mathbb{R})$ given by $(Mu)(x) = e^{i\pi S x^2} u(x)$. We compute

$$2\langle (a \circ \chi)^F u, v\rangle = \int e^{2i\pi(x-y)\xi}\big(a(x, \xi + Sx) + a(y, \xi + Sy)\big)u(y)\overline{v(x)}dydxd\xi$$

$$= \int e^{2i\pi(x-y)(\xi - Sx)} a(x, \xi) u(y)\overline{v(x)}dydxd\xi$$

$$+ \int e^{2i\pi(x-y)(\xi - Sy)} a(y, \xi) u(y)\overline{v(x)}dydxd\xi$$

$$= \int e^{2i\pi(x-y)\xi} \left(a(x,\xi)e^{-i\pi S(x-y)^2} + a(y,\xi)e^{i\pi S(x-y)^2} \right)(Mu)(y)\overline{(Mv)(x)}\,dy\,dx\,d\xi$$

so that $(a \circ \chi)^F = M^*KM$ where the kernel k of the operator K is given by

$$2k(x,y) = \hat{a}^2(x, y-x)e^{-i\pi S(x-y)^2} + \hat{a}^2(y, y-x)e^{i\pi S(x-y)^2}.$$

On the other hand the kernel l of the operator a^F is given by

$$2l(x,y) = \hat{a}^2(x, y-x) + \hat{a}^2(y, y-x).$$

Checking the case $S = 1, a(x,\xi) = e^{-\pi(x^2+\xi^2)}$, we see that

$$2k(1,0) = -e^{-2\pi} - e^{-\pi}, \quad 2l(1,0) = e^{-2\pi} + e^{-\pi},$$

proving that $K \neq a^F$ and the sought result.

2.3 Composition Formulas

We have the following composition formula $a^w b^w = (a\sharp b)^w$ with

$$(a\sharp b)(X) = 2^{2n} \iint e^{-4i\pi[X-Y, \, X-Z]} a(Y)b(Z)\,dY\,dZ, \qquad (34)$$

with an integral on $\mathbb{R}^{2n} \times \mathbb{R}^{2n}$. We can compare this with the ordinary composition formula,

$$\mathrm{Op}(a)\mathrm{Op}(b) = \mathrm{Op}(a \circ b)$$

(cf. (20)) with

$$(a \circ b)(x,\xi) = \iint e^{-2i\pi y \cdot \eta} a(x, \xi+\eta)b(y+x, \xi)\,dy\,d\eta, \qquad (35)$$

with an integral on $\mathbb{R}^n \times \mathbb{R}^n$. It is convenient to give an asymptotic version of these compositions formulae, e.g. in the semi-classical case.[5] Let m be a real number. A smooth function $a(x,\xi,\lambda)$ defined on $\mathbb{R}_x^n \times \mathbb{R}_\xi^n \times [1,+\infty)$ is in the symbol class S_{scl}^m if for any multi-indices (α,β), we have

$$\gamma_{\alpha\beta}(a) = \sup_{(x,\xi)\in\mathbb{R}^{2n}, \lambda\geq 1} |D_x^\alpha D_\xi^\beta a(x,\xi,\lambda)|\lambda^{-m+|\beta|} < \infty. \qquad (36)$$

Then one has for $a \in S_{\mathrm{scl}}^{m_1}$ and $b \in S_{\mathrm{scl}}^{m_2}$, the expansion

$$(a\sharp b)(x,\xi) = \sum_{0\leq k<N} 2^{-k} \sum_{|\alpha|+|\beta|=k} \frac{(-1)^{|\beta|}}{\alpha!\beta!} D_\xi^\alpha \partial_x^\beta a \; D_\xi^\beta \partial_x^\alpha b + r_N(a,b), \qquad (37)$$

---

[5] We use a *large* parameter λ instead of a small Planck constant h. Writing $\lambda = 1/h$ will give back the more familiar picture.

with $r_N(a, b) \in S_{\text{scl}}^{m_1+m_2-N}$. The beginning of this expansion is thus $ab + \frac{1}{2\iota}\{a, b\}$, where

$$\{a, b\} = \sum_{1 \leq j \leq n} \partial_{\xi_j} a \, \partial_{x_j} b - \partial_{x_j} a \, \partial_{\xi_j} b$$

is the Poisson bracket and $\iota = 2\pi i$. The sums inside (37) with k even are symmetric in a, b and skew-symmetric for k odd. This can be compared to the classical expansion formula

$$(a \circ b)(x, \xi) = \sum_{|\alpha| < N} \frac{1}{\alpha!} D_\xi^\alpha a \, \partial_x^\alpha b + t_N(a, b), \qquad (38)$$

with $t_N(a, b) \in S_{\text{scl}}^{m_1+m_2-N}$.

Theorem 2.1. *Let m be a real number and $a(x, \xi, \lambda)$ be in S_{scl}^m. Then the operator $a^w \lambda^{-m}$ is bounded on $L^2(\mathbb{R}^n)$ with a norm bounded above by a seminorm (36) of a.*

This theorem is a consequence of the much more general

Lemma 2.1. *Let $b(x, \xi)$ be a function defined on $\mathbb{R}^{2n}$, bounded as well as all its derivatives. Then the operators $a^w, Op(a), Op(J^t a)$ are bounded on $L^2(\mathbb{R}^n)$.*

Proof. Let us check the classical quantization with $u, v \in \mathcal{S}(\mathbb{R}^n)$, assuming that $b \in C_c^\infty(\mathbb{R}^{2n})$

$$\langle b(x, D)u, v \rangle = \int e^{2i\pi(x-y)\xi} b(x, \xi) u(y) \overline{\hat{v}(\eta)} e^{-2i\pi x\eta} \, d\eta dy dx d\xi.$$

Integrating by parts with respect to x gives with a polynomial P

$$P(D_x)\left(e^{2i\pi x(\xi-\eta)}\right) = e^{2i\pi x(\xi-\eta)} P(\xi - \eta),$$

so that with $P_k(\xi) = (1 + |\xi|^2)^{k/2}$ for $k \in 2\mathbb{N}$, we get

$$\langle b(x, D)u, v \rangle = \int e^{2i\pi(x-y)\xi} P_k(\xi-\eta)^{-1}(P_k(D_x)b)(x, \xi) u(y) \overline{\hat{v}(\eta)} e^{-2i\pi x\eta} \, d\eta dy dx d\xi.$$

Now we integrate by parts with respect to ξ so that $\langle b(x, D)u, v \rangle =$

$$\int e^{2i\pi(x-y)\xi} P_k(x-y)^{-1} P_k(D_\xi) P_k(\xi-\eta)^{-1}(P_k(D_x)b)(x, \xi) u(y) \overline{\hat{v}(\eta)} e^{-2i\pi x\eta} \, d\eta dy dx d\xi.$$

As a result we obtain that $\langle b(x, D)u, v \rangle =$

$$\sum_l \int e^{2i\pi(x-y)\xi} P_k(x-y)^{-1} \varphi_l(\xi-\eta) b_{kl}(x, \xi) u(y) \overline{\hat{v}(\eta)} e^{-2i\pi x\eta} \, d\eta dy dx d\xi,$$

where the sum is finite, $|\varphi_l(\zeta)| \le (1+|\zeta|)^{-k}$, $b_{kl} = \partial_x^\alpha \partial_\xi^\beta b$, with $|\alpha| \le k, |\beta| \le k$. We get that $\langle b(x,D)u, v \rangle$ is a finite sum of terms of type

$$\iint \underbrace{e^{2i\pi x\xi} b_{kl}(x,\xi)}_{\text{bounded}} \int e^{-2i\pi y\xi} P_k(x-y)^{-1} u(y) dy \int \varphi_l(\xi-\eta)\overline{\hat{v}(\eta)} e^{-2i\pi x\eta} d\eta \, dx d\xi. \quad (39)$$

Assuming that $k > n/2$, we get that P_k and φ_l are in $L^2(\mathbb{R}^n)$. It is then enough to check that

$$\left\| \int e^{-2i\pi y\xi} P_k(x-y)^{-1} u(y) dy \right\|_{L^2(\mathbb{R}^{2n}_{x,\xi})} \le C \|u\|_{L^2(\mathbb{R}^n)}. \quad (40)$$

In fact, using (40) we will get from the Cauchy–Schwarz inequality in $L^2(\mathbb{R}^{2n})$ that (39) is bounded above by $C' \|u\|_{L^2(\mathbb{R}^n)} \|v\|_{L^2(\mathbb{R}^n)}$, which gives the result of the lemma. Finally, we have to verify (40), which is indeed obvious since the integral inside the norm on the lhs of (40) is the partial Fourier transform of the $L^2(\mathbb{R}^{2n}_{x,y})$ function $P_k(x-y)u(y)$ whose L^2 norm is $\left\| P_k^{-1} \right\|_{L^2(\mathbb{R}^n)} \|u\|_{L^2(\mathbb{R}^n)}$.
□

Remark 2.3. The reader may have noticed that this quite original method of proof, due to I.L.Hwang [H], is also giving the sharp number of derivatives, as proven in [CM] and in particular the result

$$\|\mathrm{Op}(a)\|_{\mathcal{L}(L^2(\mathbb{R}^n))} \le C(n) \sup_{\substack{(x,\xi)\in\mathbb{R}^{2n} \\ |\alpha|\le[n/2]+1, |\beta|\le[n/2]+1}} |\partial_x^\alpha \partial_\xi^\beta a(x,\xi)|$$

where $[x]$ stands for the largest integer $\le x$ and $C(n)$ depends only on the dimension n.

3 Definition and First Properties of the Wick Quantization

3.1 Definitions

Let us consider a symplectic vector space Φ, i.e. a finite dimensional real vector space equipped with a nondegenerate alternate bilinear form σ. The form σ can be identified to an isomorphism

$$\sigma : \Phi \to \Phi^*, \quad \sigma^* = -\sigma, \quad \text{the form is } \langle \sigma X, Y \rangle_{\Phi^*, \Phi}.$$

Then the dimension of Φ is even: take e_1 a nonzero vector in Φ and define $\epsilon_1 = \sigma^{-1} e_1^*$ where $e_1^* \in \Phi^*$ is such that $\langle e_1^*, e_1 \rangle_{\Phi^*, \Phi} = 1$. We have thus

$$\langle \sigma \epsilon_1, e_1 \rangle_{\Phi^*, \Phi} = \langle e_1^*, e_1 \rangle_{\Phi^*, \Phi} = 1.$$

Let us now consider the vector space $\mathcal{V}(\epsilon_1, e_1)$: since the form is alternate, this is a plane (ϵ_1 cannot be proportional to e_1 and satisfy $\langle \sigma \epsilon_1, e_1 \rangle = 1$) and its symplectic orthogonal $\Psi = \mathcal{V}(\epsilon_1, e_1)^\sigma$ has dimension $\dim \Phi - 2$ (σ is nondegenerate); now we can restrict the form σ to Ψ. It is of course bilinear alternate; let us check that it is nondegenerate. Assuming $\sigma(X, Y) = 0$ for some $X \in \Psi$ and all $Y \in \Psi$, we get also since $X \in \mathcal{V}(\epsilon_1, e_1)^\sigma$

$$\sigma(X, e_1) = \sigma(X, \epsilon_1) = 0$$

so that X is σ-orthogonal to Φ and thus is zero. Using an induction on the dimension, we can indeed find a *symplectic basis* that is a basis $\epsilon_1, e_1, \ldots, \epsilon_n, e_n$, such that

$$\langle \sigma \epsilon_j, e_k \rangle = \delta_{j,k}, \quad \langle \sigma \epsilon_j, \epsilon_k \rangle = \langle \sigma e_j, e_k \rangle = 0.$$

Writing $X = \sum_{1 \le j \le n} \xi_j \epsilon_\mathbf{j} + x_j e_\mathbf{j}$, $Y = \sum_{1 \le j \le n} \eta_j \epsilon_\mathbf{j} + y_j e_\mathbf{j}$ we get back to the familiar

$$[X, Y] = \xi \cdot y - \eta \cdot x, \quad \text{or } \sigma = \sum_j d\xi_j \wedge dx_j.$$

Let us now assume that our symplectic vector space Φ is equipped with a positive definite quadratic form Q. The form σ can be identified to a skew-symmetric isomorphism A of the Q-Euclidean Φ, via the identity $Q(AX, Y) = \langle \sigma X, Y \rangle_{\Phi^*, \Phi}$, and this formula implies that

$$Q(X, AY) = Q(AY, X) = \langle \sigma Y, X \rangle_{\Phi^*, \Phi} = -\langle \sigma X, Y \rangle_{\Phi^*, \Phi} = -Q(AX, Y)$$

so that $A^* = -A$ (duality induced by Q). As a consequence the spectrum of A is purely imaginary and there exist $X_1, X_2 \in \Phi, \lambda \in \mathbb{R}^*$, such that $(X_1, X_2) \ne (0, 0)$ and

$$A(X_1 + iX_2) = i\lambda(X_1 + iX_2), \quad \text{i.e.} \quad AX_1 = -\lambda X_2, \quad AX_2 = \lambda X_1.$$

This implies that $\lambda Q(X_1, X_2) = Q(AX_2, X_2) = 0 = Q(X_1, X_2)$, so the vectors X_1, X_2 are Q-orthogonal and independent ($X_2 = \alpha X_1$ would imply $\alpha A X_1 = \lambda X_1$ and $\lambda Q(X_1, X_1) = 0$, i.e. $X_1 = 0 = X_2$). Moreover the plane $\mathcal{V}(X_1, X_2)$ is invariant by A and its Q-orthogonal (the X such that $Q(X, X_1) = Q(X, X_2) = 0$) coincides with its symplectic orthogonal (the X such that $Q(AX, X_1) = Q(AX, X_2) = 0$), so that $\mathcal{V}(X_1, X_2)^{\perp_Q}$ is also invariant by A. Using an induction on the dimension, we can find a symplectic basis $\epsilon_1, e_1, \ldots, \epsilon_n, e_n$, which is also orthogonal for Q, whose expression will be

$$\sum_{1 \le j \le n} \lambda_j (\xi_j^2 + x_j^2), \quad \lambda_j > 0. \tag{41}$$

A more concise argument (in fact the same) for this simultaneous diagonalization is that the Hermitian form $i\sigma$ can be diagonalized in the complex vector space equipped with the dot product Q. From this short discussion, we have to keep in mind that a positive definite quadratic form can be reduced to (41), via a suitable choice of symplectic coordinates. In the case of a semi-definite quadratic form or in the hyperbolic case, some normal forms are known but the discussion is much more involved; we refer the reader to the Sect. 21.5 of [Hör85] and to the Theorem 21.5.3 there.

Let Γ be an Euclidean norm on $\mathbb{R}^{2n}$, identified with a $2n \times 2n$ symmetric matrix; we define $\Gamma^\sigma = \sigma^* \Gamma^{-1} \sigma$, where $\sigma = \begin{pmatrix} 0 & I_n \\ -I_n & 0 \end{pmatrix}$. We shall say that Γ is a symplectic norm whenever $\Gamma = \Gamma^\sigma$. The basic examples of symplectic norms that we are going to use are

$$\Gamma_\lambda = \lambda |dx|^2 + \frac{|d\xi|^2}{\lambda} = \begin{pmatrix} \lambda I_n & 0 \\ 0 & \lambda^{-1} I_n \end{pmatrix}, \tag{42}$$

where λ is a positive parameter. Our construction of the Wick quantization could be carried out for any symplectic norm, however, for simplicity, we shall limit ourselves to the norms (42). The following definition contains also some classical properties.

Definition 3.1. Let $Y = (y, \eta)$ be a point in $\mathbb{R}^{2n}$ and $\lambda > 0$. We define first the operator

$$\Sigma_Y^\lambda = \left[2^n e^{-2\pi \Gamma_\lambda(\cdot - Y)} \right]^w. \tag{43}$$

This is a rank-one orthogonal projection: using the notations (15–16), we have

$$\Sigma_Y^\lambda u = (W_\lambda u)(Y) \varphi_Y^\lambda = \langle u, \varphi_Y^\lambda \rangle_{L^2(\mathbb{R}^n)} \varphi_Y^\lambda. \tag{44}$$

Let a be in $L^\infty(\mathbb{R}^{2n})$. The Wick($\lambda$) quantization of a is defined as

$$a^{\mathrm{Wick}(\lambda)} = \int_{\mathbb{R}^{2n}} a(Y) \Sigma_Y^\lambda dY. \tag{45}$$

To check (43), starting from (44) is an easy exercise on the Weyl quantization left to the reader.

Proposition 3.1. *Let λ be a positive number and a be in $L^\infty(\mathbb{R}^{2n})$. Then*

$$a^{Wick(\lambda)} = W_\lambda^* a^\mu W_\lambda, \quad 1^{Wick(\lambda)} = \mathrm{Id}_{L^2(\mathbb{R}^n)} \tag{46}$$

where W_λ is the isometric mapping from $L^2(\mathbb{R}^n)$ to $L^2(\mathbb{R}^{2n})$ given in (16), and a^μ the operator of multiplication by a in $L^2(\mathbb{R}^{2n})$. The operator $\pi_{H_\lambda} = W_\lambda W_\lambda^*$ is the orthogonal projection on a closed proper subspace H_λ of $L^2(\mathbb{R}^{2n})$. Moreover, we have

$$\|a^{Wick(\lambda)}\|_{\mathcal{L}(L^2(\mathbb{R}^n))} \le \|a\|_{L^\infty(\mathbb{R}^{2n})}, \tag{47}$$

$$a(X) \ge 0 \implies a^{Wick(\lambda)} \ge 0, \tag{48}$$

$$\left\| \Sigma_Y^\lambda \Sigma_Z^\lambda \right\|_{\mathcal{L}(L^2(\mathbb{R}^n))} \le 2^n e^{-\frac{\pi}{2} \Gamma_\lambda(Y - Z)}. \tag{49}$$

Moreover the kernel of $\pi_H = \pi_H^1$ is $e^{-\frac{\pi}{2}\|X - Y\|^2} e^{-i\pi[X,Y]}$, $\quad [X, Y] = \langle \sigma X, Y \rangle$.

Proof. Here we assume that $\lambda = 1$ and omit the indexation by λ. The calculations are analogous for other positive values of λ. The first properties and (48) are immediate consequences of Lemma 1.3. The operator π_H is an orthogonal projection on its range, which is the same as the range of W and the latter is closed since W is isometric. On the other hand, π_H is not onto, otherwise π_H would be the identity of $L^2(\mathbb{R}^{2n})$ and for all $u \in \mathcal{S}(\mathbb{R}^n)$, we would have

$$\|u\|^2_{L^2(\mathbb{R}^n)} = 2\operatorname{Re}\langle D_{x_1}u, ix_1u\rangle_{L^2(\mathbb{R}^n)} = 2\operatorname{Re}\langle \xi_1^{\mathrm{Wick}}u, ix_1^{\mathrm{Wick}}u\rangle_{L^2(\mathbb{R}^n)} =$$

$$= 2\operatorname{Re}\langle \xi_1 Wu, i\pi_H x_1 Wu\rangle_{L^2(\mathbb{R}^{2n})} = 2\operatorname{Re}\langle \xi_1 Wu, ix_1 Wu\rangle_{L^2(\mathbb{R}^{2n})} = 0.$$

Now, with $L^2(\mathbb{R}^n)$ dot-products, we have

$$|\langle a^{\mathrm{Wick}}u, v\rangle| = \left|\int_{\mathbb{R}^{2n}} a(Y)\langle \Sigma_Y u, v\rangle dY\right| = \left|\int_{\mathbb{R}^{2n}} a(Y)\langle \Sigma_Y u, \Sigma_Y v\rangle dY\right|$$

$$\leq \|a\|_{L^\infty(\mathbb{R}^{2n})}\int_{\mathbb{R}^{2n}} \|\Sigma_Y u\|_{L^2(\mathbb{R}^n)}\|\Sigma_Y v\|_{L^2(\mathbb{R}^n)}\, dY$$

$$\leq \|a\|_{L^\infty(\mathbb{R}^{2n})}\left(\int_{\mathbb{R}^{2n}} \|\Sigma_Y u\|^2_{L^2(\mathbb{R}^n)}\, dY\right)^{1/2}\left(\int_{\mathbb{R}^{2n}} \|\Sigma_Y v\|^2_{L^2(\mathbb{R}^n)}\, dY\right)^{1/2}$$

$$= \|a\|_{L^\infty(\mathbb{R}^{2n})}\|u\|_{L^2(\mathbb{R}^n)}\|v\|_{L^2(\mathbb{R}^n)},$$

yielding (47). For $Y, Z \in \mathbb{R}^{2n}$ a straightforward computation shows that the Weyl symbol of $\Sigma_Y \Sigma_Z$ is, as a function of the variable $X \in \mathbb{R}^{2n}$, setting $\Gamma_1(T) = |T|^2$

$$e^{-\frac{\pi}{2}|Y-Z|^2}e^{-2i\pi[X-Y, X-Z]}2^n e^{-2\pi|X-\frac{Y+Z}{2}|^2}.$$

Since for the Weyl quantization, one has $\|a^w\|_{\mathcal{L}(L^2(\mathbb{R}^n))} \leq 2^n \|a\|_{L^1(\mathbb{R}^{2n})}$, we get the result (49). The very last assertion is left as an (easy) exercise for the reader.   □

Remark 3.1. The positivity property (48) is not satisfied for the Weyl quantization since the Wigner function $\mathcal{H}(u, u)$ (see (22)) is not always non-negative, although it is actually positive if u is a Gaussian function. We leave to the reader the computation of

$$\mathcal{H}(u_1, u_1)(x, \xi) = 4\pi 2^n e^{-2\pi(|x|^2+|\xi|^2)}\left(\xi_1^2 + x_1^2 - \frac{1}{4\pi}\right)$$

$$u_1(x) = 2^{n/4}x_1 2\pi^{1/2}e^{-\pi|x|^2}$$

which is negative in a neighborhood V of the origin. Now, choosing a nonnegative $a(x, \xi) \in C_c^\infty(V)$ and using (21) we get $\langle a^w u_1, u_1\rangle < 0$. On the other hand we have the familiar

$$\mathcal{H}(u_0, u_0)(x, \xi) = 2^n e^{-2\pi(|x|^2+|\xi|^2)}, \quad u_0(x) = 2^{n/4}e^{-\pi|x|^2}.$$

3.2 The Gårding Inequality with Gain of One Derivative

Proposition 3.2. *Let m be a real number and $p(x, \xi, \lambda)$ be a symbol in S_{scl}^m (see (36)). Then*

$$p^{Wick(\lambda)} = p^w + r(p)^w, \tag{50}$$

with $r(p) \in S_{scl}^{m-1}$ so that the mapping $p \mapsto r(p)$ is continuous. Moreover, $r(p) = 0$ if p is a linear form or a constant.

Proof. From the Definition 3.1.1, one has $p^{Wick(\lambda)} = \widetilde{p}^w$, with

$$\widetilde{p}(X) = \int_{\mathbb{R}^{2n}} p(X + Y) \, e^{-2\pi \Gamma_\lambda(Y)} 2^n dY$$

$$= p(X) + \underbrace{\int_0^1 \int_{\mathbb{R}^{2n}} (1 - \theta) p''(X + \theta Y) Y^2 e^{-2\pi \Gamma_\lambda(Y)} 2^n dY \, d\theta}_{r(p)(X)}. \tag{51}$$

We note now that the estimates (36) of S_{scl}^m on p are equivalent to

$$|p^{(k)}(X) T^k| \le C_k \lambda^{m - \frac{k}{2}} \Gamma_\lambda(T)^{\frac{k}{2}} \quad \text{or} \quad |p^{(k)}(X)|_{\Gamma_\lambda} \le C_k \lambda^{m - \frac{k}{2}}.$$

Thus we get

$$|r(p)^{(k)}(X)|_{\Gamma_\lambda} \le C_{k+2} \lambda^{m - \frac{k+2}{2}} \int_{\mathbb{R}^{2n}} \Gamma_\lambda(Y) e^{-2\pi \Gamma_\lambda(Y)} 2^{n-1} dY,$$

and since $\det(\Gamma_\lambda) = 1$, the integral above is a constant and this implies that $r \in S_{scl}^{m-1}$. The last point in the proposition follows from the formula (51) showing that $r(p)$ depends linearly on p''. $\qquad\qquad \square$

Remark 3.2. For further understanding of our results, it would be better to use symbol classes defined by a metric in the phase space, as introduced in the Chap. 18 of [Hör85]. As we have seen above,

$$S_{scl}^m = S(\lambda^m, \lambda^{-1} \Gamma_\lambda),$$

that is symbols such that

$$|a^{(k)}(X) T^k| \le \gamma_k(a) \lambda^{m - \frac{k}{2}} \Gamma_\lambda(T)^{\frac{k}{2}},$$

or more accurately, for all $k \in \mathbb{N}$,

$$\gamma_k(a) = \sup_{\substack{X \in \mathbb{R}^{2n}, \lambda \ge 1, \\ T \in \mathbb{R}^{2n}, \Gamma_\lambda(T) = 1}} |a^{(k)}(X) T^k| \lambda^{-m + \frac{k}{2}} < +\infty.$$

The following theorem was proven in 1966 by L.Hörmander and a generalization to systems was given the same year by P.Lax and L.Nirenberg. The reader can check the Theorem 18.6.14 in [Hör85] for a (much) wider generalization of this statement. The name given to this inequality by the aforementioned authors was "Sharp Gårding inequality", a terminology that may look inappropriate nowadays since, in the scalar case, a drastic improvement of that sharpness was given in 1978 by C.Fefferman and D.H.Phong in [FP] (see our Sect. 5 below). However, in the vector-valued case, the Fefferman–Phong inequality is not true in general, as observed in [Br90]; a class of counterexamples were studied more systematically in [P].

Theorem 3.1. *Let $a(x, \xi, \lambda)$ be a symbol in S^1_{scl} (cf. (36)), taking nonnegative values. Then the operator a^w is semi-bounded from below in $L^2(\mathbb{R}^n)$, and more precisely, there exists a semi-norm $\gamma_{\alpha\beta}(a)$ of a and a constant C_n, depending only on the dimension such that*

$$a^w + C_n \gamma_{\alpha\beta}(a) \geq 0, \quad \text{as an operator.} \tag{52}$$

Proof. It appears as an immediate consequence of the Proposition 3.2 and of (48): we have from (50)

$$a^w = a^{\text{Wick}(\lambda)} - r(a)^w, \qquad r(a) \in S^0_{scl} \text{ thus } -r(a)^w \in \mathcal{L}(L^2(\mathbb{R}^n)),$$

and from (48) $a^{\text{Wick}(\lambda)} \geq 0$, yielding the result. $\qquad \square$

3.3 Variations

In this section, we show that using the method of proof of the Theorem 18.1.14 in [Hör85], we can in fact obtain a stronger result than the Theorem 3.1.

Let $\varphi(x, \xi)$ an even L^2 function on $\mathbb{R}^{2n}$ with L^2 norm 1. We define, using (35), $\Psi = \varphi^* \circ \varphi$, with φ^* standing for the standard symbol of the adjoint. We note that Ψ is even and that

$$\iint \Psi(x, \xi) dx = 1. \tag{53}$$

In fact, we have

$$\Psi(x, \xi) = \iint \varphi^*(x, \xi + \eta) \varphi(x + y, \xi) e^{-2i\pi y\eta} dy d\eta$$

$$= \iiiint \bar{\varphi}(x + z, \zeta + \xi + \eta) e^{-2i\pi z\zeta} \varphi(x + y, \xi) e^{-2i\pi y\eta} dz dy d\zeta d\eta$$

$$= \iiiint \bar{\varphi}(x + z, \xi + \eta) e^{-2i\pi z\zeta} \varphi(x + y, \xi) e^{-2i\pi y(\eta - \zeta)} dz dy d\zeta d\eta$$

$$= \iiiint \bar{\varphi}(z,\eta)e^{-2i\pi(z-x)\zeta}\varphi(y,\xi)e^{-2i\pi(y-x)(\eta-\zeta-\xi)}dzdyd\zeta d\eta$$

$$= \iint \bar{\varphi}(y,\eta)\varphi(y,\xi)e^{-2i\pi(y-x)(\eta-\xi)}dyd\eta,$$

and since φ is even we obtain

$$\Psi(-x,-\xi) = \iint \bar{\varphi}(y,\eta)\varphi(y,-\xi)e^{-2i\pi(y+x)(\eta+\xi)}dyd\eta$$

$$= \iint \bar{\varphi}(-y,-\eta)\varphi(-y,-\xi)e^{-2i\pi(-y+x)(-\eta+\xi)}dyd\eta$$

$$= \iint \bar{\varphi}(y,\eta)\varphi(y,\xi)e^{-2i\pi(y-x)(\eta-\xi)}dyd\eta = \Psi(x,\xi).$$

Moreover we have

$$\iint \Psi(x,\xi)dxd\xi = \iiiint \bar{\varphi}(y,\eta)\varphi(y,\xi)e^{-2i\pi(y-x)(\eta-\xi)}dydxd\eta d\xi =$$

$$= \iint \bar{\varphi}(y,\xi)\varphi(y,\xi)dyd\xi = 1.$$

We consider now the symbol a^b defined by

$$a^b(x,\xi) = \iint a(x+y,\xi+\eta)\Psi(y,\eta)dyd\eta = \iint a(y,\eta)\Psi(x-y,\xi-\eta)dyd\eta. \quad (54)$$

Recalling the definition of Sect. 2.2, we use the phase translation

$$(\tau_{y,\eta}u)(x) = u(x-y)e^{2i\pi(x-\frac{y}{2})\eta}, \quad \text{so that} \quad \tau_{y,\eta}^* = \tau_{-y,-\eta}$$

and we get

$$0 \leq \tau_{y,\eta}\mathrm{Op}(\varphi^*(x,\xi))\mathrm{Op}(\varphi(x,\xi))\tau_{y,\eta}^* = \tau_{y,\eta}\mathrm{Op}(\Psi(x,\xi))\tau_{y,\eta}^* =$$

$$= \mathrm{Op}((x,\xi) \mapsto \Psi(x-y,\xi-\eta)), \quad (55)$$

since

$$(\tau_{y,\eta}\mathrm{Op}(\Psi(x,\xi))\tau_{y,\eta}^*u)(x) = e^{2i\pi(x-\frac{y}{2})\eta}\int \Psi(x-y,\xi)e^{2i\pi(x-y)\xi}\widehat{\tau_{y,\eta}^*u}(\xi)d\xi$$

$$= e^{2i\pi(x-\frac{y}{2})\eta}\iint \Psi(x-y,\xi)e^{2i\pi(x-y-z)\xi}(\tau_{y,\eta}^*u)(z)dzd\xi$$

$$= e^{2i\pi(x-\frac{y}{2})\eta}\iint \Psi(x-y,\xi)e^{2i\pi(x-y-z)\xi}u(z+y)e^{-2i\pi(z+\frac{y}{2})\eta}dzd\xi$$

$$= e^{2i\pi(x-\frac{y}{2})\eta}\iint \Psi(x-y,\xi)e^{2i\pi(x-z)\xi}u(z)e^{-2i\pi(z-\frac{y}{2})\eta}dzd\xi$$

$$= \iint \Psi(x - y, \xi) e^{2i\pi(x-z)(\xi+\eta)} u(z) dz d\xi$$

$$= \iint \Psi(x - y, \xi - \eta) e^{2i\pi(x-z)\xi} u(z) dz d\xi$$

$$= \int \Psi(x - y, \xi - \eta) e^{2i\pi x\xi} \hat{u}(\xi) d\xi$$

$$= (\text{Op}((x, \xi) \mapsto \Psi(x - y, \xi - \eta)) u)(x).$$

From (54) and $a \geq 0$, we get

$$\text{Op}(a^b) = \iint a(y, \eta) \text{Op}(\Psi(x - y, \xi - \eta)) dy d\eta \geq 0, \quad \text{as an operator.} \quad (56)$$

Lemma 3.1. *Let a be a function defined on $\mathbb{R}^{2n}$ such that $a'' \in S_{0,0}^0$ (smooth bounded functions as well as all their derivatives). Then with a^b defined in (54), $a - a^b$ belongs to $S_{0,0}^0$.*

Proof. We have

$$2a^b(x, \xi) - 2a(x, \xi) = \iint (a(x+y, \xi+\eta) + a(x-y, \xi-\eta) - 2a(x, \xi)) \Psi(y, \eta) dy d\eta$$

$$= \iint a_{y,\eta}^{[2]}(x, \xi) \Psi(y, \eta) dy d\eta,$$

with

$$a_Y^{[2]}(X) = \int_{-1}^1 (1 - |\theta|) a''(X + \theta Y) Y^2 d\theta, \qquad Y = (y, \eta), \ X = (x, \xi),$$

a symbol which belongs to $S_{0,0}^0$ with semi-norms controlled by semi-norms of $a'' \times |Y|^2$. We have

$$2a^b(X) - 2a(X) = \int_{-1}^1 \int_{\mathbb{R}^{2n}} (1 - |\theta|) a''(X + \theta Y) Y^2 \Psi(Y) dY d\theta.$$

We may also assume that Ψ belongs to $L^1(\mathbb{R}^{2n})$ and is rapidly decreasing, entailing that for all semi-norms γ in $S_{0,0}^0$,

$$\gamma(a^b - a) \leq \iint \gamma(a_{y,\eta}^{[2]}) |\Psi(y, \eta)| dy d\eta < \infty.$$

$\square$

As a consequence, we get the

Theorem 3.2. *Let a be a nonnegative function defined on $\mathbb{R}^{2n}$ such that the Hessian $a'' \in S_{0,0}^0$. Then the operators a^w and $\text{Re}(a(x, D))$ are semi-bounded from below.*

Proof. We have from the Lemma 3.1 and the Lemma 2.1

$$\mathrm{Op}(a) \in \mathrm{Op}(a^b) + \mathrm{Op}(S^0_{0,0}) \subset \mathrm{Op}(a^b) + \mathcal{L}(L^2(\mathbb{R}^n)),$$

so that using (56), we get that $\mathrm{Re}\,\mathrm{Op}(a)$ is semi-bounded from below. We can also change the quantization and consider a symbol a satisfying the assumptions of the Theorem 3.2: with $a^w = \mathrm{Op}(J^{1/2}a)$, we see from our Sect. 2.1 that $J^{1/2}a \in a + i\pi D_x \cdot D_\xi a + S^0_{0,0}$ so that

$$a^w \in \mathrm{Op}(a) + \mathrm{Op}(i\pi D_x \cdot D_\xi a) + \mathcal{L}(L^2(\mathbb{R}^n))$$

and taking real parts of the operators we have

$$a^w \in \mathrm{Re}\,\mathrm{Op}(a) + \mathrm{Re}\,\mathrm{Op}(i\pi D_x \cdot D_\xi a) + \mathcal{L}(L^2(\mathbb{R}^n)). \tag{57}$$

We check now, using that a is real-valued (entailing $i\pi D_x \cdot D_\xi a \in i\mathbb{R}$),

$$
\begin{aligned}
2\,\mathrm{Re}\,\mathrm{Op}(i\pi D_x \cdot D_\xi a) &= \mathrm{Op}(i\pi D_x \cdot D_\xi a + \overline{J i\pi D_x \cdot D_\xi a}) \\
&= \mathrm{Op}((\mathrm{Id} - J)i\pi D_x \cdot D_\xi a)
\end{aligned}
$$

which belongs to $\mathrm{Op}(S^0_{0,0})$. As a consequence, (57) is giving that a^w is semi-bounded from below since $\mathrm{Re}\,\mathrm{Op}(a)$ is already proven so. $\qquad\square$

It is interesting to see that the non-asymptotic result of Theorem 3.3.2 implies the asymptotic statement of Theorem 3.1; as a matter of fact, if we consider a nonnegative symbol $a(x, \xi, \lambda)$ in S^1_{scl}, the operator $a(x, \xi, \lambda)^w$ is unitarily equivalent to $a(x\lambda^{-1/2}, \xi\lambda^{1/2}, \lambda)^w$ and from the estimates (36), we get that the nonnegative symbol

$$b(x, \xi) = a(x\lambda^{-1/2}, \xi\lambda^{1/2}, \lambda)$$

satisfies indeed the assumptions of the theorem. As a result, the operator b^w (and thus the unitarily equivalent a^w) is semi-bounded from below.

As said above, that proof of the sharp Gårding inequality is borrowed from [Hör85]. This is a general idea of mollifying the symbol by a normalized function of type $\varphi^* \circ \varphi$. Nothing else at this stage is really needed and the classical so-called *coherent states method* is simply dealing with a Gaussian function φ, with the only but no crucial advantage that the computations of $\varphi^* \circ \varphi$ can be made explicitly with other Gaussians. This point of view is precisely the most synthetic and seems suitable to tackle a group situation.

Our last remark is dealing with the standard classes of pseudodifferential operators and with the asymptotic point of view, which plays an important role for PDE. The standard classes of symbols $S^m_{\rho,\delta}$ on $\mathbb{R}^n$ are well-known. Sticking for simplicity with the case $\rho = 1, \delta = 0$ and calling $S^m = S^m_{1,0}$, one can get standard continuity and composition results. However the statement of the sharp Gårding inequality is: let a be a nonnegative symbol in S^1, then

$$\mathrm{Re}\langle \mathrm{Op}(a)u, u \rangle + C\,\|u\|^2_{L^2} \geq 0. \tag{58}$$

This result is in fact a consequence of Theorem 3.2. Let us sketch the proof of the non-semi-classical (58), using Theorem 3.2:

1. Using a Littlewood–Paley decomposition, one writes

$$a(x,\xi) = \sum_{\nu \in \mathbb{N}} \varphi_\nu(\xi) a(x,\xi), \quad \mathrm{Op}(a) = \sum_{\nu \in \mathbb{N}} \psi_\nu(D)\mathrm{Op}(\varphi_\nu a)\psi_\nu(D) + R$$

with $\psi_\nu = 1$ on the support of φ_ν; R belongs to $S^{-\infty}$ (it is not completely obvious because of the summation in ν).
2. We are reduced to prove a semiclassical statement, since the conditions on the derivatives of $\varphi_\nu a$ are expressed in terms of the frequency $2^\nu = 1/h$. Essentially, the statement to be proven is: for b smooth nonnegative bounded with bounded derivatives the operator $h^{-1}\mathrm{Op}(b(x,h\xi))$ is semi-bounded from below (uniformly in $h \in\,]0,1]$).
3. We note that the previous operator is unitarily equivalent to

$$h^{-1}\mathrm{Op}(b(h^{1/2}x, h^{1/2}\xi))$$

and that the seminorms in $S_{0,0}^0$ of A'' where

$$A(x,\xi) = h^{-1}b(h^{1/2}x, h^{1/2}\xi)$$

are bounded for $h \in\,]0,1]$. It is indeed the case and we can use Theorem 3.2 to conclude.

4 Energy Estimates via the Wick Quantization

4.1 Subelliptic Operators Satisfying Condition (P)

We intend to illustrate in this section the usefulness of Wick quantization to prove energy estimates. We want to give a simple proof of a well-known theorem on subellipticity for differential operators. In this case, the proof of the theorem conjectured in the papers of Egorov [Eg] is known since 1971 with the work of Treves [Tr]. Our method here falls short of giving a proof in the general case (i.e. for pseudo-differential operators). Hörmander gave a complete proof in 1979 of the general theorem on subellipticity which can be found in Chap. 27 of [Hör85]. We will go a little beyond the differential case, proving the theorem when the zero set of the imaginary part is included in its critical set (see Theorem 4.2) below. Moreover, we believe that this elementary proof, reducing actually the problem to simple ordinary differential equations, is a good example of the wave-packet technique.

Let P be a principal type pseudodifferential operator on a manifold $\mathcal{M}$ whose principal symbol p satisfies condition $(\overline{\psi})$:

$$(\overline{\psi}) \quad \begin{cases} \forall z \in \mathbb{C}, \operatorname{Im}(zp) \text{ does not change sign from } + \text{ to } - \\ \text{along the oriented bicharacteristic curves of } \operatorname{Re}(zp). \end{cases} \quad (59)$$

Assume also that for some $z \in \mathbb{C}$ and in some conic open set $\Omega \subset T^*(\mathcal{M})\backslash 0$ and some integer k

$$H^k_{\operatorname{Re}(zp)}(\operatorname{Im}(zp)) \neq 0. \quad (60)$$

The following condition is equivalent to $(\overline{\psi})$ for differential operators

$$(P) \quad \begin{cases} \forall z \in \mathbb{C}, \operatorname{Im}(zp) \text{ does not change sign} \\ \text{along the oriented bicharacteristic curves of } \operatorname{Re}(zp). \end{cases} \quad (61)$$

Theorem 4.1. *Let P be a properly supported principal type pseudo-differential operator of order m on a manifold $\mathcal{M}$ satisfying $(60-61)$. Then P is subelliptic on Ω with loss of $k/(k+1)$ derivatives: for any real s*

$$u \in \mathcal{D}'(\mathcal{M}), \quad Pu \in H^s(\Omega) \Longrightarrow u \in H^{s+m-\frac{k}{k+1}}(\Omega). \quad (62)$$

Here, k is necessarily an even integer.

After classical reductions, Theorem 4.1 in $n+1$ dimensions will follow from Theorem 4.2 below. We need first to state a

Definition 4.1. Let n be an integer and $A_\Lambda(t,x,\xi)$ be a family of smooth functions on $\mathbb{R}_t \times \mathbb{R}_x^n \times \mathbb{R}_\xi^n$, depending on a parameter $\Lambda \geq 1$. Let m be a real number. We shall say that $(A_\Lambda) \in S_n^m$ if for each $(l, \alpha, \beta) \in \mathbb{N} \times \mathbb{N}^n \times \mathbb{N}^n$,

$$\sup\nolimits_{\Lambda \geq 1, (t,x,\xi) \in \mathbb{R}^{2n+1}} |(D_t^l D_x^\alpha D_\xi^\beta A_\Lambda)(t,x,\xi)|\Lambda^{-m+\frac{|\alpha|+|\beta|}{2}} = C_{l\alpha\beta} < \infty, (63)$$

$$\operatorname{supp} A_\Lambda \subset [-1,1] \times \{x \in \mathbb{R}^n, |x| \leq \Lambda^{1/2}\} \times \{\xi \in \mathbb{R}^n, |\xi| \leq \Lambda^{1/2}\}. \quad (64)$$

The semi-norms of the family (A_Λ) are defined as the constants $C_{l\alpha\beta}$ in (63). Note that for each t, the function $(x,\xi) \mapsto A_\Lambda(t, \Lambda^{1/2}x, \Lambda^{-1/2}\xi)$ belongs to S_{scl}^m as defined by (36).

We can now state the

Theorem 4.2. *Let n be an integer and $Q_\Lambda(t,x,\xi)$ be a family of smooth functions such that $(Q_\Lambda) \in S_n^1$. Assume moreover that*

$$Q_\Lambda(t,x,\xi) = 0 \implies d_{x,\xi}Q_\Lambda(t,x,\xi) = 0. \quad (65)$$

$$\inf_{\substack{\Lambda \geq 1, |2t| \leq 1, \\ |2x| \leq \Lambda^{1/2}, |2\xi| \leq \Lambda^{1/2}}} |D_t^k Q_\Lambda(t,x,\xi)|\Lambda^{-1} = \delta_0 > 0. \quad (66)$$

Let $\chi_\Lambda(t,x,\xi)$ be a family of smooth functions such that the family $(\chi_\Lambda(\frac{t}{2},\frac{x}{2},\frac{\xi}{2}))_\Lambda \in S_n^0$. There exists a positive constant δ_1, such that, for any $u(t,x) \in C_c^\infty((-\frac{1}{2},\frac{1}{2})_t, \mathcal{S}(\mathbb{R}_x^n))$,

$$\|D_t u + iQ_\Lambda(t,x,D_x)u\|_{L^2(\mathbb{R}^{n+1})} + \|u\|_{L^2(\mathbb{R}^{n+1})} \geq \delta_1 \Lambda^{\frac{1}{k+1}} \|\chi_\Lambda(t,x,D_x)u\|_{L^2(\mathbb{R}^{n+1})}.$$

$$(67)$$

Note that δ_1 depends only on the dimension n, δ_0 and the semi-norms of the families $(Q_\Lambda), (\chi_\Lambda)$.

Let us note right now that condition (65) is satisfied by nonnegative (and nonpositive) functions. Properties (60–61) imply that, $q \geq 0$ or $q \leq 0$ in some conic open neighborhood of a point where (60) is satisfied.

4.2 Polynomial Behaviour of Some Functions

Lemma 4.1. *Let $k \in \mathbb{N}^*, \delta > 0$ and $C > 0$ be given. Let I be an interval of $\mathbb{R}$ and $q : I \longrightarrow \mathbb{R}$ be a C^k function satisfying*

$$\inf_{t \in I} |q^{(k)}(t)| \geq \delta. \tag{68}$$

Then, for any $h > 0$, the set

$$\{t \in I, |q(t)| \leq Ch^k\} \subset \cup_{1 \leq l \leq k} J_l, \tag{69}$$

where J_l is an interval with length $h(\alpha_k C\delta^{-1})^{1/k}, \alpha_k = 2^{2k}k!$. As a consequence the Lebesgue measure of $\{t \in I, |q(t)| \leq Ch^k\}$ is smaller than

$$h(\frac{C}{\delta})^{1/k}4k(k!)^{1/k} \leq h\left(\frac{C}{\delta}\right)^{1/k}4k^2.$$

Proof. Let $k \in \mathbb{N}^*$, h a positive number and set $E_k(h,C,q) = \{t \in I, |q(t)| \leq Ch^k\}$. Let us first assume $k = 1$. Assume that $t, t_0 \in E_1(h,C,q)$; then the mean value theorem and (68) imply

$$2Ch \geq |q(t) - q(t_0)| \geq \delta|t - t_0|$$

so that $E_1(h,C,q) \cap \{t, |t - t_0| > h2C/\delta\} = \emptyset$: otherwise we would have $2Ch > h\delta 2C/\delta$. As a result for any $t_0, t \in E_1(h,C,q), |t - t_0| \leq h2C/\delta$. Either $E_1(h,C,q)$ is empty or it is not empty and then included in an interval with length $\leq h4C/\delta$.

Let us now assume that $k \geq 2$. If $E_k(h,C,q) = \emptyset$, (69) is true. We assume that there exists $t_0 \in E_k(h,C,q)$, and we write for $t \in I$,

$$q(t) = q(t_0) + \underbrace{\int_0^1 q'(t_0 + \theta(t - t_0))d\theta(t - t_0)}_{Q(t)}. \tag{70}$$

Then, if $t \in E_k(h, C, q)$, we have $2Ch^k \geq |Q(t)(t - t_0)|$. Now, for a given $\omega > 0$, either $|t - t_0| \leq \omega h/2$ and $t \in [t_0 - \omega h/2, t_0 + \omega h/2]$, or $|t - t_0| > \omega h/2$ and from the previous inequality we infer $|Q(t)| \leq \omega^{-1} 4Ch^{k-1}$, i.e. we get that

$$E_k(h, C, q) \subset [t_0 - \omega h/2, t_0 + \omega h/2] \cup E_{k-1}(h, \omega^{-1} 4C, Q). \tag{71}$$

But the function Q satisfies the assumptions of the lemma with $k - 1$, δ replaced by δ/k: in fact for $t \in I$, $Q^{(k-1)}(t) = \int_0^1 q^{(k)}(t_0 + \theta(t - t_0))\theta^{k-1}d\theta$, and if $q^{(k)}(t) \geq \delta$ on I, we get $Q^{(k-1)}(t) \geq \delta/k$. By induction on k and using (71), we get that

$$E_k(h, C, q) \subset [t_0 - \omega h/2, t_0 + \omega h/2] \cup_{1 \leq l \leq k-1} J_l,$$
$$|J_l| \leq h(4C\omega^{-1}k\delta^{-1}\alpha_{k-1})^{1/(k-1)}. \tag{72}$$

We choose now $\omega = (4C\omega^{-1}k\delta^{-1}\alpha_{k-1})^{1/(k-1)}$, i.e. $\omega^k = 4C\delta^{-1}k\alpha_{k-1}$, that is

$$\omega = (C\delta^{-1}4k\alpha_{k-1})^{1/k}$$

yielding the result if $\alpha_k = 4k\alpha_{k-1}$, i.e. $\alpha_k = (4k)(4(k - 1))\dots(4 \times 2)\alpha_1 = 4^{k-1}k!2^2 = 2^{2k}k!$. The proof of the lemma is complete. $\quad\square$

Lemma 4.2. *Let $f : \mathbb{R} \mapsto [0, +\infty)$ be a C^1 function so that (distribution derivative) $f'' \in L^\infty(\mathbb{R})$. Then for all $x \in \mathbb{R}$,*

$$f'(x)^2 \leq 2f(x)\|f''\|_{L^\infty(\mathbb{R})}. \tag{73}$$

Proof. The following formula is true for f distribution, $h \in \mathbb{R}$ given :

$$f(x + h) = f(x) + f'(x)h + \int_0^1 (1 - \theta)f''(x + \theta h)d\theta h^2.$$

Then, since $f \in C^1$, for all $h \in \mathbb{R}$, we get $0 \leq f(x) + f'(x)h + \frac{1}{2}\|f''\|_{L^\infty(\mathbb{R})}h^2$. The nonpositivity of the discriminant of this polynomial is given by (73). $\quad\square$

It is easy to extend this lemma to functions whose zero set is included in the critical set :

Remark 4.1. If $f : \mathbb{R} \mapsto \mathbb{R}$ is twice differentiable, $f'' \in L^\infty$ and $f'(x) = 0$ when $f(x) = 0$, then

$$f'(x)^2 \leq 2|f(x)|\|f''\|_{L^\infty(\mathbb{R})}. \tag{74}$$

In fact, for $f \in C^1$, $F(x) = |f(x)|$ is such that $F'(x) = f'(x)f(x)/|f(x)|$ on $f(x) \neq 0$. Moreover, if $f(x) = 0$ then $f'(x) = 0$ so that

$$F(x + h) - F(x) = |f(x + h)| = |f(x) + f'(x)h + o(h)| = o(h),$$

so that $F'(x) = 0$ there. We get then that F is C^1 with

$$F'(x) = s(x)f'(x), \quad s(x) = \frac{f(x)}{|f(x)|} \text{ for } f(x) \neq 0, s(x) = 0 \text{ elsewhere.}$$

Moreover, if $f(x) = 0$,

$$F'(x + h) - F'(x) = s(x + h)[f'(x) + f''(x)h + o(h)] = s(x + h)[f''(x)h + o(h)]. \tag{75}$$

If $f''(x) \neq 0$, $s(x+h) = \text{sign}[f(x) + f'(x)h + f''(x)h^2/2 + o(h^2)] = \text{sign}(f''(x))$ for h small enough, we get $F''(x) = |f''(x)|$ there. If $f''(x) = 0$, we get from (75) that $F''(x) = 0$. We can apply the Lemma 4.2 to F and obtain (74).

Lemma 4.3. *Let q be a smooth real-valued function defined on $(-1, 1)$ such that*

$$q(t) > 0 \text{ and } s > t \implies q(s) \geq 0. \tag{76}$$

Let $\Phi \in C_c^\infty((-1,1))$ be given. There exists a function $S : (-1,1) \mapsto \{\pm\frac{1}{2}, \pm\frac{3}{2}\}$ such that for any $\rho > 0$ and $\Lambda \geq 1$,

$$2\,\text{Re}\langle \rho D_t \Phi + i\Lambda q\Phi, iS\Phi \rangle_{L^2(\mathbb{R})} \geq \rho \|\Phi\|_{L^\infty(\mathbb{R})}^2 + \int \Lambda |q(t)||\Phi(t)|^2 dt. \tag{77}$$

If in addition q satisfies (68)

$$[\frac{\gamma(k,\delta)}{\rho} + 2]\, 2\text{Re}\langle \rho D_t \Phi + i\Lambda q\Phi, iS\Phi \rangle_{L^2(\mathbb{R})} \geq \int [\, \Lambda|q(t)| + \Lambda^{\frac{1}{k+1}} \,] |\Phi(t)|^2 dt, \tag{78}$$

where $\gamma(k, \delta)$ is a positive constant depending only on k, δ. As a consequence we have

$$2[\frac{\gamma(k,\delta)}{\rho} + 2]\, \|\rho D_t \Phi + i\Lambda q\Phi\|_{L^2(\mathbb{R})} \|\Phi\|_{L^2(\mathbb{R})} \geq \int [\, \Lambda|q(t)| + \Lambda^{\frac{1}{k+1}} \,] |\Phi(t)|^2 dt.$$

Proof. We define $\theta = \sup\{t \in (-1,1), q(t) < 0\}$ and $\theta = -1$ if this set is empty. The condition (76) implies readily

$$q(t)\,\text{sign}(t - \theta) = |q(t)|.$$

We compute then, with a given $T \in (-1,1)$, H the characteristic function of $\mathbb{R}_+$,

$$2\,\text{Re}\langle \rho D_t\Phi + i\Lambda q\Phi, i[H(t-T)H(T-\theta) - H(T-t)H(\theta-T) + \frac{1}{2}\,\text{sign}(t-\theta)]\Phi\rangle_{L^2(\mathbb{R})}$$

$$= \rho|\Phi(T)|^2 + 2H(T-\theta)\int_T^1 \Lambda|q(t)|\,|\Phi(t)|^2dt + 2H(\theta-T)\int_{-1}^T \Lambda|q(t)|\,|\Phi(t)|^2dt +$$

$$+ \int_{-1}^1 \Lambda|q(t)|\,|\Phi(t)|^2dt + \rho|\Phi(\theta)|^2. \tag{79}$$

This implies (77). To get (79), we notice first, applying Lemma 4.1 ($\mathcal{L}$ stands for the Lebesgue measure) and (77), that the following inequalities hold:

$$\Lambda^{\frac{1}{k+1}}\int |\Phi(t)|^2dt = \Lambda^{\frac{1}{k+1}}\int_{\{|\Lambda q(t)| < \Lambda^{\frac{1}{k+1}}\}} |\Phi(t)|^2dt$$

$$+ \Lambda^{\frac{1}{k+1}}\int_{\{|\Lambda q(t)| \ge \Lambda^{\frac{1}{k+1}}\}} |\Phi(t)|^2dt$$

$$\le \Lambda^{\frac{1}{k+1}}\|\Phi\|^2_{L^\infty(\mathbb{R})} \times \mathcal{L}[\{t \in (-1,1), |q(t)|$$

$$\le \Lambda^{-\frac{k}{k+1}}\}] + \int |\Lambda q(t)|\,|\Phi(t)|^2dt$$

$$\le \gamma(k,\delta)\|\Phi\|^2_{L^\infty(\mathbb{R})} + 2\mathrm{Re}\langle \rho D_t\Phi + i\Lambda q\Phi, iS\Phi\rangle_{L^2(\mathbb{R})}$$

$$\le [\frac{2\gamma(k,\delta)}{\rho} + 2]\,\mathrm{Re}\langle \rho D_t\Phi + i\Lambda q\Phi, iS\Phi\rangle_{L^2(\mathbb{R})},$$

which gives (79). The proof of the Lemma 4.3 is complete. $\square$

Lemma 4.4. *Let (Q_Λ) be a family of smooth functions in $S_n^1(def.4.1.2)$. Assume that (66) is satisfied as well as (76) for each $q(t) = Q_\Lambda(t,x,\xi)$. There exists a constant C, such that for any $\Phi \in C_c^\infty(\mathbb{R}_t \times \mathbb{R}_x^n \times \mathbb{R}_\xi^n)$ supported in*

$$\max\{|2t|, |2x\Lambda^{-1/2}|, |2\xi\Lambda^{-1/2}|\} \le 1,$$

the following inequality holds (here $\mathbb{R}_X^d = \mathbb{R}_x^n \times \mathbb{R}_\xi^n$ and the norms are $L^2(\mathbb{R}_t \times \mathbb{R}_X^d)$)

$$C\|D_t\Phi + iQ_\Lambda\Phi\|\,\|\Phi\| \ge \Lambda^{\frac{1}{k+1}}\|\Phi\|^2 + \iint |Q_\Lambda(t,X)|\,|\Phi(t,X)|^2 dtdX. \tag{80}$$

Moreover, for $\Phi \in C_c^\infty((-1,1),\mathcal{S}(\mathbb{R}^{2n}))$

$$C\|D_t\Phi + iQ_\Lambda\Phi\|\,\|\Phi\| \ge \iint |Q_\Lambda(t,X)|\,|\Phi(t,X)|^2 dtdX. \tag{81}$$

Note that C depends only on the dimension, the semi-norms of the family (Q_Λ) and δ_0 in (66).

Proof. We want to apply the Lemma 4.3 to

$$q(t) = \Lambda^{-1}Q_\Lambda(\frac{t}{2}, X), \text{ whenever } X = (x,\xi) \text{ such that } \max\{|2x|, |2\xi|\} \le \Lambda^{1/2} \tag{82}$$

with δ in (68) given by $2^{-k}\delta_0$, δ_0 defined in (66). From (79), we get, for each $X = (x, \xi)$ and $\Psi(t, X)$ smooth supported in

$$|t| < 1 \times \{|2x\Lambda^{-1/2}| \leq 1\} \times \{|2\xi\Lambda^{-1/2}| \leq 1\},$$

the inequality

$$2\left[\frac{\gamma(k, \delta)}{\rho} + 2\right] \left\| \rho D_t \Psi(t, X) + iQ_\Lambda(\frac{t}{2}, X)\Psi(t, X) \right\|_{L^2(\mathbb{R}_t)} \|\Psi(t, X)\|_{L^2(\mathbb{R}_t)}$$

$$\geq \int [|Q_\Lambda(\frac{t}{2}, X)| + \Lambda^{\frac{1}{k+1}}] |\Psi(t, X)|^2 dt. \tag{83}$$

Integrating (83) with respect to X, the Cauchy–Schwarz inequality gives the result in (80), with $\Psi(t, X) = \Phi(\frac{t}{2}, X)$ and $\rho = 2$. To get (81), we only need to use (77) and integrate with respect to X. □

4.3 Energy Identities

Lemma 4.5. *Let (Q_Λ) be a family of smooth functions satisfying the assumptions of the Lemma 4.4. We write Q_Λ^μ for the operator of multiplication by $Q_\Lambda(t, X)$ on $L^2(\mathbb{R}^{2n})$. If π_H is defined in Proposition 3.1 (for $\lambda = 1$), we have the following estimate for the commutator: for any $\Psi \in L^2(\mathbb{R}^{2n})$*

$$\|[\pi_H, Q_\Lambda^\mu]\Psi\|_{L^2(\mathbb{R}^{2n})}^2 \leq c_n C \int |Q_\Lambda(t, X)| |\Psi(X)|^2 dX + c_n C \|\Psi\|_{L^2(\mathbb{R}^{2n})}^2, \tag{84}$$

where c_n depends only on the dimension and $C = \max_{|\alpha|+|\beta|=2} C_{0\alpha\beta}$ (these constants are the semi-norms of the family (Q_Λ) defined in (63)).

Proof. From the Proposition 3.1, the kernel of the commutator is

$$e^{-\frac{\pi}{2}|X-Y|^2} e^{-i\pi[X,Y]} [Q_\Lambda(t, Y) - Q_\Lambda(t, X)] =$$

$$e^{-\frac{\pi}{2}|X-Y|^2} e^{-i\pi[X,Y]} [d_X Q_\Lambda(t, Y)(Y - X) - \int_0^1 Q_\Lambda''(t, Y + \theta(X - Y)) d\theta (Y - X)^2]. \tag{85}$$

The second term in the bracket gives rise to a bounded operator, thanks to (63). Since the multiplication by $d_X Q(t, X)$ can be estimated by (74), we get the result of the lemma. □

Proof of Theorem 4.2. We apply now the Lemma 4.4 to a function

$$\Phi(t, X) = \chi(X)(Wu)(t, X)$$

where W is given in the Proposition 3.1, $u(t,x)$ is in $C_c^\infty((-1,1), \mathcal{S}(\mathbb{R}^n))$, $\chi_\Lambda(X)$ satisfies (63) with $m = 0$ and

$$\text{supp}\,\chi_\Lambda \subset \{X = (x, \xi), 2\Lambda^{-1/2}|x| \le 1, 2\Lambda^{-1/2}|\xi| \le 1\}. \qquad (86)$$

We get, with $L^2(\mathbb{R}_t \times \mathbb{R}^{2n}_{x,\xi})$ norms

$$C\,\|D_t \chi_\Lambda W u + iQ_\Lambda \chi_\Lambda W u\|\,\|\chi_\Lambda W u\|$$

$$\ge \Lambda^{\frac{1}{k+1}}\,\|\chi_\Lambda W u\|^2 + \iint |Q_\Lambda(t,X)|\,|\chi_\Lambda W u(t,X)|^2 dtdX, \qquad (87)$$

which implies, with $\pi_K = \text{Id} - \pi_H$,

$$C\Big(\|\pi_H[D_t W u + iQ_\Lambda W u]\|^2 + \|\pi_K[D_t W u + iQ_\Lambda W u]\|^2\Big)^{1/2}\|W u\| \ge$$

$$\Lambda^{\frac{1}{k+1}}\,\|\chi_\Lambda W u\|^2 + \iint |Q_\Lambda(t,X)|\,|\chi_\Lambda W u(t,X)|^2 dtdX. \qquad (88)$$

Moreover, from the Lemma 4.4 we have

$$C\,\|D_t W u + iQ_\Lambda W u\|\,\|W u\| \ge \iint |Q_\Lambda(t,X)||W u(t,X)|^2 dtdX, \qquad (89)$$

We obtain

$$C\,\|W[D_t u + iW^* Q_\Lambda W u]\|\,\|W u\| + C\,\|[\pi_H, Q_\Lambda] W u]\|\,\|W u\| \ge \qquad (90)$$

$$\Lambda^{\frac{1}{k+1}}\,\|\chi_\Lambda W u\|^2 + \iint |Q_\Lambda(t,X)||W u(t,X)|^2 dtdX.$$

Using now the Lemma 4.5, we estimate the bracket in (90):

$$C\,\|W[D_t u + iW^* Q_\Lambda W u]\|\,\|W u\|$$

$$+ C_1\varepsilon \iint |Q_\Lambda(t,X)||W u(t,X)|^2 dtdX + C_1\varepsilon^{-1}\,\|W u\|^2 \qquad (91)$$

$$\ge \Lambda^{\frac{1}{k+1}}\,\|\chi_\Lambda W u\|^2 + \iint |Q_\Lambda(t,X)||W u(t,X)|^2 dtdX.$$

We get (67), choosing ε small enough, using the fact that W is isometric (Proposition 3.1), and that

$$W^* Q_\Lambda W - Q_\Lambda(t, x, D_x)$$

is uniformly bounded on $L^2(\mathbb{R}^n)$(Proposition 3.1). The proof of Theorem 4.2 is complete.

5 The Fefferman–Phong Inequality

5.1 The Semi-Classical Inequality

We consider a function $a \in C^\infty(\mathbb{R}^{2n})$ bounded as well as all its derivatives. The (semi-classical) Fefferman–Phong inequality states that, if a is a nonnegative function, there exists C such that, for all $u \in L^2(\mathbb{R}^n)$ and all $h \in (0,1)$

$$\mathrm{Re}\langle a(x, hD)u, u\rangle_{L^2(\mathbb{R}^n)} + Ch^2 \|u\|_{L^2}^2 \geq 0,$$

or equivalently (with an a priori different constant C)

$$a(x, h\xi)^w + Ch^2 \geq 0.$$

The constants C above depend only a finite number of derivatives of a. Let us ask our first question:

Q1: *How many derivatives of a are needed to control C?*

From the proof by Fefferman and Phong ([FP]), it is clear that the number N of derivatives of a needed to control C should be

$$N = 4 + \nu(n).$$

Since the proof is using an induction on the dimension, it is not completely obvious to answer to our question with a reasonably simple ν. We remark that, with a unitary equivalence,

$$h^{-2}a(x, h\xi)^w \equiv h^{-2}a(xh^{1/2}, h^{1/2}\xi)^w.$$

Defining $A(x, \xi) = h^{-2}a(xh^{1/2}, h^{1/2}\xi)$, we see that the following property holds:

$$A(x, \xi) \geq 0, \qquad A^{(k)} \text{ is bounded for } k \geq 4. \qquad (\sharp)$$

Bony proved in 1998 ([B99]) that

$$(\sharp) \implies A^w + C \geq 0.$$

Naturally, from the above identities, this implies the Fefferman–Phong inequality. This result shows a twofold phenomenon:

1. Only derivatives with order ≥ 4 are needed.
2. The control of these derivatives is quite weak, of type $S_{0,0}^0$. In particular, the derivatives of large order do not get small (the class $S_{0,0}^0$ does not have an asymptotic calculus).

Our second question is

Q2: *Is it possible to relax $(\sharp)$ by asking only $A^{(4)} \in \mathcal{A}$,* where $\mathcal{A}$ is a suitable Banach algebra containing $S^0_{0,0}$? We shall in fact prove a result involving a Wiener-type algebra introduced by Sjöstrand in [Sjö94]. To formulate this, we need first to introduce that algebra.

5.2 The Sjöstrand Algebra

Let $\mathbb{Z}^{2n}$ be the standard lattice in $\mathbb{R}^{2n}_X$ and let $1 = \sum_{j \in \mathbb{Z}^{2n}} \chi_0(X - j), \chi_0 \in C^\infty_c(\mathbb{R}^{2n})$, be a partition of unity. We note $\chi_j(X) = \chi_0(X - j)$.

Definition 5.1. Let $a \in \mathcal{S}'(\mathbb{R}^{2n})$. We shall say that a belongs to $\mathcal{A}$ whenever $\omega_a \in L^1(\mathbb{R}^{2n})$, with $\omega_a(\Xi) = \sup_{j \in \mathbb{Z}^{2n}} |\mathcal{F}(\chi_j a)(\Xi)|$. $\mathcal{A}$ is a Banach algebra for the multiplication with the norm $\|a\|_{\mathcal{A}} = \|\omega_a\|_{L^1(\mathbb{R}^{2n})}$.

The next three lemmas are Propositions 1.2.1, 1.2.3 and Lemma A.2.1 in [LM].

Lemma 5.1. We have $S^0_{0,0} \subset S^0_{0,0;2n+1} \subset \mathcal{A} \subset C^0(\mathbb{R}^{2n}) \cap L^\infty(\mathbb{R}^{2n})$, where $S_{0,0;2n+1}$ is the set of functions defined on $\mathbb{R}^{2n}$ such that $|(\partial^\alpha_\xi \partial^\beta_x a)(x, \xi)| \le C_{\alpha\beta}$ for $|\alpha| + |\beta| \le 2n + 1$. The algebra $\mathcal{A}$ is stable by change of quantization, i.e. for all t real, $a \in \mathcal{A} \Longleftrightarrow J^t a = \exp(2i\pi t D_x \cdot D_\xi) a \in \mathcal{A}$.

We recall that $(a_1 \sharp a_2)^w = a_1^w a_2^w$ with

$$(a_1 \sharp a_2)(X) = 2^{2n} \iint_{\mathbb{R}^{2n} \times \mathbb{R}^{2n}} a_1(Y_1) a_2(Y_2) e^{-4i\pi[X - Y_1, X - Y_2]} dY_1 dY_2.$$

Lemma 5.2. The bilinear map $a_1, a_2 \mapsto a_1 \sharp a_2$ is defined on $\mathcal{A} \times \mathcal{A}$ and continuous valued in $\mathcal{A}$, which is a (noncommutative) Banach algebra for $\sharp$. The maps $a \mapsto a^w, a(x, D)$ are continuous from $\mathcal{A}$ to $\mathcal{L}(L^2(\mathbb{R}^n))$.

Lemma 5.3. Let b be a function in $\mathcal{A}$ and $T \in \mathbb{R}^{2n}, t \in \mathbb{R}$. Then the functions $\tau_T b$, b_t defined by $\tau_T b(X) = b(X - T), b_t(X) = b(tX)$ belong to $\mathcal{A}$ and

$$\sup_{T \in \mathbb{R}^{2n}} \|\tau_T b\|_{\mathcal{A}} \le C \|b\|_{\mathcal{A}}, \qquad \|b_t\|_{\mathcal{A}} \le (1 + |t|)^{2n} C \|b\|_{\mathcal{A}}.$$

Remark 5.1 (Comments on the Wiener Lemma). The standard Wiener's lemma states that if $a \in \ell^1(\mathbb{Z}^d)$ is such that $u \mapsto a * u = C_a u$ is invertible as an operator on $\ell^2(\mathbb{Z}^d)$, then the inverse operator is of the form C_b for some $b \in \ell^1(\mathbb{Z}^d)$. Sjöstrand has proven several types of Wiener lemmas for $\mathcal{A}$ ([Sjö95]). First a commutative version, saying that if $a \in \mathcal{A}$ and $1/a$ is a bounded function, then $1/a$ belongs to $\mathcal{A}$. Next, a noncommutative version of the Wiener lemma for the algebra $\mathcal{A}$: if an operator a^w with $a \in \mathcal{A}$ is invertible as a continuous operator on L^2, then the inverse operator is b^w with $b \in \mathcal{A}$. In a paper by Gröchenig and Leinert ([GL]), the authors prove several versions of the noncommutative Wiener lemma, and their definition of the twisted convolution is indeed very close to (a discrete version of) the composition formula above.

The main result of this chapter is the following

Theorem 5.1. *There exists a constant C such that, for all nonnegative functions a defined on $\mathbb{R}^{2n}$ satisfying $a^{(4)} \in \mathcal{A}$, the operator a^w is semi-bounded from below and, more precisely, satisfies*

$$a^w + C\|a^{(4)}\|_{\mathcal{A}} \geq 0.$$

The constant C depends only on the dimension n.

Note that this answers positively to our question (about relaxing the assumption on $a^{(4)}$), and as a byproduct gives the answer $4 + 2n + \epsilon$ for the number of derivatives needed to control C in the Fefferman–Phong inequality.[6] Some results of this type were proven by Sjöstrand in [Sjö95], namely the standard Gårding inequality with gain of one derivative for his class, $a \geq 0, a'' \in \mathcal{A} \implies a(x, h\xi)^w + Ch \geq 0$. A version of the Hörmander–Melin inequality with gain of $6/5$ of derivatives (see [Hör79]) was given by Hérau ([H01]) who used a limited regularity on the symbol a, only such that $a^{(3)} \in \mathcal{A}$.

The Chap. 2 implies readily the improvement of the Gårding inequality with gain of one derivative. Take $a \geq 0$ such that $a'' \in \mathcal{A}$: then $a^w = a^{\text{Wick}} - r(a)^w \geq -r(a)^w$, with $r(a)(X) = \int_0^1 \int_{\mathbb{R}^{2n}} (1 - \theta) a''(X + \theta Y) Y^2 e^{-2\pi|Y|^2} 2^n \, dY \, d\theta$. Since $\mathcal{A}$ is stable by translation (see the Lemma 5.3), we see that $r(a) \in \mathcal{A}$ and thus $r(a)^w$ is bounded on $L^2(\mathbb{R}^n)$ from the Lemma 5.2.

5.3 Composition Formulas

The next three lemmas are Lemmas 2.2.1, 2.3.1, 2.3.3 in [LM].

Lemma 5.4. *Let a be a function defined on $\mathbb{R}^{2n}$ such that the fourth derivatives $a^{(4)}$ belong to $\mathcal{A}$. Then we have*

$$a^w = \left(a - \frac{1}{8\pi} \operatorname{trace} a''\right)^{\text{Wick}} + \rho_0(a^{(4)})^w,$$

with $\rho_0(a^{(4)}) \in \mathcal{A}$: more precisely $\|\rho_0(a^{(4)})\|_{\mathcal{A}} \leq C_n \|a^{(4)}\|_{\mathcal{A}}$.

One should not expect the quantity $a - \frac{1}{8\pi} \operatorname{trace} a''$ to be nonnegative: this quantity will take negative values even in the simplest case $a(x, \xi) = x^2 + \xi^2$,

---

[6] This threshold was improved recently by A.Boulkhemair [Bo06] who proved that only $4 + n + \epsilon$ derivatives were needed.

so that the positivity of the quantization expressed by the Lemma 4 is far from enough to get our result.

Remark 5.2. We note that, from the Lemma 5.4 and the L^2 boundedness of operators with symbols in $\mathcal{A}$, the theorem is reduced to proving

$$a \geq 0, a^{(4)} \in \mathcal{A} \Longrightarrow \left(a - \frac{1}{8\pi} \operatorname{trace} a''\right)^{\text{Wick}} + C \geq 0.$$

Lemma 5.5. *For $p, q \in L^\infty(\mathbb{R}^{2n})$ real-valued with $p'' \in L^\infty(\mathbb{R}^{2n})$, we have*

$$\operatorname{Re}\left(p^{\text{Wick}} q^{\text{Wick}}\right) = \left(pq - \frac{1}{4\pi} \nabla p \cdot \nabla q\right)^{\text{Wick}} + R,$$

with $\|R\|_{\mathcal{L}(L^2(\mathbb{R}^n))} \leq C(n) \|p''\|_{L^\infty} \|q\|_{L^\infty}$.

Lemma 5.6. *For p measurable real-valued function such that p'', $(p'p'')'$, $(pp'')'' \in L^\infty$, we have*

$$p^{\text{Wick}} p^{\text{Wick}} = \int \left[p(Z)^2 - \frac{1}{4\pi} |\nabla p(Z)|^2\right] \Sigma_Z dZ + S,$$

$$\|S\|_{\mathcal{L}(L^2(\mathbb{R}^n))} \leq C(n) \left(\|p''\|_{L^\infty}^2 + \|(p''p')'\|_{L^\infty} + \|(pp'')''\|_{L^\infty}\right).$$

Remark 5.3 (Further reduction). To get our theorem, we shall prove

$$a \geq 0, a^{(4)} \in L^\infty(\mathbb{R}^{2n}) \Longrightarrow \left(a - \frac{1}{8\pi} \operatorname{trace} a''\right)^{\text{Wick}} + C \geq 0.$$

We leave now the arguments of harmonic analysis and we will use a structure theorem on nonnegative $C^{3,1}$ functions as sum of squares of $C^{1,1}$ functions to write the operator $\left(a - \frac{1}{8\pi} \operatorname{trace} a''\right)^{\text{Wick}}$ as a sum of squares of operators, up to L^2-bounded operators, thanks to the last two lemmas.

5.4 Sketching the Proof

Our main argument relies on a decomposition theorem for nonnegative functions as sum of squares.

Theorem 5.2. *Let $m \in \mathbb{N}$. There exists an integer N and a positive constant C such that the following property holds. Let a be a nonnegative $C^{3,1}$ function defined on $\mathbb{R}^m$ such that $a^{(4)} \in L^\infty$; then we can write*

$$a = \sum_{1 \leq j \leq N} b_j^2$$

where the b_j are $C^{1,1}$ functions such that $b''_j, (b'_j b''_j)', (b_j b''_j)'' \in L^\infty$. More precisely, we have

$$\|b''_j\|^2_{L^\infty} + \|(b'_j b''_j)'\|_{L^\infty} + \|(b_j b''_j)''\|_{L^\infty} \leq C \|a^{(4)}\|_{L^\infty}.$$

Note that this implies that each function b_j is such that b^2_j is $C^{3,1}$ and that N and C depend only on the dimension m.

Part of this theorem is a consequence of the classical proof of the Fefferman–Phong inequality in [FP] and of the more refined analysis of Bony ([B99]) (see also the papers by Guan [Gu97] and Tataru [Ta]). However the control of the L^∞ norm of the quantities $(b'_j b''_j)', (b_j b''_j)''$ seems to be new and is important for us.

Sketching the proof. We use a Calderón–Zygmund method and define

$$\rho(x) = \big(|a(x)| + |a''(x)|^2\big)^{1/4}, \quad \Omega = \{x, \rho(x) > 0\},$$

assuming as we may $\|a^{(4)}\|_{L^\infty} \leq 1$. Note that, since ρ is continuous, the set Ω is open. The metric $|dx|^2/\rho(x)^2$ is slowly varying in Ω: $\exists r_0 > 0, C_0 \geq 1$ such that

$$x \in \Omega, |y - x| \leq r_0 \rho(x) \implies y \in \Omega, C_0^{-1} \leq \frac{\rho(x)}{\rho(y)} \leq C_0.$$

The constants r_0, C_0 can be chosen as "universal" constants, thanks to the normalization on $a^{(4)}$ above. Moreover the nonnegativity of a implies with $\gamma_j = 1$ for $j = 0, 2, 4$, $\gamma_1 = 3, \gamma_3 = 4$,

$$|a^{(j)}(x)| \leq \gamma_j \rho(x)^{4-j}, \quad 1 \leq j \leq 4.$$

Remark 5.4. We shall use the following notation: let A be a symmetric k-linear form on real normed vector space V. We define the norm of A by

$$\|A\| = \sup_{\|T\|=1} |AT^k|.$$

Since the symmetrized products of $T_1 \otimes \cdots \otimes T_k$ can be written as a linear combination of k-th powers, that norm is equivalent to the natural norm

$$\|A\| = \sup_{\substack{\|T_j\|=1, \\ 1 \leq j \leq k}} |AT_1 \ldots T_k|$$

and in fact, when V is Euclidean, we have the equality $\|A\| = \|A\|$ (see [K]). For an arbitrary normed space, the best estimate is $\|A\| \leq \frac{k^k}{k!} \|A\|$ (see the Remark 3.1.2 in [LM]).

The basic properties of slowly varying metrics are summarized in the following lemma (see, e.g. Sect. 1.4 in [Hör85]).

Lemma 5.7. *Let* a, ρ, Ω, r_0 *be as above. There exists a positive number* $r_0' \leq r_0$, *such that for all* $r \in]0, r_0']$, *there exists a sequence* $(x_\nu)_{\nu \in \mathbb{N}}$ *of points in* Ω *and a positive number* M_r, *such that the following properties are satisfied. We define* $U_\nu, U_\nu^*, U_\nu^{**}$ *as the closed Euclidean balls with center* x_ν *and radius* $r\rho_\nu, 2r\rho_\nu, 4r\rho_\nu$ *with* $\rho_\nu = \rho(x_\nu)$. *There exist two families of nonnegative smooth functions on* $\mathbb{R}^m$, $(\varphi_\nu)_{\nu \in \mathbb{N}}, (\psi_\nu)_{\nu \in \mathbb{N}}$ *such that*

$$\sum_\nu \varphi_\nu^2(x) = 1_\Omega(x), \ \operatorname{supp} \varphi_\nu \subset U_\nu, \psi_\nu \equiv 1 \ \text{ on } U_\nu^*,$$

$\operatorname{supp} \psi_\nu \subset U_\nu^{**} \subset \Omega$. *Moreover, for all integers* l, *we have*

$$\sup_{x \in \Omega, \nu \in \mathbb{N}} \|\varphi_\nu^{(l)}(x)\| \rho_\nu^l + \sup_{x \in \Omega, \nu \in \mathbb{N}} \|\psi_\nu^{(l)}(x)\| \rho_\nu^l < \infty.$$

The overlap of the balls U_ν^{**} *is bounded, i.e.*

$$\bigcap_{\nu \in \mathcal{N}} U_\nu^{**} \neq \emptyset \quad \Longrightarrow \quad \#\mathcal{N} \leq M_r.$$

Moreover, $\rho(x) \sim \rho_\nu$ *all over* U_ν^{**} *(i.e. the ratios* $\rho(x)/\rho_\nu$ *are bounded above and below by a fixed constant, provided that* $x \in U_\nu^{**}$*).*

Since a is vanishing on Ω^c, we obtain

$$a(x) = \sum_{\nu \in \mathbb{N}} a(x)\varphi_\nu^2(x).$$

Definition 5.2. Let a, ρ, Ω be as above. Let θ be a positive number $\leq \theta_0$, where $\theta_0 < 1/2$ is a fixed constant. A point $x \in \Omega$ is said to be

(1) θ-elliptic whenever $a(x) \geq \theta\rho(x)^4$,

(2) θ-nondegenerate whenever $a(x) < \theta\rho(x)^4$: we have then $\|a''(x)\|^2 \geq \rho(x)^4/2$.

Let us first consider the "elliptic" indices ν such that x_ν is θ-elliptic. For $x \in U_\nu^{**}$, we have $a(x) \sim \rho_\nu^4$, so that with

$$b_\nu(x) = a(x)^{1/2}\psi_\nu(x), \quad b_\nu^2 = a\psi_\nu^2, \quad \varphi_\nu^2 b_\nu^2 = a\varphi_\nu^2$$

and on $\operatorname{supp} \varphi_\nu$ (where $\psi_\nu \equiv 1$),

$$\begin{cases} b_\nu' & = 2^{-1}a^{-1/2}a', \\ b_\nu'' & = -2^{-2}a^{-3/2}a'^2 + 2^{-1}a^{-1/2}a'', \\ b_\nu''' & = 3 \times 2^{-3}a^{-5/2}a'^3 - \frac{3}{4}a^{-3/2}a'a'' + 2^{-1}a^{-1/2}a''', \\ b_\nu^{(4)} & = -\frac{15}{16}a^{-7/2}a'^4 + \frac{9}{4}a^{-5/2}a'^2a'' - \frac{3}{4}a^{-3/2}a''^2 \\ & \qquad -a^{-3/2}a'a''' + \frac{1}{2}a^{-1/2}a^{(4)}, \end{cases}$$

yielding easily the result. The whole difficulty is concentrated on the next case.

The nondegenerate indices ν *are those for which* x_ν *is* θ-*nondegenerate.* Since a'' is large, according to our scaling, we may choose the coordinates on U_ν such that

$$\partial_1^2 a(x) \geq \rho_\nu^2/2 \text{ for } |x - x_\nu| \lesssim \rho_\nu.$$

Since we know also that a is small at some point in U_ν (if the constant θ_0 is suitably chosen, cf. the Lemma A.1.5 in [LM]), we get that $\partial_1 a$ vanishes somewhere in U_ν. From the implicit function theorem, there exists α such that $\partial_1 a(\alpha(x'), x') = 0$ and thus, with $\beta = x_1 - \alpha(x')$, $R = \left(\int_0^1 (1-t)\partial_1^2 a(\alpha(x') + t(x_1 - \alpha(x')), x')dt\right)^{1/2}$,

$$a(x) = a(x_1, x') = R^2 \beta^2 + a(\alpha(x'), x') =$$

$$= \int_0^1 (1-t)\partial_1^2 a\big(\alpha(x') + t(x_1 - \alpha(x')), x'\big)dt\big(x_1 - \alpha(x')\big)^2 + a(\alpha(x'), x').$$

We find easily $|\alpha(x') - x_{\nu 1}| \lesssim \rho_\nu$, $|\alpha'(x')| \lesssim 1$, $|\alpha''(x')| \lesssim \rho_\nu^{-1}$, $|\alpha'''(x')|$ $\lesssim \rho_\nu^{-2}$. Following Bony's argument, we compute the derivatives of

$$x' \mapsto a(\alpha(x'), x') = c(x').$$

We have, denoting by ∂_2 the x'-partial derivative,

$$c' = \alpha'\partial_1 a + \partial_2 a = \partial_2 a,$$

(here we have used the identity $\partial_1 a(\alpha(x'), x') \equiv 0)$,

$$c'' = \alpha'\partial_1\partial_2 a + \partial_2^2 a,$$

$$c''' = \alpha''\partial_1\partial_2 a + \alpha'^2\partial_1^2\partial_2 a + 2\alpha'\partial_1\partial_2^2 a + \partial_2^3 a,$$

$$c'''' = \alpha'''\partial_1\partial_2 a + 3\alpha''\alpha'\partial_1^2\partial_2 a + 3\alpha''\partial_1\partial_2^2 a$$

$$+ \alpha'^3\partial_1^3\partial_2 a + 3\alpha'^2\partial_1^2\partial_2^2 a + 3\alpha'\partial_1\partial_2^3 a + \partial_2^4 a,$$

so that $|c'| \lesssim \rho^3$, $|c''| \lesssim \rho^2$, $|c'''| \lesssim \rho$, $|c''''| \lesssim 1$.

This forces the function $B(x) = R(x)^2(x_1 - \alpha)^2$ to be $C^{3,1}$ with a j-th derivative bounded above by ρ_ν^{4-j} ($0 \leq j \leq 4$), since it is the case for a and c. Defining $b(x) = R(x)(x_1 - \alpha(x'))$ we see that

$$a = b^2 + c, \quad |(b^2)^{(j)}| = |B^{(j)}| \lesssim \rho_\nu^{4-j}, \ 0 \leq j \leq 4.$$

As a consequence, we have

$$R^2\beta^2 = \overbrace{B(\alpha(x'), x')}^{=0} + \overbrace{\int_0^1 \partial_1 B(\alpha(x') + \theta(x_1 - \alpha(x')), x')d\theta}^{\in C^{2,1}} \beta,$$

$$|\beta^{(j)}| \lesssim \rho^{1-j}, 0 \leq j \leq 3,$$

and since the open set $\{\beta \neq 0\}$ is dense,

$$R^2\beta = \int_0^1 \partial_1 B(\alpha(x') + \theta(x_1 - \alpha(x')), x')d\theta \in C^{2,1},$$

$$|(R^2\beta)^{(j)}| \lesssim \rho_\nu^{3-j}, \ 0 \leq j \leq 3.$$

Also we have $0 < R^2 = \omega \in C^{1,1}, \omega \sim \rho_\nu^2$ and

$$|\omega^{(j)}| \lesssim \rho_\nu^{2-j}, 0 \leq j \leq 2,$$

entailing that with $R = \omega^{1/2}$,

$$|R' = \frac{1}{2}\omega^{-1/2}\omega'| \lesssim 1, \ |R'' = -\frac{1}{4}\omega^{-3/2}\omega'^2 + \frac{1}{2}\omega^{-1/2}\omega''| \lesssim \rho_\nu^{-1}.$$

Using Leibniz' formula, we get

$$(R^2\beta)''' = (\omega\beta)''' = \omega'''\beta + 3\omega''\beta' + 3\omega'\beta'' + \omega\beta''',$$

which makes sense since ω''' is a distribution of order 1 and β is $C^{2,1}$. We know that $(\omega\beta)'''$ is L^∞, and since it is also the case of $\omega''\beta', \omega'\beta'', \omega\beta'''$, we get that $\omega'''\beta$ is bounded. On the other hand we have, since $\omega = R^2$,

$$\omega''' = 6R'R'' + 2\underbrace{R}_{C^{1,1}}\underbrace{R'''}_{\substack{\text{distribution} \\ \text{of order 1}}}$$

entailing that $\beta(6R'R'' + 2RR''')$ is L^∞ and since it is the case of $\beta R'R''$, we get that $\beta RR'''$ is L^∞. With $b = R\beta$, we get $b'b'' = (R'\beta + R\beta')(R''\beta + 2R'\beta' + R\beta'')$ and to check that $(b'b'')'$ is in L^∞, it is enough to check the derivatives of $R''\beta R'\beta$, $R''\beta R\beta'$ which are, up to bounded terms,

$$R'''\beta R'\beta = R'''\beta RR'\frac{\beta}{R}, \quad R'''\beta R\beta'$$

which are bounded according to the estimates above. Note that b'' is bounded. We want also to verify that $(bb'')''$ is bounded. We use that $(b^2)^{(4)}$ is bounded and since we have

$$\underbrace{(b^2)''''}_{\text{bounded}} = 2(b' \otimes b' + bb'')'' = 2\underbrace{(b' \otimes b'' + b'' \otimes b')'}_{\text{bounded}} + 2(bb'')'',$$

we obtain the boundedness of $(bb'')''$. We can conclude by using an induction on the dimension (c is defined on $\mathbb{R}^{m-1}$) and a standard argument due to Guan ([Gu97]) on slowly varying metrics.

Lemma 5.8. *Let a be a nonnegative function defined on $\mathbb{R}^{2n}$ such that $a^{(4)}$ belongs to $L^\infty(\mathbb{R}^{2n})$. We have from the Theorem 5.2 the identity $a = \sum_{1 \leq j \leq N} b_j^2$ along with some estimates on each b_j and its derivatives. Then we have*

$$\left(a - \frac{1}{8\pi} \operatorname{trace} a'' \right)^{Wick} = \sum_{1 \le j \le N} \left[\left(b_j - \frac{1}{8\pi} \operatorname{trace} b_j'' \right)^{Wick} \right]^2 + R$$

where R is a L^2-bounded operator such that $\|R\|_{\mathcal{L}(L^2(\mathbb{R}^n))} \le C\|a^{(4)}\|_{L^\infty(\mathbb{R}^{2n})}$, C depending only on the dimension n.

This lemma is Lemma 3.2.1 in [LM] and is a direct consequence of Sect. 5.3 and of the Theorem 5.2. It allows us to obtain the reduction of Remark 5.3 and to get the proof of the Theorem 5.1.

5.5 A Final Comment

One may ask the following question: why did we not apply the induction argument on the Sjöstrand algebra $\mathcal{A}$ directly, and avoid that complicated detour with the Wick calculus? The answer to that interrogation is simple: as seen above the Fefferman–Phong induction procedure requires a cutting process (this is the metric $dX^2/\rho(X)^2$) and also a bending of the phase space (the function α is not linear). Although the cutting part may respect $\mathcal{A}$, it is not very likely that the rigid affine structure of $\mathcal{A}$ would survive the bending. We were somehow forced to push the induction procedure in some other corner, far away from the quantization business, and our theorem on nonnegative functions, although proven by induction on the dimension, is collecting all the information on lower dimensions.

Finally, as additional referenes to the contents of our lectures, we quote [AM], [BF], [Be], [B04], [BC], [Bo97], [Bo99], [CF], [Fo], [G], [L97], [L98], [L02], [L90], [LN], [Sh].

6 Appendix

6.1 Cotlar's Lemma

We recall the statement of the celebrated Cotlar's lemma in a version given in the paper [BL](Lemma 4.2.3′) (see also [Hör79], [U]).

Lemma 6.1 (Cotlar's lemma). *Let* $(\Omega, \mathcal{M}, \mu)$ *be a σ-finite measured space where μ is a positive σ-finite measure and let $\mathbf{H}$ be a Hilbert space. Let $\omega \mapsto A_\omega$ be a weakly measurable mapping from Ω into $\mathcal{L}(\mathbf{H})$. We assume that*

$$M = \max \left(\sup_{\omega \in \Omega} \int_\Omega \|A_\omega^* A_{\omega'}\|^{1/2} \, d\mu(\omega'), \ \sup_{\omega \in \Omega} \int_\Omega \|A_\omega A_{\omega'}^*\|^{1/2} \, d\mu(\omega') \right) < +\infty.$$

Then the operator $A = \int_\Omega A_\omega d\mu(\omega)$ *is bounded on $\mathbf{H}$ with norm less than M.*

Lemma 6.2. *Let ω be a measurable function defined on $\mathbb{R}^{2n} \times \mathbb{R}^{2n}$ such that*

$$|\omega(Y, Z)| \leq \gamma_0 \big(1 + |Y - Z|\big)^{N_0}.$$

Then the operator $\iint \omega(Y, Z) \Sigma_Y \Sigma_Z dY dZ$ is bounded on $L^2(\mathbb{R}^n)$ with $\mathcal{L}(L^2(\mathbb{R}^n))$ norm bounded above by a constant depending on γ_0, N_0.

Proof. Writing

$$\Sigma_Y \Sigma_Z \Sigma_{Y'} \Sigma_{Z'} = \Sigma_Y \Sigma_Z \; \Sigma_Z \Sigma_{Y'} \; \Sigma_{Y'} \Sigma_{Z'}$$

we see that it is an immediate consequence of the Lemma 6.1 and of the formula (49). $\qquad\square$

References

[AM]   H. Ando, Y. Morimoto, *Wick calculus and the Cauchy problem for some dispersive equations*,Osaka J. Math., 39, (2002), 1, 123–147

[BF]   R.Beals, C.Fefferman, *On local solvability of linear partial differential equations*, Ann. of Math., 97, (1973), 482–498

[Be]   F.A. Berezin, *Quantization*, Math.USSR, Izvest., 8,(1974), 1109-1165

[B99]   J.M. Bony, *Sur l'inégalité de fefferman-phong*, Séminaire EDP, Ecole Polytechnique (1998-99), Exposé 3.

[B04]   ———, *Décomposition des fonctions positives en sommes de carrés*, Journées Equations aux Dérivées Partielles, (2004), Exposé 3, Ecole Polytech., Palaiseau

[BC]   J.M. Bony, J.Y. Chemin, *Espaces fonctionnels associés, au calcul de Weyl-Hörmander*, Bull. Soc. Math. France, 122, (1994), 77-118.

[BL]   J.M. Bony, N. Lerner, *Quantification asymptotique et microlocalisations d'ordre supérieur*, Ann. Ec.Norm.Sup., 22, (1989), 337-433.

[Bo97] A. Boulkhemair, *Remarks on a Wiener type pseudodifferential algebra and Fourier integral operators*, Math.Res.Lett., 4, (1997), 53-67.

[Bo99] ———, *L^2 estimates for Weyl quantization*, J.Func.Anal., 165, (1999), 173–204

[Bo06] ———, *Private communication*, March 2006.

[Br90] R. Brummelhuis, *A counterexample to the fefferman–phong inequality for systems*, C.R. Acad. Sci. Paris **310**, (1990), série I, 95–98.

[CF]   A. Cordoba, C. Fefferman, *Wave packets and Fourier integral operators*, Comm. PDE, 3, (1978), (11), 979–1005.

[CM]   R.D. Coifman, Y. Meyer, *Au delà des opérateurs pseudo-différentiels*, vol. 57, Astérisque, Société Mathématique de France, 1978.

[Eg]   Y. V. Egorov, *Subelliptic pseudodifferential operators*, Soviet Math. Dok., 10, (1969), 1056-1059.

[FP]   C. Fefferman, D.H. Phong, *On positivity of pseudo-differential equations*, Proc.Nat.Acad.Sci. **75** (1978), 4673–4674.

[Fo]   G.B. Folland, *Harmonic analysis in phase space*, Princeton University Press, Annals of Math.Studies, 122, (1989).

[G]   G. Glaeser, *Racine carrée d'une fonction différentiable*, Ann.Inst.Fourier **13** (1963), 2, 203–210.

[Gu97] P. Guan, *C^2 A priori estimates for degenerate monge-ampère equations*, Duke Math. J. **86** (1997), (2), 323–346.

[GL]   K.Gröchenig, M.Leinert, *Wiener's lemma for twisted convolution and Gabor frames*, J. Amer. Math. Soc., 17 (2004), (1), 1–18.

[H01]   F. Hérau, *Melin-hörmander inequality in a wiener type pseudo-differential algebra*, Ark. Mat. **39** (2001), 2, 311–338.

[Hör79]   L. Hörmander, *The weyl calculus of pseudodifferential operators*, Comm. Pure Appl. Math. **32** (1979), 3, 360–444.

[Hör85]   _____, *The analysis of linear partial differential operators i-iv*, Springer Verlag, 1983-85.

[H]   I.L. Hwang, *The l^2 boundedness of pseudo-differential operators*, Trans. Amer. Math. Soc. **302** (1987), 55–76.

[K]   O.D. Kellogg, *On bounded polynomials in several variables*, Math.Z. **27** (1928), 55–64.

[L97]   N. Lerner, *Energy methods via coherent states and advanced pseudo-differential calculus*, Multidimensional complex analysis and partial differential equations, ed. P.Cordaro, H.Jacobowitz, S.Gindikin, Contemporary Mathematics, 205, (1997), 177–201.

[L98]   _____, *Perturbation and energy estimates*, Ann.Sci.ENS, (1998), 31, 843–886.

[L02]   _____, *Solving pseudo-differential equations*, Proceedings of the ICM 2002 in Beijing, (2002), Higher Education Press, 711–720, II.

[L90]   _____, *Wick-Wigner functions and tomographic methods*, SIAM Journal of Mathematical Analysis, 21, 1990, (4), 1083–1092.

[LM]   N. Lerner, Y. Morimoto, *On the fefferman–phong inequality and a wiener-type algebra of pseudodifferential operators*, preprint (october 2005), http://perso.univ-rennes1.fr/nicolas.lerner/.

[LN]   N.Lerner, J.Nourrigat, *Lower bounds for pseudo-differential operators*, Ann. Inst. Fourier, (1990), 3, 40, 657–682.

[P]   A. Parmeggiani, *A class of counterexamples to the fefferman–phong inequality for systems*, Comm. Partial Differential Equations **29** (2004), 9–10, 1281–1303.

[S]   I. Segal, *Transforms for operators and asymptotic automorphisms over a locally compact abelian group*, Math.Scand. (1963), 31– 43.

[Sjö94]   J. Sjöstrand, *An algebra of pseudodifferential operators*, Math.Res.Lett. **1** (1994), 2, 189–192.

[Sjö95]   _____, *Wiener type algebras of pseudodifferential operators*, Séminaire EDP, École Polytechnique (1994-95), Exposé 4.

[Sh]   M. Shubin, *Pseudo-differential operators and spectral theory*, Springer-Verlag, 1985.

[Ta]   D. Tataru, *On the fefferman–phong inequality and related problems*, Comm. Partial Differential Equations **27** (2002), (11–12), 2101–2138.

[Tr]   F. Treves, *A new method of proof of subelliptic estimates*, Comm.Pure Appl. Math., 24, (1971), 71–115.

[U]   A. Unterberger, *Oscillateur harmonique et opérateurs pseudo-différentiels*, Ann.Inst.Fourier **29** (1979), 3, 201–221.

[Wi]   A. Weil, *Sur certains groupes d'opérateurs unitaires*, Acta Math. **111** (1964), 143–211.

[Wy]   H. Weyl, *Gruppentheorie und quantenmechanik*, Verlag von S.Hirzel, Leipzig, 1928.

Schatten Properties
for Pseudo-Differential Operators
on Modulation Spaces

J. Toft

Abstract Let $M^{p,q}_{(\omega)}$ be the modulation space with parameters p, q and weight function ω. Also let $t \in \mathbf{R}$ and assume that $a \in M^{p,q}_{(\omega)}$. We establish sufficient and necessary conditions on $p, q \in [1, \infty]$, ω_1, ω_2 and ω in order to the pseudo-differential operator $a_t(x, D)$ should be a Schatten–von Neumann operator from $M^{2,2}_{(\omega_1)}$ to $M^{2,2}_{(\omega_2)}$ of certain order.

1 Introduction

In [21, 22], Gröchenig and Heil present a method, based on time–frequency analysis when investigating pseudo-differential operators with non-smooth symbols belonging to non-weighted modulation spaces. Here they make suitable Gabor expansions of the symbols, which in some extent reduce the problems in such way that the symbols are translations and modulations of a fix and well-known function. In that end, they are able to make a somewhat detailed study of compactness, and prove embedding properties between Schatten–von Neumann classes of pseudo-differential operators acting on L^2, and modulation spaces.

Furthermore, they prove that any pseudo-differential operator with symbol in the modulation space $M^{\infty,1}$ (denoted by S_w in [32] by Sjöstrand) is continuous on any non-weighted modulation space $M^{p,q}$. Since $L^2 = M^{2,2}$, it follows in particular that such operators are continuous on L^2, a property which was proved by Sjöstrand in [31], where modulation spaces were used as symbol classes for the first time. Furthermore, since S^0_0, the set of functions which are bounded together with all their derivatives, is contained in

Joachim Toft

Department of Mathematics and Systems Engineering, Växjö University, Vejdes plats 6,7
35195 Växjö, Sweden
e-mail: joachim.toft@vxu.se

L. Rodino, M.W. Wong (eds.) *Pseudo-Differential Operators*. Lecture Notes in Mathematics 1949.
© Springer-Verlag Berlin Heidelberg 2008

$M^{\infty,1}$, it follows from these investigations that any pseudo-differential operator with symbol in S_0^0 is continuous on $M^{p,q}$. The latter result was remarked in the L^2-case in [31], and for general p and q, the result is a special case of Theorem 2.1 in [33] by Tachizawa.

Some further improvements and extensions of the results above have been done since [21, 22, 31–33]. In [5], Boulkemair extend the L^2 continuity to Fourier integral operators with symbols in $M^{\infty,1}$ and phase functions of rather general types. In the independent papers [23,39], continuity for pseudo-differential operators with symbol class $M^{p,q}$ acting on modulation spaces, are considered. In [40] these results were further extended in the case of Weyl operators where the symbols belong to weighted modulation spaces. Some further properties concerning embeddings between Schatten–von Neumann classes in the pseudo-differential calculus and modulation spaces can also be found in [39].

Important parts in this context concern modulation spaces, and their properties. These spaces were introduced by Feichtinger in [10,12] during the period 1980–1983 as an appropriate family of function and distribution spaces to have in background when discussing certain problems within time–frequency analysis. The basic theory of such spaces were thereafter established and extended by Feichtinger and Gröchenig (see, e.g. [11,15,16,20], and the references therein). Roughly speaking, for an appropriate weight function ω, the modulation space $M_{(\omega)}^{p,q}$ is obtained by imposing a mixed $L_{(\omega)}^{p,q}$-norm on the short-time Fourier transform of a tempered distribution. The non-weighted modulation space $M^{p,q}$ is then obtained by choosing $\omega = 1$. In terms of modulation spaces it is sometimes easy to obtain information concerning growth and decay properties, as well as certain localization and regularity properties for distributions.

As additional references in the above contexts, we quote [1,3,6–8,25–27,34].

In this paper we make a brief review of the discussions in [42] concerning continuity for pseudo-differential operators in background of modulation space theory. More precisely, we consider pseudo-differential operators where the corresponding symbols belong to appropriate modulation spaces, and discuss continuity for such operators when acting on modulation spaces. Especially we are concerned with a somewhat detailed study of continuity and compactness for pseudo-differential operators acting between modulation spaces of Hilbert type in terms of Schatten–von Neumann classes. In particular we investigate trace-class and Hilbert–Schmidt properties.

Except for the Hilbert–Schmidt case it is in general a hard task to find complete characterizations of Schatten–von Neumann classes. One is therefore forced to find embeddings between such classes and other spaces which are more convenient. In Sect. 5 we discuss embeddings between such classes and modulation spaces, and generalize certain results in [22, 36, 39]. In contrast to the latter papers, the situation in Sect. 5 is more complicated depending on the fact that we consider operators acting on modulation spaces which involve weight functions of general types, instead of operators acting on L^2. In particular, by choosing the involved weight functions in appropriate ways, we may use our results to discuss Schatten–von Neumann properties for pseudo-differential operators acting between weighted Lebesgue spaces and/or Sobolev spaces of Hilbert type.

The general types of modulation spaces which are involved in the continuity investigations cause new problems comparing to [22, 39, 40]. These problems are overcome by using a related Gabor technique as in [22], leading to a convenient way to expand the symbols, and discretization of certain parts of the problems. The requested results are thereafter obtained by using techniques in modulation space theory, which are well known within time–frequency analysis, in combination with certain duality properties for Schatten–von Neumann classes in pseudo-differential calculus, presented in Sect. 4, and harmonic analysis.

In order to describe our results in more details we recall the definition of modulation spaces. Assume that $\chi \in \mathscr{S}(\mathbf{R}^m)\backslash 0$, $p, q \in [1, \infty]$ and that ω is an appropriate function on $\mathbf{R}^{2m}$, and let $\tau_x\chi(y) = \chi(y-x)$ when $x, y \in \mathbf{R}^m$. (We use the same notation for the usual functions and distribution spaces as in, e.g. [24].) Then the modulation space $M^{p,q}_{(\omega)}(\mathbf{R}^m)$ consists of all $f \in \mathscr{S}'(\mathbf{R}^m)$ such that

$$\|f\|_{M^{p,q}_{(\omega)}} = \|f\|_{M^{p,q,\chi}_{(\omega)}}$$

$$\equiv \left(\int \left(\int |\mathscr{F}(f\,\tau_x\chi)(\xi)\omega(x,\xi)|^p\,dx \right)^{q/p} d\xi \right)^{1/q} < \infty, \tag{1}$$

with the obvious modifications when $p = \infty$ and/or $q = \infty$. Here $\mathscr{F}$ denotes the Fourier transform on $\mathscr{S}'(\mathbf{R}^m)$, which takes the form

$$\mathscr{F}f(\xi) = \hat{f}(\xi) = (2\pi)^{-m/2} \int f(x)e^{-i\langle x,\xi\rangle}\,dx$$

when $f \in \mathscr{S}(\mathbf{R}^m)$. Moreover, the function $(x, \xi) \mapsto \mathscr{F}(f\,\tau_x\chi)(\xi)$ is called the *short-time Fourier transform* with *window function*, or just *window*, χ for f in the literature. (In the literature, the terms coherent state transform and coherent state also occur.)

Next assume that $t \in \mathbf{R}$ is fixed and that $a \in \mathscr{S}(\mathbf{R}^{2m})$. Then the pseudo-differential operator $a_t(x, D)$ is the continuous operator on $\mathscr{S}(\mathbf{R}^m)$ which is defined by the formula

$$(a_t(x, D)f)(x) = (\mathrm{Op}_{(t)}(a)f)(x)$$

$$= (2\pi)^{-m} \iint a((1-t)x + ty, \xi)f(y)e^{i\langle x-y,\xi\rangle}\,dyd\xi. \tag{2}$$

The definition of $a_t(x, D)$ extends to any $a \in \mathscr{S}'(\mathbf{R}^{2m})$, and then $a_t(x, D)$ is continuous from $\mathscr{S}(\mathbf{R}^m)$ to $\mathscr{S}'(\mathbf{R}^m)$. (See, e.g. [24], or Sect. 2.) If $t = 1/2$, then $a_t(x, D)$ is equal to the Weyl operator $a^w(x, D)$ for a. If instead $t = 0$, then the standard (Kohn–Nirenberg) representation $a(x, D)$ is obtained.

In Sect. 5 we discuss continuity and Schatten–von Neumann properties for pseudo-differential operators acting on modulation spaces of Hilbert type when the operator symbols belong to modulation spaces. In particular we find appropriate conditions on ω, ω_1, ω_2 and p, q in order for

$$a_t(x, D) : M^{2,2}_{(\omega_1)} \to M^{2,2}_{(\omega_2)}$$

to be a Schatten–von Neumann operator of certain degree when $a \in M^{p,q}_{(\omega)}$.

Recall that an operator T from $M^{2,2}_{(\omega_1)}$ to $M^{2,2}_{(\omega_2)}$ belongs to $\mathscr{I}_p$, the set of Schatten–von Neumann operators of order $p \in [1, \infty]$, if and only if

$$\sup \left(\sum |(Tf_j, g_j)_{M^2_{(\omega_2)}}|^p \right)^{1/p} < \infty,$$

where the supremum is taken over all orthonormal sequences (f_j) in $M^{2,2}_{(\omega_1)}$ and (g_j) in $M^{2,2}_{(\omega_2)}$. In particular, this implies that $\mathscr{I}_\infty$ is the set of linear and continuous operators, and that T is compact when $T \in \mathscr{I}_p$ and $p < \infty$ (cf. [29,30]). We are then concerned with classification and embedding properties for the set $s_{t,p}(\omega_1, \omega_2)$ which consists of all $a \in \mathscr{S}'(\mathbf{R}^{2m})$ such that $a_t(x, D) \in \mathscr{I}_p$. In Sect. 5 we prove that

$$M^{p,q_1}_{(\omega)} \subseteq s_{t,p}(\omega_1, \omega_2) \subseteq M^{p,q_2}_{(\omega)}, \tag{3}$$

for appropriate choices of q_1, q_2 and ω. In particular, our investigations concern Schatten–von Neumann properties for pseudo-differential operators which map the Sobolev space $H^2_{s_1}$ or the weighted Lebesgue space $L^2_{s_1}$ to $H^2_{s_2}$ or $L^2_{s_2}$, since each one of these spaces agrees with $M^2_{(\omega)}$ if ω is chosen in an appropriate way. Here $H^p_s(\mathbf{R}^m)$ is the Sobolev space of distributions with s derivatives in L^p, i.e. the set of all $f \in \mathscr{S}'$ such that $(1 - \Delta)^{s/2} f \in L^p(\mathbf{R}^m)$.

2 Preliminaries

In this section we discuss basic properties for modulation spaces. The proofs are in many cases omitted since they can be found in [4,10–13,15–17,21,35–40].

We start by recalling some properties of the weight functions which are involved. We say that the function $\omega \in L^\infty_{loc}(\mathbf{R}^m)$ is v-*moderate* for some appropriate positive function $v \in L^\infty_{loc}(\mathbf{R}^m)$, if there is a constant $C > 0$ such that

$$\omega(x_1 + x_2) \leq C\omega(x_1)v(x_2), \quad x_1, x_2 \in \mathbf{R}^m. \tag{4}$$

The function v is then said to *moderate* ω. If in addition (4) holds for $\omega = v$, then v is said to be a *moderate* or *submultiplicative function*.

As in [40] we let $\mathscr{P}(\mathbf{R}^m)$ denote the cone which consists of all $0 < \omega \in L^\infty_{loc}(\mathbf{R}^m)$ such that ω is v-moderate, for some polynomial v on $\mathbf{R}^m$.

Note that if $\omega \in \mathscr{P}(\mathbf{R}^m)$, then $\omega(x) + \omega(x)^{-1} \leq P(x)$, $x \in \mathbf{R}^m$ for some polynomial P on $\mathbf{R}^m$.

Let ω_1 and ω_2 be positive functions. If ω_2/ω_1 is a bounded, then we write $\omega_2 \prec \omega_1$. They are called equivalent when $\omega_1 \prec \omega_2 \prec \omega_1$, and then and we write $\omega_1 \sim \omega_2$.

If $\mathscr{H}$ is a Hilbert space, then its scalar product is denoted by $(\,\cdot\,,\,\cdot\,)_{\mathscr{H}}$, or $(\,\cdot\,,\,\cdot\,)$ when there are no confusions about the Hilbert space structure.

The duality between a topological vector space and its dual is denoted by $\langle\,\cdot\,,\,\cdot\,\rangle$. For admissible a and b in $\mathscr{S}'(\mathbf{R}^m)$, we set $(a,b) = \langle a, \bar{b}\rangle$, and it is obvious that $(\,\cdot\,,\,\cdot\,)$ on L^2 is the usual scalar product.

Next assume that $\mathscr{B}_1$ and $\mathscr{B}_2$ are topological spaces. Then $\mathscr{B}_1 \hookrightarrow \mathscr{B}_2$ means that $\mathscr{B}_1$ is continuously embedded in $\mathscr{B}_2$. In the case that $\mathscr{B}_1$ and $\mathscr{B}_2$ are Banach spaces, $\mathscr{B}_1 \hookrightarrow \mathscr{B}_2$ is equivalent to $\mathscr{B}_1 \subseteq \mathscr{B}_2$ and $\|x\|_{\mathscr{B}_2} \leq C\|x\|_{\mathscr{B}_1}$, for some constant $C > 0$ which is independent of $x \in \mathscr{B}_1$.

Next let V_1 and V_2 be vector spaces such that $V_1 \oplus V_2 = \mathbf{R}^m$ and $V_2 = V_1^\perp$, and assume that $v_0 \in \mathscr{S}'(V_1)$ and that $v(x_1, x_2) = (v_0 \otimes 1)(x_1, x_2)$, where $x_j \in V_j$ for $j = 1, 2$. Then $v(x_1, x_2)$ is identified with $v_0(x_1)$, and we set $v(x_1, x_2) = v(x_1)$.

Assume that $\omega \in \mathscr{P}(\mathbf{R}^{2m})$, $p, q \in [1, \infty]$, and that $\chi \in \mathscr{S}(\mathbf{R}^m) \setminus 0$. Then recall that the *modulation space* $M^{p,q}_{(\omega)}(\mathbf{R}^m)$ is the set of all $f \in \mathscr{S}'(\mathbf{R}^m)$ such that (1) holds. We note that the definition of $M^{p,q}_{(\omega)}(\mathbf{R}^m)$ is independent of the choice of window χ, and that different choices of χ give rise to equivalent norms. (See Proposition 2.1 below.)

If $\omega = 1$, then the notation $M^{p,q}$ is used instead of $M^{p,q}_{(\omega)}$. Moreover we set $M^p_{(\omega)} = M^{p,p}_{(\omega)}$ and $M^p = M^{p,p}$.

Remark 1 *We are also concerned with the following family of function and distribution spaces which are related to the Wiener amalgam spaces. Assume that $p, q \in [1, \infty]$ and that $\omega \in \mathscr{P}(\mathbf{R}^{2m})$. Then the space $W^{p,q}_{(\omega)}(\mathbf{R}^m)$ consists of all $a \in \mathscr{S}'(\mathbf{R}^m)$ such that*

$$\|a\|_{W^{p,q}_{(\omega)}} = \left(\int \left(\int |\mathscr{F}(a\,\tau_x\chi)(\xi)\omega(x,\xi)|^p \, d\xi \right)^{q/p} dx \right)^{1/q}$$

is finite (cf. Definition 4 in [17]).

We recall that $W^{p,q}_{(\omega)} = \mathscr{F}M^{p,q}_{(\omega_0)}$ when $\omega_0(x,\xi) = \omega(-\xi, x) \in \mathscr{P}(\mathbf{R}^{2m})$. In fact, let $\check{\chi}(x) = \chi(-x)$ as usual. Then Parseval's formula and a change of the order of integration shows that

$$|\mathscr{F}^{-1}(\widehat{a}\,\tau_\xi\widehat{\chi})(x)| = |\mathscr{F}(a\,\tau_x\check{\chi})(\xi)|, \tag{5}$$

and the assertion follows. We refer to [12,17] for more facts about the $W^{p,q}_{(\omega)}$-spaces.

The convention of indicating weight functions with parenthesis is used also in other situations. For example, if $\omega \in \mathscr{P}(\mathbf{R}^m)$, then $L^p_{(\omega)}(\mathbf{R}^m)$ is the set of all measurable functions f on $\mathbf{R}^m$ such that $f\omega \in L^p(\mathbf{R}^m)$, i.e. such that $\|f\|_{L^p_{(\omega)}} \equiv \|f\omega\|_{L^p}$ is finite.

Next we consider the Fourier transform of functions and distributions defined on $\mathbf{R}^{2m}$. By interpreting $\mathbf{R}^{2m}$ as the phase space with dual variables (y, η), we let the phase space Fourier transform, $\widetilde{\mathscr{F}}$, be defined by the formula

$$(\widetilde{\mathscr{F}}f)(y,\eta) = \widehat{f}(y,\eta) \equiv (2\pi)^{-m} \iint f(x,\xi)e^{-i(\langle x,\eta\rangle + \langle y,\xi\rangle)}\,dx d\xi, \qquad (6)$$

when $f \in L^1(\mathbf{R}^{2m})$, i.e. $(\widetilde{\mathscr{F}}f)(x,\xi) = (\mathscr{F}f)(\xi,x)$. Then $\widetilde{\mathscr{F}}$ is a homeomorphism on $\mathscr{S}(\mathbf{R}^{2m})$ which extends to a homeomorphism on $\mathscr{S}'(\mathbf{R}^{2m})$ and to a unitary map on $L^2(\mathbf{R}^{2m})$, since similar facts hold for $\mathscr{F}$.

We use the notation $\widetilde{M}^{p,q}_{(\omega)}$, $\widetilde{M}^{p}_{(\omega)}$, $\widetilde{M}^{p,q}$ and $\widetilde{M}^{p}$ instead of $M^{p,q}_{(\omega)}$, $M^{p}_{(\omega)}$, $M^{p,q}$ and M^p respectively, when $\widetilde{\mathscr{F}}$ is used instead of $\mathscr{F}$, in the definition of modulation spaces.

The following proposition is a consequence of well-known facts in [12] or [21]. Here and in what follows, we let p' denote the conjugate exponent of p, i.e. $1/p + 1/p' = 1$.

Proposition 2.1 *Assume that* $p, q, p_j, q_j \in [1, \infty]$ *for* $j = 1, 2$, *and that* $\omega, \omega_1, \omega_2, v \in \mathscr{P}(\mathbf{R}^{2m})$ *are such that* ω *is* v-*moderate. Then the following are true:*

(1) *If* $\chi \in \mathscr{S}(\mathbf{R}^m) \setminus 0$, *then* $f \in M^{p,q}_{(\omega)}(\mathbf{R}^m)$ *if and only if* (1) *holds, i.e.* $M^{p,q}_{(\omega)}(\mathbf{R}^m)$ *is independent of the choice of* χ. *Moreover,* $M^{p,q}_{(\omega)}$ *is a Banach space under the norm in* (1), *and different choices of* χ *give rise to equivalent norms.*

(2) *If* $p_1 \leq p_2$, $q_1 \leq q_2$ *and* $\omega_2 \prec \omega_1$, *then*

$$\mathscr{S}(\mathbf{R}^m) \hookrightarrow M^{p_1,q_1}_{(\omega_1)}(\mathbf{R}^m) \hookrightarrow M^{p_2,q_2}_{(\omega_2)}(\mathbf{R}^m) \hookrightarrow \mathscr{S}'(\mathbf{R}^m).$$

(3) *The sesqui-linear form* $(\,\cdot\,,\,\cdot\,)$ *on* $\mathscr{S}$ *extends to a continuous map from* $M^{p,q}_{(\omega)}(\mathbf{R}^m) \times M^{p',q'}_{(1/\omega)}(\mathbf{R}^m)$ *to* $\mathbf{C}$. *On the other hand, if* $\|a\| = \sup|(a,b)|$, *where the supremum is taken over all* $b \in M^{p',q'}_{(1/\omega)}(\mathbf{R}^m)$ *such that* $\|b\|_{M^{p',q'}_{(1/\omega)}} \leq 1$, *then* $\|\cdot\|$ *and* $\|\cdot\|_{M^{p,q}_{(\omega)}}$ *are equivalent norms.*

(4) *If* $p, q < \infty$, *then* $\mathscr{S}(\mathbf{R}^m)$ *is dense in* $M^{p,q}_{(\omega)}(\mathbf{R}^m)$. *The dual space of* $M^{p,q}_{(\omega)}(\mathbf{R}^m)$ *can be identified with* $M^{p',q'}_{(1/\omega)}(\mathbf{R}^m)$, *through the form* $(\,\cdot\,,\,\cdot\,)_{L^2}$. *Moreover,* $\mathscr{S}(\mathbf{R}^m)$ *is weakly dense in* $M^{\infty}_{(\omega)}(\mathbf{R}^m)$.

Proposition 2.1 (1) permits us to be rather vague about to the choice of $\chi \in M^1_{(v)} \setminus 0$ in (1). For example, if $C > 0$ is a constant and Ω is a subset of $\mathscr{S}'$, then $\|a\|_{M^{p,q}_{(\omega)}} \leq C$ for every $a \in \Omega$, means that the inequality holds for some choice of $\chi \in M^1_{(v)} \setminus 0$ and every $a \in \Omega$. Evidently, for any other choice of $\chi \in M^1_{(v)} \setminus 0$, a similar inequality is true although C may have to be replaced by a larger constant, if necessary.

Next we discuss weight functions which are common in the applications. For any $s, t \in \mathbf{R}$, set

$$\sigma_t(x) = \langle x \rangle^t, \quad \sigma_{s,t}(x, \xi) = \langle \xi \rangle^s \langle x \rangle^t, \tag{7}$$

when $x, \xi \in \mathbf{R}^m$. Then it follows that $\sigma_t \in \mathscr{P}_0(\mathbf{R}^m)$ and $\sigma_{s,t} \in \mathscr{P}_0(\mathbf{R}^{2m})$ for every $s, t \in \mathbf{R}$, and σ_t is $\sigma_{|t|}$-moderate and $\sigma_{s,t}$ is $\sigma_{|s|,|t|}$-moderate. Obviously, $\sigma_s(x, \xi) = (1 + |x|^2 + |\xi|^2)^{s/2}$, and $\sigma_{s,t} = \sigma_t \otimes \sigma_s$. Moreover, if $\omega \in \mathscr{P}(\mathbf{R}^m)$, then ω is σ_t-moderate provided t is chosen large enough.

For convenience, we use the notations L^p_s, $M^{p,q}_s$ and $M^{p,q}_{s,t}$ instead of $L^p_{(\sigma_s)}$, $M^{p,q}_{(\sigma_s)}$ and $M^{p,q}_{(\sigma_{s,t})}$ respectively.

It is also convenient to let $\mathcal{M}^{p,q}_{(\omega)}(\mathbf{R}^m)$ be the completion of $\mathscr{S}(\mathbf{R}^m)$ under the norm $\|\cdot\|_{M^{p,q}_{(\omega)}}$. Then $\mathcal{M}^{p,q}_{(\omega)} \subseteq M^{p,q}_{(\omega)}$ with equality if and only if $p < \infty$ and $q < \infty$. It follows that most of the properties which are valid for $M^{p,q}_{(\omega)}(\mathbf{R}^m)$, also hold for $\mathcal{M}^{p,q}_{(\omega)}(\mathbf{R}^m)$.

Remark 2 *Assume that $p, q, q_1, q_2 \in [1, \infty]$. Then the following properties for modulation spaces hold:*

(1) *If $q_1 \leq \min(p, p')$ and $q_2 \geq \max(p, p')$, then $M^{p,q_1} \subseteq L^p \subseteq M^{p,q_2}$. In particular, $M^2 = L^2$.*

(2) $S^0_0 = \cap_{s \in \mathbf{R}} M^{\infty,1}_{s,0}$.

(3) *If $\omega \in \mathscr{P}(\mathbf{R}^{2m})$ is such that $\omega(x, \xi) = \omega(x)$, then $M^{p,q}_{(\omega)}(\mathbf{R}^m) \hookrightarrow C(\mathbf{R}^m)$ if and only if $q = 1$.*

(4) $M^{1,\infty}$ *is a convolution algebra which contains all measures on $\mathbf{R}^m$ with bounded mass.*

(5) *If Ω is a subset of $\mathscr{P}(\mathbf{R}^{2m})$ such that for any polynomial P on $\mathbf{R}^{2m}$, there is an element $\omega \in \Omega$ such that P/ω is bounded, then*

$$\mathscr{S}(\mathbf{R}^m) = \cap_{\omega \in \Omega} M^{p,q}_{(\omega)}(\mathbf{R}^m), \quad \mathscr{S}'(\mathbf{R}^m) = \cup_{\omega \in \Omega} M^{p,q}_{(1/\omega)}(\mathbf{R}^m).$$

(6) *If $s, t \in \mathbf{R}$ are such that $t \geq 0$, then*

$$M^2_{s,0} = H^2_s, \quad M^2_{0,s} = L^2_s, \quad \text{and} \quad M^2_t = L^2_t \cap H^2_t.$$

(See, e.g. [10–12, 14–17, 21, 38–40].)

Next we recall some facts from Chaps. 12 and 13 in [21] concerning extension of the Gabor theory to modulation spaces. Let $(x_j)_{j \in I}$ and $(\xi_k)_{k \in I}$ be lattices in $\mathbf{R}^m$, and assume that $\chi \in \mathscr{S}(\mathbf{R}^m)$ is fixed and satisfies

$$C^{-1} \leq \sum_{j \in I} |\chi(x - x_j)|^2 \leq C, \quad x \in \mathbf{R}^m, \tag{8}$$

for some constant C. If (ξ_k) is sufficiently dense, then it follows from [21] and Sect. 7.3 in [24] that there exists a function $\psi \in \mathscr{S}(\mathbf{R}^m)$ such that

$$f(x) = \sum_{j,k \in I} c_{j,k}(f) e^{i\langle x, \xi_k \rangle} \chi(x - x_j) \tag{9}$$

$$= \sum_{j,k \in I} d_{j,k}(f) e^{i\langle x, \xi_k \rangle} \psi(x - x_j), \tag{9'}$$

for every $f \in \mathscr{S}'(\mathbf{R}^m)$, where $c_{j,k}(f)$ and $d_{j,k}(f)$ are the "Fourier coefficients" for f, given by the formulas

$$c_{j,k}(f) = c_{j,k} = \mathscr{F}(f\tau_{x_j}\psi)(\xi_k) \quad \text{and} \tag{11}$$

$$d_{j,k}(f) = d_{j,k} = \mathscr{F}(f\tau_{x_j}\chi)(\xi_k). \tag{11'}$$

Here the sums converge in $\mathscr{S}'(\mathbf{R}^m)$.

In order to present some further properties in the case of modulation spaces, it is convenient to consider the following sequence spaces. Assume that $\lambda = (\lambda_{j,k})_{j,k \in I}$ is a (fix) sequence of non-negative numbers, and that $p, q \in [1, \infty]$. Then let $l_{(\lambda)}^{p,q}$ be the Banach space of sequences $(c_{j,k})_{j,k \in I}$ of complex numbers such that

$$\|(c_{j,k})_{j,k \in I}\|_{l_{(\lambda)}^{p,q}} \equiv \left(\sum_k \left(\sum_j |c_{j,k} \lambda_{j,k}|^p \right)^{q/p} \right)^{1/q}$$

is finite. Also let $\ell_{(\lambda)}^{p,q}$ be the completion of all finite sequences (i.e. all sequences $(c_{j,k})_{j,k \in I}$ such that only a finite numbers of $c_{j,k}$ are non-zero) under the norm $\| \cdot \|_{l_{(\lambda)}^{p,q}}$. Furthermore we set $l^{p,q} = l_{(\lambda)}^{p,q}$ when $\lambda_j = 1$ for every j, $l_{(\lambda)}^p = l_{(\lambda)}^{p,p}$, and similarly for the $\ell_{(\lambda)}^{p,q}$-spaces.

The following proposition shows that essential parts of the Gabor theory, in the context of L^2-spaces, carry over to modulation spaces. This shows in particular that there is a convenient way to discretize modulation spaces. The proof is omitted, since the result follows from Chaps. 12 and 13 in [21].

Proposition 2.2 *Let $(x_j)_{j \in I}$ be a lattice, $\omega \in \mathscr{P}(\mathbf{R}^{2m})$, and assume that $\chi \in \mathscr{S}(\mathbf{R}^m)$ satisfies (8). Then there is a lattice $(\xi_k)_{k \in I}$ and a function $\psi \in \mathscr{S}(\mathbf{R}^m)$ such that the following are true:*

(1) *If $f \in \mathscr{S}'(\mathbf{R}^m)$, and $c_{j,k}(f)$ and $d_{j,k}(f)$ are given by (11) and (11') respectively, then (9) and (9') hold.*

(2) *If $\lambda = (\omega(x_j, \xi_k))_{j,k}$ and $f \in \mathscr{S}'(\mathbf{R}^m)$, then the following conditions are equivalent:*

(a) $f \in M^{p,q}_{(\omega)}(\mathbf{R}^m)$

(b) $(c_{j,k}(f))_{j,k} \in l^{p,q}_{(\lambda)}$

(c) $(d_{j,k}(f))_{j,k} \in l^{p,q}_{(\lambda)}$

Moreover, if (a)–(c) *are fulfilled and in addition* $p, q < \infty$, *then the sums in* (9) *and* (9') *converge to* f *in* $M^{p,q}_{(\omega)}$. *Furthermore, the norms* $f \mapsto \|(c_{j,k}(f))_{j,k}\|_{l^{p,q}_{(\lambda)}}$, $f \mapsto \|(d_{j,k}(f))_{j,k}\|_{l^{p,q}_{(\lambda)}}$ *and* $\| \cdot \|_{M^{p,q}_{(\omega)}}$ *are equivalent.*

Remark 3 *We note that the equivalence* (a)–(c) *in Proposition 2.2 holds, after* $M^{p,q}_{(\omega)}$ *and* $l^{p,q}_{(\lambda)}$ *have been replaced by* $\mathcal{M}^{p,q}_{(\omega)}$ *and* $\ell^{p,q}_{(\lambda)}$ *respectively.*

Next we discuss (complex) interpolation properties for modulation spaces. Such properties were carefully investigated in [12] for non-weighted modulation spaces, and thereafter extended in several directions in [16], were interpolation properties for coorbit spaces were established. As a consequence of [16] we have the following proposition.

Proposition 2.3 *Assume that* $0 < \theta < 1$ *and that* $p, q, p_j, q_j \in [1, \infty]$ *satisfy*

$$\frac{1}{p} = \frac{1-\theta}{p_1} + \frac{\theta}{p_2} \quad and \quad \frac{1}{q} = \frac{1-\theta}{q_1} + \frac{\theta}{q_2}.$$

Also assume that $\omega_1, \omega_2 \in \mathscr{P}(\mathbf{R}^{2m})$ *and let* $\omega = \omega_1^{1-\theta} \omega_2^{\theta}$. *Then*

$$(\mathcal{M}^{p_1,q_1}_{(\omega_1)}(\mathbf{R}^m), \mathcal{M}^{p_2,q_2}_{(\omega_2)}(\mathbf{R}^m))_{[\theta]} = \mathcal{M}^{p,q}_{(\omega)}(\mathbf{R}^m).$$

Next we recall some facts in Chap. XVIII in [24] concerning pseudo-differential operators. Assume that $a \in \mathscr{S}(\mathbf{R}^{2m})$, and that $t \in \mathbf{R}$ is fixed. Then the pseudo-differential operator $a_t(x, D)$ in (2) is a linear and continuous operator on $\mathscr{S}(\mathbf{R}^m)$, as remarked in the introduction. For general $a \in \mathscr{S}'(\mathbf{R}^{2m})$, the pseudo-differential operator $a_t(x, D)$ is defined as the continuous operator from $\mathscr{S}(\mathbf{R}^m)$ to $\mathscr{S}'(\mathbf{R}^m)$ with distribution kernel

$$K_{t,a}(x, y) = (2\pi)^{-m/2}(\mathscr{F}_2^{-1}a)((1 - t)x + ty, y - x), \tag{13}$$

Here $\mathscr{F}_2 F$ is the partial Fourier transform of $F(x, y) \in \mathscr{S}(\mathbf{R}^{2m})$ with respect to the y-variable. This definition makes sense, since the mappings $\mathscr{F}_2$ and $F(x, y) \mapsto F((1 - t)x + ty, y - x)$ are homeomorphisms on $\mathscr{S}'(\mathbf{R}^{2m})$. We also note that this definition of $a_t(x, D)$ agrees with the operator in (2) when $a \in \mathscr{S}(\mathbf{R}^{2m})$.

Furthermore, any linear and continuous operator T from $\mathscr{S}(\mathbf{R}^m)$ to $\mathscr{S}'(\mathbf{R}^m)$ has a distribution kernel K in $\mathscr{S}'(\mathbf{R}^{2m})$ in view of kernel theorem of Schwartz. By Fourier's inversion formula we may then find a unique $a \in \mathscr{S}'$ such that (13) is fulfilled with $K = K_{t,a}$. Consequently, for every fixed $t \in \mathbf{R}$, there is a one to one correspondence between linear and continuous operators from $\mathscr{S}(\mathbf{R}^m)$ to $\mathscr{S}'(\mathbf{R}^m)$, and $\mathrm{Op}_{(t)}(\mathscr{S}'(\mathbf{R}^{2m}))$, the set of all $a_t(x, D)$ such that $a \in \mathscr{S}'(\mathbf{R}^{2m})$.

In particular, if $a \in \mathscr{S}'(\mathbf{R}^{2m})$ and $s, t \in \mathbf{R}$, then there is a unique $b \in \mathscr{S}'(\mathbf{R}^{2m})$ such that $a_s(x, D) = b_t(x, D)$. By straight-forward applications of Fourier's inversion formula, it follows that

$$a_s(x, D) = b_t(x, D) \quad \Leftrightarrow \quad b(x, \xi) = e^{i(t-s)\langle D_x, D_\xi \rangle} a(x, \xi). \tag{14}$$

(Cf. Sect. 18.5 in [24].)

We may also express the relations between a and b here above in terms of convolution operators. In fact, by Fourier's inversion formula it follows that if $t \neq 0$ and $\Phi(x, \xi) = \langle x, \xi \rangle / t$, then there is a constant c such that $e^{it\langle D_x, D_\xi \rangle} a = c\, e^{i\Phi} * a$. If instead $t = 0$, then $e^{it\langle D_x, D_\xi \rangle} a = a = \delta * a$. These relations motivate us to consider continuity properties for operators of the form

$$f \mapsto S_\Phi f \equiv (e^{i\Phi} \otimes \delta_{V_2}) * f, \tag{15}$$

where δ_{V_2} is the delta function on the vector space $V_2 \subseteq \mathbf{R}^m$ and Φ is a real-valued and non-degenerate quadratic form on $V_1 = V_2^\perp$. The operator S_Φ is essentially a composition of a partial Fourier transform with respect to the variables in V_1 and a non-degenerate matrix $A_\Phi / 2$. In particular, Fourier's inversion formula gives that S_Φ is a homeomorphism on $\mathscr{S}(\mathbf{R}^m)$ which extends uniquely to a homeomorphism on $\mathscr{S}'(\mathbf{R}^m)$.

A brief study of the operator S_Φ when acting on modulation spaces can be found in [36, 38, 40, 42]. The following proposition is a restatement of Proposition 1.7 in [42] and seems to be, in this context, the most general result. We omit the proof.

Proposition 2.4 *Assume that* $\chi \in \mathscr{S}(\mathbf{R}^m)$, $\omega \in \mathscr{P}(\mathbf{R}^{2m})$, $p, q \in [1, \infty]$, *and that* $V_1, V_2 \subseteq \mathbf{R}^m$ *are vector spaces such that* $V_2 = V_1^\perp$. *Also assume that* Φ *is a real-valued and non-degenerate quadratic form on* V_1, *and let* $A_\Phi / 2$ *be the corresponding matrix. If* $\xi = (\xi_1, \xi_2)$ *where* $\xi_j \in V_j$ *for* $j = 1, 2$, *then*

$$\|S_\Phi f\|_{M^{p,q,\chi}_{(\omega_\Phi)}} = \|f\|_{M^{p,q,\psi}_{(\omega)}}, \quad where \quad f \in \mathscr{S}'(\mathbf{R}^m),$$

$$\omega_\Phi(x, \xi) = \omega(x - A_\Phi^{-1}\xi_1, \xi) \quad and \quad \psi = S_\Phi \chi. \tag{16}$$

In particular, the following are true:

(1) *The map* (15) *on* $\mathscr{S}'(\mathbf{R}^m)$ *restricts to a homeomorphism from* $M^{p,q}_{(\omega)}(\mathbf{R}^m)$ *to* $M^{p,q}_{(\omega_\Phi)}(\mathbf{R}^m)$.

(2) *If* $t \in \mathbf{R}$, $\omega_0 \in \mathscr{P}(\mathbf{R}^{2m} \oplus \mathbf{R}^{2m})$, *and*

$$\omega_t(x, \xi, y, \eta) = \omega_0(x - ty, \xi - t\eta, y, \eta),$$

then the map $e^{it\langle D_x, D_\xi \rangle}$ *on* $\mathscr{S}'(\mathbf{R}^{2m})$ *restricts to a homeomorphism from* $\widetilde{M}^{p,q}_{(\omega_0)}(\mathbf{R}^{2m})$ *to* $\widetilde{M}^{p,q}_{(\omega_t)}(\mathbf{R}^{2m})$.

We finish the section by giving some remarks on Wigner distributions and Weyl operators of rank one. The Wigner distribution for $f \in \mathscr{S}(\mathbf{R}^m)$ and $g \in \mathscr{S}(\mathbf{R}^m)$ is defined by the formula

$$W_{f,g}(x, \xi) = (2\pi)^{-m/2} \int f(x - y/2)\overline{g(x + y/2)}e^{i\langle y, \xi\rangle} \, dy. \qquad (17)$$

For future references we note that the map $(f, g) \mapsto W_{f,g}$ is continuous from $\mathscr{S}(\mathbf{R}^m) \times \mathscr{S}(\mathbf{R}^m)$ to $\mathscr{S}(\mathbf{R}^{2m})$ which extends uniquely to continuous mappings from $\mathscr{S}'(\mathbf{R}^m) \times \mathscr{S}'(\mathbf{R}^m)$ to $\mathscr{S}'(\mathbf{R}^{2m})$, and from $L^2(\mathbf{R}^m) \times L^2(\mathbf{R}^m)$ to $L^2(\mathbf{R}^{2m})$. (See [18, 35, 37].)

Next assume that $a \in \mathscr{S}(\mathbf{R}^{2m})$ and that $f, g \in \mathscr{S}(\mathbf{R}^m)$. Then it follows from (2) with $t = 1/2$ and straight-forward computations that

$$(a^w(x, D)f, g) = (2\pi)^{-m/2}(a, W_{g,f}), \qquad (18)$$

In particular, (18) and Fourier's inversion formula imply that if $f_1, f_2 \in \mathscr{S}(\mathbf{R}^m)$, then

$$a^w(x, D)f(x) = (2\pi)^{-m/2}(f, f_2)f_1(x) \quad \Leftrightarrow \quad a = W_{f_1, f_2}. \qquad (19)$$

Consequently, a Weyl operator is a rank one operator if and only if its symbol is a Wigner distribution.

3 Schatten–Von Neumann Classes for Operators Acting on Hilbert Spaces

In this section we discuss Schatten–von Neumann classes of linear operators from a Hilbert space $\mathscr{H}_1$ to another Hilbert space $\mathscr{H}_2$. Such operator classes were introduced by R. Schatten in [28] in the case $\mathscr{H}_1 = \mathscr{H}_2$ (see also [19, 30]). The general situation when $\mathscr{H}_1$ is not necessarily equal to $\mathscr{H}_2$ have thereafter been discussed in [2, 29]. Here we give a short introduction, based on an argument which essentially reduces the situation to the case $\mathscr{H}_1 = \mathscr{H}_2$.

For any Hilbert space $\mathscr{H}$, we let $\mathrm{ON}(\mathscr{H})$ be the set of orthonormal sequences in $\mathscr{H}$. Assume that $T : \mathscr{H}_1 \to \mathscr{H}_2$ is linear, and that $p \in [1, \infty]$. Then set

$$\|T\|_{\mathscr{I}_p} = \|T\|_{\mathscr{I}_p(\mathscr{H}_1, \mathscr{H}_2)} \equiv \sup \left(\sum |(Tf_j, g_j)_{\mathscr{H}_2}|^p \right)^{1/p} \qquad (20)$$

(with obvious modifications when $p = \infty$). Here the supremum is taken over all $(f_j) \in \mathrm{ON}(\mathscr{H}_1)$ and $(g_j) \in \mathrm{ON}(\mathscr{H}_2)$. Now recall that $\mathscr{I}_p = \mathscr{I}_p(\mathscr{H}_1, \mathscr{H}_2)$, the Schatten–von Neumann class of order p, consists of all linear operators T from $\mathscr{H}_1$ to $\mathscr{H}_2$ such that $\|T\|_{\mathscr{I}_p(\mathscr{H}_1, \mathscr{H}_2)}$ is finite. Obviously, $\mathscr{I}_\infty(\mathscr{H}_1, \mathscr{H}_2)$ consists of all continuous operators from $\mathscr{H}_1$ to $\mathscr{H}_2$. If $\mathscr{H}_1 = \mathscr{H}_2$, then the shorter notation $\mathscr{I}_p(\mathscr{H}_1)$ is used instead of $\mathscr{I}_p(\mathscr{H}_1, \mathscr{H}_2)$. We also let $\mathscr{I}_\sharp(\mathscr{H}_1, \mathscr{H}_2)$ be the set of all linear and compact operators from $\mathscr{H}_1$ to $\mathscr{H}_2$, and equip this space with the norm $\| \cdot \|_{\mathscr{I}_\infty}$ as usual. The spaces $\mathscr{I}_1(\mathscr{H}_1, \mathscr{H}_2)$ and $\mathscr{I}_2(\mathscr{H}_1, \mathscr{H}_2)$ are called the sets of trace-class operators and Hilbert–Schmidt operators respectively. These definitions agree with the old ones when $\mathscr{H}_1 = \mathscr{H}_2$, and in this case the norms $\| \cdot \|_{\mathscr{I}_1}$ and $\| \cdot \|_{\mathscr{I}_2}$ agree with the trace-class norm and Hilbert–Schmidt norm respectively.

Another description of Schatten–von Neumann classes can be obtained in terms of singular numbers. Assume first that T above is compact. Then by the spectral theorem it follows that

$$Tf = \sum_{j=1}^{\infty} \lambda_j (f, g_j)_{\mathscr{H}_1} f_j, \quad f \in \mathscr{H}_1, \tag{21}$$

for some sequences $(f_j)_{j=1}^{\infty} \in \mathrm{ON}(\mathscr{H}_2)$ and $(g_j)_{j=1}^{\infty} \in \mathrm{ON}(\mathscr{H}_1)$, and some sequence $\lambda_1 \geq \lambda_2 \geq \cdots \geq 0$. Here the numbers λ_j are called the singular numbers for T, and we use the notation $\sigma_j(T)$ for these numbers, i.e. $\sigma_j(T) = \lambda_j$.

There is a canonical way to extend the definition of singular values to bounded operators, which are not necessarily compact. More precisely, for any closed subspace V of $\mathscr{H}_1$, set

$$\mu_V(T) = \sup_{f \in V, \, \|f\|_{\mathscr{H}_1} \leq 1} \|Tf\|_{\mathscr{H}_2},$$

Then let $\sigma_j(T)$ be defined by the formula

$$\sigma_j(T) = \sigma_{j, \mathscr{H}_1, \mathscr{H}_2}(T) \equiv \inf_{\dim V^\perp = j-1} \mu_V(T). \tag{22}$$

It is straight-forward to verify that $\sigma_j(T)$ agrees with the earlier definition when T is compact. Moreover, $T \in \mathscr{I}_p$ if and only if $(\sigma_j(T)) \in l^p$, and

$$\|T\|_{\mathscr{I}_p} = \|(\sigma_j(T))\|_{l^p}. \tag{23}$$

From now on we assume that the involved Hilbert spaces are separable. Then without loss of generality we may in many situations concerning $\mathscr{I}_p(\mathscr{H}_1, \mathscr{H}_2)$ reduce ourself to the case $\mathscr{H}_1 = \mathscr{H}_2$.

In fact, assume that (f_j^0) and (g_j^0) are fixed orthonormal bases for $\mathscr{H}_1$ and $\mathscr{H}_2$ respectively, and let T_0 be the linear map, defined by the formula

$$T_0 \left(\sum \alpha_j f_j^0 \right) = \sum \alpha_j g_j^0.$$

Here $(\alpha_j) \in l^2$ is arbitrary. Then T_0 is an isometric bijection from $\mathscr{H}_1$ to $\mathscr{H}_2$, $T_0^* \circ T_0 = \mathrm{Id}_{\mathscr{H}_1}$ and $T_0 \circ T_0^* = \mathrm{Id}_{\mathscr{H}_2}$. Consequently, $(f_j) \mapsto (T_0 f_j)$ is a bijection from $\mathrm{ON}(\mathscr{H}_1)$ to $\mathrm{ON}(\mathscr{H}_2)$. This in turn implies that $T \mapsto T_0 \circ T$

is an isometric homeomorphism from $\mathscr{I}_p(\mathscr{H}_1)$ to $\mathscr{I}_p(\mathscr{H}_1, \mathscr{H}_2)$, i.e. there is a canonical identification between $\mathscr{I}_p(\mathscr{H}_1)$ and $\mathscr{I}_p(\mathscr{H}_1, \mathscr{H}_2)$.

In Propositions 3.1–3.6 below we list some properties for spaces of the type $\mathscr{I}_p(\mathscr{H}_1, \mathscr{H}_2)$, which are well known in the case $\mathscr{H}_1 = \mathscr{H}_2$ (see [30]). The general case, when $\mathscr{H}_1$ is not necessarily equal to $\mathscr{H}_2$ is now a consequence of the identification here above and the corresponding results in [30]. In the general case, the results can also be found in [2,19,29].

Proposition 3.1 *Assume that* $p, p_j \in [1, \infty]$ *for* $1 \leq j \leq 2$ *such that* $p_1 < p_2 < \infty$. *Also assume that* $\mathscr{H}_1$ *and* $\mathscr{H}_2$ *are separable Hilbert spaces. Then* $\mathscr{I}_p$ *is a Banach space,*

$$\mathscr{I}_{p_1}(\mathscr{H}_1, \mathscr{H}_2) \subseteq \mathscr{I}_{p_2}(\mathscr{H}_1, \mathscr{H}_2) \subseteq \mathscr{I}_\sharp(\mathscr{H}_1, \mathscr{H}_2) \subseteq \mathscr{I}_\infty(\mathscr{H}_1, \mathscr{H}_2), \quad (24)$$

and

$$\|T\|_{\mathscr{I}_\infty} \leq \|T\|_{\mathscr{I}_{p_2}} \leq \|T\|_{\mathscr{I}_{p_1}}, \quad T \in \mathscr{I}_\infty(\mathscr{H}_1, \mathscr{H}_2). \quad (25)$$

Moreover, equalities in (25) *occur if and only if* T *is a rank one operator, i.e.* $Tf = (f, g_1)_{\mathscr{H}_1} g_2$ *for some* $g_1 \in \mathscr{H}_1$ *and* $g_2 \in \mathscr{H}_2$, *and then* $\|T\|_{\mathscr{I}_p} = \|g_1\|_{\mathscr{H}_1} \|g_2\|_{\mathscr{H}_2}$ *for every* $p \in [1, \infty]$.

The next result concerns algebraic properties for Schatten–von Neumann classes.

Proposition 3.2 *Assume that* $p, q, r \in [1, \infty]$ *such that* $1/p + 1/q = 1/r$. *Also assume that* $\mathscr{H}_j$ *for* $1 \leq j \leq 3$ *are separable Hilbert spaces. If* $T_1 \in \mathscr{I}_p(\mathscr{H}_1, \mathscr{H}_2)$ *and* $T_2 \in \mathscr{I}_q(\mathscr{H}_2, \mathscr{H}_3)$, *then* $T = T_2 \circ T_1 \in \mathscr{I}_r(\mathscr{H}_1, \mathscr{H}_3)$, *and*

$$\|T_2 \circ T_1\|_{\mathscr{I}_r} \leq \|T_1\|_{\mathscr{I}_p} \|T_2\|_{\mathscr{I}_q}. \quad (26)$$

On the other hand, for any $T \in \mathscr{I}_r(\mathscr{H}_1, \mathscr{H}_3)$, *there are operators* $T_1 \in \mathscr{I}_p(\mathscr{H}_1, \mathscr{H}_2)$ *and* $T_2 \in \mathscr{I}_q(\mathscr{H}_2, \mathscr{H}_3)$ *such that* $T = T_2 \circ T_1$ *and equality is attained in* (26).

Remark 4 *Assume that* $p \in [1, \infty]$ *and that* $\mathscr{H}_j$ *for* $j = 1, \ldots 4$ *are Hilbert spaces such that* $\mathscr{H}_1 \hookrightarrow \mathscr{H}_2$, *and* $\mathscr{H}_3 \hookrightarrow \mathscr{H}_4$. *Then it follows from Proposition 3.2 that these embeddings induce the embedding* $\mathscr{I}_p(\mathscr{H}_2, \mathscr{H}_3) \hookrightarrow \mathscr{I}_p(\mathscr{H}_1, \mathscr{H}_4)$. *This is also a consequence of* (23), *since it follows from the assumptions that there exists a constant* C *such that if* T *is linear from* $\mathscr{H}_2$ *to* $\mathscr{H}_3$ *and* $j \geq 1$, *then* $\sigma_{j, \mathscr{H}_1, \mathscr{H}_4}(T) \leq C \sigma_{j, \mathscr{H}_2, \mathscr{H}_3}(T)$.

We note that $T \in \mathscr{I}_p(\mathscr{H}_1, \mathscr{H}_2)$ if and only if $T^* \in \mathscr{I}_p(\mathscr{H}_2, \mathscr{H}_1)$, in view of (20). The next proposition deals with duality properties. Here recall that $p' \in [1, \infty]$ denotes the conjugate exponent for $p \in [1, \infty]$, i.e. $1/p + 1/p' = 1$.

Proposition 3.3 *Assume that* $p \in [1, \infty]$, *and that* $\mathscr{H}_1$ *and* $\mathscr{H}_2$ *are separable Hilbert spaces. Then the form*

$$(T_1, T_2) = (T_1, T_2)_{\mathscr{H}_1, \mathscr{H}_2} \equiv \mathrm{tr}_{\mathscr{H}_1}(T_2^* \circ T_1)$$

on $\mathscr{I}_1(\mathscr{H}_1, \mathscr{H}_2)$ extends uniquely to a continuous and sesqui-linear form on $\mathscr{I}_p(\mathscr{H}_1, \mathscr{H}_2) \times \mathscr{I}_{p'}(\mathscr{H}_1, \mathscr{H}_2)$, and for every $T_1 \in \mathscr{I}_p$ and $T_2 \in \mathscr{I}_{p'}$, then

$$(T_1, T_2)_{\mathscr{H}_1, \mathscr{H}_2} = \overline{(T_2, T_1)_{\mathscr{H}_2, \mathscr{H}_1}}$$

$$|(T_1, T_2)_{\mathscr{H}_1, \mathscr{H}_2}| \leq \|T_1\|_{\mathscr{I}_p} \|T_2\|_{\mathscr{I}_{p'}}, \quad \|T_1\|_{\mathscr{I}_p} = \sup |(T_1, S)_{\mathscr{H}_1, \mathscr{H}_2}|,$$

where the supremum is taken over all $S \in \mathscr{I}_{p'}$ such that $\|S\|_{\mathscr{I}_{p'}} \leq 1$. If in addition $p < \infty$, then the dual space for $\mathscr{I}_p(\mathscr{H}_1, \mathscr{H}_2)$ can be identified with $\mathscr{I}_{p'}(\mathscr{H}_1, \mathscr{H}_2)$ through this form.

In view of Proposition 3.3 we note that $\mathscr{I}_2(\mathscr{H}_1, \mathscr{H}_2)$ is a Hilbert space with scalar product $(\cdot, \cdot)_{\mathscr{H}_1, \mathscr{H}_2}$, and that the corresponding norm agrees with the Hilbert–Schmidt norm $\| \cdot \|_{\mathscr{I}_2}$.

The first part of the next proposition follows immediately from Proposition 3.1, since $\mathscr{I}_p \subseteq \mathscr{I}_\sharp$ when $p < \infty$, and $\mathscr{I}_1(\mathscr{H}_1, \mathscr{H}_2)$ contains all operators of finite rank.

Proposition 3.4 Assume that $p \in [1, \infty)$, and that $\mathscr{H}_1$ and $\mathscr{H}_2$ are separable Hilbert spaces. Then $\mathscr{I}_1(\mathscr{H}_1, \mathscr{H}_2)$ is dense in $\mathscr{I}_\sharp(\mathscr{H}_1, \mathscr{H}_2)$ and in $\mathscr{I}_p(\mathscr{H}_1, \mathscr{H}_2)$. It is dense in $\mathscr{I}_\infty(\mathscr{H}_1, \mathscr{H}_2)$ with respect to the weak* topology.

The next proposition deals with spectral properties. Here recall that ℓ^∞ is the set of all bounded sequences $(\lambda_j)_{j\geq 1}$ such that $\lambda_j \to \infty$ as j tends to infinity.

Proposition 3.5 Assume that $T \in \mathscr{I}_\sharp(\mathscr{H}_1, \mathscr{H}_2)$. Then (21) holds for some choice of sequences $(f_j)_{j=1}^\infty \in \mathrm{ON}(\mathscr{H}_1)$, $(g_j)_{j=1}^\infty \in \mathrm{ON}(\mathscr{H}_2)$ and $\lambda = (\lambda_j)_{j=1}^\infty \in \ell^\infty$, where the sum on the right-hand side in (21) convergences with respect to the operator norm. Moreover, if $1 \leq p < \infty$ then $T \in \mathscr{I}_p(\mathscr{H}_1, \mathscr{H}_2)$, if and only if $\lambda \in l^p$, and then $\|T\|_{\mathscr{I}_p} = \|\lambda\|_{l^p}$ and the sum on the right-hand side of (21) converges with respect to the norm $\| \cdot \|_{\mathscr{I}_p}$.

The next result concerns interpolation properties.

Proposition 3.6 Assume that $p, p_1, p_2 \in [1, \infty]$ and $0 \leq \theta \leq 1$ such that $1/p = (1-\theta)/p_1 + \theta/p_2$. Assume also that $\mathscr{H}_1$ and $\mathscr{H}_2$ are separable Hilbert spaces. Then the (complex) interpolation space $(\mathscr{I}_{p_1}, \mathscr{I}_{p_2})_{[\theta]}$ is equal to $\mathscr{I}_p$ with equality in norms.

4 Schatten–Von Neumann Classes for Operators Acting on Modulation Spaces

In this section we discuss some properties of Schatten–von Neumann classes of linear operators acting from $M^2_{(\omega_1)}(\mathbf{R}^m)$ to $M^2_{(\omega_2)}(\mathbf{R}^m)$, where $\omega_1, \omega_2 \in$

$\mathscr{P}(\mathbf{R}^{2m})$. We are especially concerned with finding appropriate identifications of the dual of $s_{t,p}(\omega_1, \omega_2)$ when $p < \infty$ (see the introduction or below). In the first part we prove that there is a canonical way to identify the dual of $s_{t,p}(\omega_1, \omega_2)$ with $s_{t,p'}(\omega_1, \omega_2)$. In the second part we use this property to prove that the dual of $s_{t,p}(\omega_1, \omega_2)$ can also be identified with $s_{t,p'}(1/\omega_1, 1/\omega_2)$ through a unique extension of the L^2 product from $\mathscr{S}$.

We start by considering Schatten–von Neumann classes in the context of pseudo-differential calculus. Let $t \in \mathbf{R}$, $p \in [1, \infty]$ and $\omega_1, \omega_2 \in \mathscr{P}(\mathbf{R}^{2m})$ be fixed. From the introduction we recall that $s_{t,p}(\omega_1, \omega_2)$ consists of all $a \in \mathscr{S}'(\mathbf{R}^{2m})$ such that $a_t(x, D) \in \mathscr{I}_p(M^2_{(\omega_1)}, M^2_{(\omega_2)})$. Also let $s_{t,\sharp}(\omega_1, \omega_2)$ be the set of all $a \in \mathscr{S}'(\mathbf{R}^{2m})$ such that $a_t(x, D) \in \mathscr{I}_\sharp(M^2_{(\omega_1)}, M^2_{(\omega_2)})$, We let $s_{t,p}(\omega_1, \omega_2)$ and $s_{t,\sharp}(\omega_1, \omega_2)$ be equipped by the norms

$$\|a\|_{s_{t,p}} = \|a\|_{s_{t,p}(\omega_1,\omega_2)} \equiv \|a_t(x, D)\|_{\mathscr{I}_p(M^2_{(\omega_1)}, M^2_{(\omega_2)})}$$

and $\| \cdot \|_{s_{t,\infty}}$ respectively. Since the Weyl quantization is important in our investigations we also use the notations s_p^w and $s_\sharp^w$ instead of $s_{t,p}$ and $s_{t,\sharp}$ respectively when $t = 1/2$ and $p \in [1, \infty]$. In the case $\omega_1 = \omega_2 = \omega$, then we use the notation $s_{t,p}(\omega)$ and $s_p^w(\omega)$ instead of $s_{t,p}(\omega_1, \omega_2)$ and $s_p^w(\omega_1, \omega_2)$ respectively.

Remark 5 *Let $\omega_j \in \mathscr{P}(\mathbf{R}^{2m})$ for $j = 1, 2$, $t \in \mathbf{R}$ and $p \in [1, \infty]$. Then recall that different windows ψ_j and χ_j in $\mathscr{S}(\mathbf{R}^m) \setminus 0$ give rise to different norms $\| \cdot \|_{(j,1)}$ and $\| \cdot \|_{(j,2)}$ for $M^2_{(\omega_j)}$, $j = 1, 2$. This in turn implies that the definition of $s_{t,p}(\omega_1, \omega_2)$ norm depends on the chosen windows for $M^2_{(\omega_j)}$. However, as for $M^2_{(\omega_j)}$, we claim that $s_{t,p}(\omega_1, \omega_2)$ is independent of the choice of windows and that different windows give rise to equivalent norms.*

In fact, let $\| \cdot \|_{(1)}$ be the $s_{t,p}(\omega_1, \omega_2)$-norm when the window for $M^2_{(\omega_j)}$ is ψ_j for $j = 1, 2$, and let $\| \cdot \|_{(2)}$ be the $s_{t,p}(\omega_1, \omega_2)$ norm when the window for $M^2_{(\omega_j)}$ is χ_j for $j = 1, 2$. Then

$$C_j^{-1}\|f\|_{(j,1)} \le \|f\|_{(j,2)} \le C_j\|f\|_{(j,1)}$$

for $j = 1, 2$, and some positive constants C_1 and C_2 which are independent of $f \in \mathscr{S}'(\mathbf{R}^m)$. Hence (22) and (23) show that if $C = C_1 C_2$, then

$$C^{-1}\|a\|_{(1)} \le \|a\|_{(2)} \le C\|a\|_{(1)}, \quad a \in \mathscr{S}'(\mathbf{R}^{2m}).$$

This proves the assertion.

In order to avoid ambiguity of the $s_{t,p}(\omega_1, \omega_2)$ norm, we assume from now on that the windows for $M^2_{(\omega_1)}$ and $M^2_{(\omega_2)}$ are fixed.

From the fact that each linear and continuous operator from $\mathscr{S}(\mathbf{R}^m)$ to $\mathscr{S}'(\mathbf{R}^m)$ is equal to $a_t(x, D)$, for a unique $a \in \mathscr{S}'(\mathbf{R}^{2m})$, it follows that the map $a \mapsto a_t(x, D)$ is an isometric homeomorphism from $s_{t,p}(\omega_1, \omega_2)$ to

$\mathscr{I}_p(M^2_{(\omega_1)}, M^2_{(\omega_2)})$ when $p \in [1, \infty]$, and from $s_{t,\sharp}(\omega_1, \omega_2)$ to $\mathscr{I}_\sharp(M^2_{(\omega_1)}, M^2_{(\omega_2)})$. Consequently, most of the properties which are listed in Sect. 3 carry over to the $s_{t,p}$-spaces. Hence Proposition 3.1 shows that $s_{t,p}(\omega_1, \omega_2)$ is a scale of Banach spaces which increase with the parameter $p \in [1, \infty]$. Moreover, Proposition 3.3 shows that the norm $\| \cdot \|_{s_{t,2}(\omega_1,\omega_2)}$ in $s_{t,2}(\omega_1, \omega_2)$ induces a scalar product

$$(\cdot, \cdot)_{s_{t,2}} = (\cdot, \cdot)_{s_{t,2}(\omega_1,\omega_2)}$$

and that if $p < \infty$, then the dual for $s_{t,p}(\omega_1, \omega_2)$ can be identified with $s_{t,p'}(\omega_1, \omega_2)$ through a uniquely extension of the $s_{t,2}(\omega_1, \omega_2)$ product from $s_{t,1}(\omega_1, \omega_2)$.

A problem in this context is the somewhat complicated structure of the form $(\cdot, \cdot)_{s_{t,2}(\omega_1,\omega_2)}$, compared to, e.g. the scalar product on L^2, which in general fits pseudo-differential calculus well. In the remaining part of the section we therefore focus on a possible replacement of the form $(\cdot, \cdot)_{s_{t,2}}$ with L^2 when discussing duality. From our investigations it turns out that indeed the dual space of $s_{t,p}(\omega_1, \omega_2)$ may be identified with $s_{t,p'}(1/\omega_1, 1/\omega_2)$, by a unique and continuous extension of the L^2 product from $\mathscr{S}$.

We start by making some preparations. Assume that $\omega \in \mathscr{P}(\mathbf{R}^{2m})$. By Theorem 2.1 it follows that the dual for $M^2_{(\omega)}(\mathbf{R}^m)$ can be identified with $M^2_{(1/\omega)}(\mathbf{R}^m)$ through the scalar product on $L^2(\mathbf{R}^m)$. On the other hand, since $M^2_{(\omega)}$ is a Hilbert space, its dual can also be identified with $M^2_{(\omega)}$ through the scalar product on $M^2_{(\omega)}$. Consequently, there exist unique homeomorphisms

$$\mathcal{T}_\omega : \left(M^2_{(\omega)}\right)^* \to M^2_{(\omega)} \quad \text{and} \quad \mathcal{R}_\omega : M^2_{(\omega)} \to M^2_{(1/\omega)}$$

such that if $\ell \in \left(M^2_{(\omega)}\right)^*$ and $h = \mathcal{T}_\omega \ell \in M^2_{(\omega)}$, then

$$\ell(f) = (f, h)_{M^2_{(\omega)}} = (f, \mathcal{R}_\omega h)_{L^2}, \quad f \in M^2_{(\omega)} \tag{27}$$

and

$$C^{-1} \| \mathcal{R}_\omega h \|_{M^2_{(1/\omega)}} \le \| \ell \| = \| h \|_{M^2_{(\omega)}} \le C \| \mathcal{R}_\omega h \|_{M^2_{(1/\omega)}} \tag{28}$$

for some constant C which is independent of h. We observe that (27) and (28) imply that if $(f_j)_{j=1}^\infty \in \mathrm{ON}(M^2_{(\omega)})$ and $g_k = \mathcal{R}_\omega f_k$, then

$$(f_j, g_k)_{L^2} = \delta_{j,k} \quad \text{and} \quad C^{-1} \le \| g_k \|_{M^2_{(1/\omega)}} \le C. \tag{29}$$

For convenience we set $\mathrm{ON}_\omega = \mathrm{ON}(M^2_{(\omega)})$, and we let ON^*_ω be the set of all sequences $(g_j)_{j=1}^\infty$ in $M^2_{(1/\omega)}$ such that $g_j = \mathcal{R}_\omega f_j$ for some $(f_j)_{j=1}^\infty \in \mathrm{ON}(M^2_{(\omega)})$. The following characterization of $s_{t,p}(\omega_1, \omega_2)$ follows immediately from (18), Proposition 3.5 and the homoeomorphism property of $\mathcal{R}_\omega$.

Proposition 4.1 *Assume that $t \in \mathbf{R}$ and $\omega_1, \omega_2 \in \mathscr{P}(\mathbf{R}^{2m})$, Also assume that $a \in s_{t,\sharp}(\omega_1, \omega_2)$. Then for some $(g_j)_{j=1}^\infty \in \mathrm{ON}^*_{\omega_1}$, $(h_j)_{j=1}^\infty \in \mathrm{ON}_{\omega_2}$ and $\lambda = (\lambda_j)_{j=1}^\infty \in \ell^\infty$ it holds*

$$a = \sum_{j=1}^{\infty} \lambda_j W_{h_j,g_j} \tag{30}$$

(with convergence with respect to the norm $\| \cdot \|_{s_{t,\infty}}$). Moreover, if $1 \leq p < \infty$ then $a \in s_{t,p}(\omega_1, \omega_2)$, if and only if $\lambda \in l^p$, and then

$$\|a\|_{s_{t,p}(\omega_1,\omega_2)} = \|\lambda\|_{l^p}$$

for some constant C which is independent of a.

On the other hand, if a is given by (30) for some $(g_j)_{j=1}^{\infty} \in \mathrm{ON}_{\omega_1}^*$, $(h_j)_{j=1}^{\infty} \in \mathrm{ON}_{\omega_2}$ and $\lambda = (\lambda_j)_{j=1}^{\infty} \in l^p$, then $a \in s_{t,p}(\omega_1, \omega_2)$.

As an application of Proposition 4.1 we have the following result. We omit the proof since it can be found in [42].

Theorem 6 *Assume that $t \in \mathbf{R}$, $p \in [1, \infty)$ and that $\omega_1, \omega_2 \in \mathscr{P}(\mathbf{R}^{2m})$. Then the scalar product on $L^2(\mathbf{R}^{2m})$ extends uniquely to a duality between $s_{t,p}(\omega_1, \omega_2)$ and $s_{t,p'}(1/\omega_1, 1/\omega_2)$, and the dual space for $s_{t,p}(\omega_1, \omega_2)$ can be identified with $s_{t,p'}(1/\omega_1, 1/\omega_2)$ through this form. Moreover, if $\ell \in s_{t,p}(\omega_1, \omega_2)^*$*   *and $a \in s_{t,p'}(1/\omega_1, 1/\omega_2)$ such that $\overline{\ell(b)} = (a, b)_{L^2}$ when $b \in s_{t,p}(\omega_1, \omega_2)$, then*

$$C^{-1}\|a\|_{s_{t,p'}(1/\omega_1,1/\omega_2)} \leq \|\ell\| \leq C\|a\|_{s_{t,p'}(1/\omega_1,1/\omega_2)}$$

for some constant C which only depends on ω_1 and ω_2.

Remark 7 *Theorem 6 appeared after fruitful discussions with Paolo Boggiatto.*

5 Continuity and Schatten–Von Neumann Properties for Pseudo-Differential Operators

In this section, we discuss continuity and Schatten–von Neumann properties for pseudo-differential operators, when the operator symbols belong to appropriate classes of modulation spaces. In particular we extend some of the continuity properties of Sect. 5 in [40]. An important ingredient in these investigations concern continuity properties for the Wigner distributions in context of modulation spaces, as well as the central role for Wigner distributions in the Weyl calculus of pseudo-differential operators.

First assume that $\mathscr{B}_j$ for $j = 1, 2, 3$ are Frechét spaces such that

$$\mathscr{S}(\mathbf{R}^{2m}) \hookrightarrow \mathscr{B}_1 \hookrightarrow \mathscr{S}'(\mathbf{R}^{2m}), \quad \mathscr{S}(\mathbf{R}^m) \hookrightarrow \mathscr{B}_2, \mathscr{B}_3 \hookrightarrow \mathscr{S}'(\mathbf{R}^m),$$

and that $(a, f, g) \mapsto (a, W_{g,f})$ is well-defined and sequently continuous from $\mathscr{B}_1 \times \mathscr{B}_2 \times \mathscr{B}_3$ to $\mathbf{C}$. Then (18) is taken as the definition of $a^w(x, D)f$ as an element in $\mathscr{B}_3'$ when $f \in \mathscr{B}_2$, and it follows that $a^w(x, D)$ is a continuous operator from $\mathscr{B}_2$ to $\mathscr{B}_3'$.

Next we discuss continuity properties for pseudo-differential operator, and prove in a moment that if $t \in \mathbf{R}$,

$$1/p_1 - 1/p_2 = 1/q_1 - 1/q_2 = 1 - 1/p - 1/q, \quad q \leq p_2, q_2 \leq p, \qquad (31)$$

ω_1, ω_2 and ω are appropriate weight functions and $a \in M_{(\omega)}^{p,q}$, then $a_t(x, D)$ is continuous from $M_{(\omega_1)}^{p_1,q_1}$ to $M_{(\omega_2)}^{p_2,q_2}$. As a first step we consider continuity properties for Wigner distributions in background of modulation space theory.

Proposition 5.1 *Assume that $p_j, q_j, p, q \in [1, \infty]$ are such that $p \leq p_j, q_j \leq q$, for $j = 1, 2$, and that*

$$1/p_1 + 1/p_2 = 1/q_1 + 1/q_2 = 1/p + 1/q. \qquad (32)$$

holds. Also assume that $\omega_1, \omega_2 \in \mathscr{P}(\mathbf{R}^{2m})$, and that $\omega \in \mathscr{P}(\mathbf{R}^{2m} \oplus \mathbf{R}^{2m})$ satisfy

$$\omega(x, \xi, y, \eta) \leq C\omega_1(x - y/2, \xi + \eta/2)\omega_2(x + y/2, \xi - \eta/2). \qquad (33)$$

Then the map $(f_1, f_2) \mapsto W_{f_1,f_2}$ from $\mathscr{S}'(\mathbf{R}^m) \times \mathscr{S}'(\mathbf{R}^m)$ to $\mathscr{S}'(\mathbf{R}^{2m})$ restricts to a continuous mapping from $M_{(\omega_1)}^{p_1,q_1}(\mathbf{R}^m) \times M_{(\omega_2)}^{p_2,q_2}(\mathbf{R}^m)$ to $\widetilde{M}_{(\omega)}^{p,q}(\mathbf{R}^{2m})$, and for some constant C,

$$\|W_{f_1,f_2}\|_{\widetilde{M}_{(\omega)}^{p,q}} \leq C\|f_1\|_{M_{(\omega_1)}^{p_1,q_1}}\|f_2\|_{M_{(\omega_2)}^{p_2,q_2}} \qquad (34)$$

when $f_1, f_2 \in \mathscr{S}'(\mathbf{R}^m)$.

We omit the proof since the result with proofs can be found in [40, 42].

Proposition 5.1 can be used to obtain continuity for pseudo-differential operators when acting on modulation spaces. More precisely, in [40, 42], Proposition 5.1 is used prove the following result. We omit the proof.

Theorem 8 *Assume that $t \in \mathbf{R}$ and $p, q, p_j, q_j \in [1, \infty]$ for $j = 1, 2$, satisfy (31). Also assume that $\omega \in \mathscr{P}(\mathbf{R}^{2m} \oplus \mathbf{R}^{2m})$ and $\omega_1, \omega_2 \in \mathscr{P}(\mathbf{R}^{2m})$ satisfy*

$$\frac{\omega_2(x - ty, \xi + (1-t)\eta)}{\omega_1(x + (1-t)y, \xi - t\eta)} \leq C\omega(x, \xi, y, \eta) \qquad (35)$$

for some constant C. If $a \in \widetilde{M}_{(\omega)}^{p,q}(\mathbf{R}^{2m})$, then $a_t(x, D)$ from $\mathscr{S}(\mathbf{R}^m)$ to $\mathscr{S}'(\mathbf{R}^m)$ extends uniquely to a continuous mapping from $M_{(\omega_1)}^{p_1,q_1}(\mathbf{R}^m)$ to $M_{(\omega_2)}^{p_2,q_2}(\mathbf{R}^m)$.

Moreover, if in addition $a \in \widetilde{\mathcal{M}}_{(\omega)}^{p,q}$, *then* $a_t(x, D) : M_{(\omega_1)}^{p_1,q_1} \to M_{(\omega_2)}^{p_2,q_2}$ *is compact.*

Remark 9 *If* $p = p_1 = p_2 = \infty$ *and* $q = q_1 = q_2 = 1$, *or* $p = q_1 = q_2 = \infty$ *and* $q = p_1 = p_2 = 1$ *in Theorem 8, then* $\mathcal{S}$ *is not dense in any of the involving spaces. However, in spite of these facts, there are no ambiguity in order to define* $a_t(x, D)f$ *when* $f \in M^{p_1,q_1}$ *(see, e.g. [42]).*

In the case $p = q = \infty$ in Theorem 8, the converse is also true, i.e. we have the following result.

Theorem 10 *Assume that* $t \in \mathbf{R}$, $a \in \mathcal{S}'(\mathbf{R}^{2m})$, $\omega \in \mathcal{P}(\mathbf{R}^{2m} \oplus \mathbf{R}^{2m})$, *and* $\omega_1, \omega_2 \in \mathcal{P}(\mathbf{R}^{2m})$ *such that* (35) *holds. Then the operator* $a_t(x, D)$ *from* $\mathcal{S}(\mathbf{R}^m)$ *to* $\mathcal{S}'(\mathbf{R}^m)$ *extends to a continuous mapping from* $M_{(\omega_1)}^1(\mathbf{R}^m)$ *to* $M_{(\omega_2)}^\infty(\mathbf{R}^m)$, *if and only if* $a \in \widetilde{M}_{(\omega)}^\infty(\mathbf{R}^{2m})$.

For the proof we need the following two propositions of independent interest, where the first proposition is a slight generalization of Feichtinger–Gröchenig's kernel theorem, named as Schwartz–Gröchenig's kernel theorem in Theorem 4.1 in [40]. (See also Theorem 14.4.1 in [21].)

Proposition 5.2 *Assume that* $m = m_1 + m_2$, $\omega_j \in \mathcal{P}(\mathbf{R}^{2m_j})$ *for* $j = 1, 2$ *and* $\omega \in \mathcal{P}(\mathbf{R}^m \oplus \mathbf{R}^m)$ *such that*

$$\omega(x, y, \xi, \eta) = \omega_2(x, \xi)/\omega_1(y, -\eta). \tag{36}$$

Also assume that T *is a linear and continuous map from* $\mathcal{S}(\mathbf{R}^{m_1})$ *to* $\mathcal{S}'(\mathbf{R}^{m_2})$. *Then* T *extends to a continuous mapping from* $M_{(\omega_1)}^\infty(\mathbf{R}^{m_1})$ *to* $M_{(\omega_2)}^\infty(\mathbf{R}^{m_2})$, *if and only if it exists an element* $K \in M_{(\omega)}^\infty(\mathbf{R}^m)$ *such that*

$$(Tf)(x) = \langle K(x, \cdot), f \rangle. \tag{37}$$

Here the right-hand side in (37) should be interpreted as the distribution u, given by the formula $\langle u, g \rangle = \langle K, g \otimes f \rangle$, or alternatively, by the formula $(u, g) = (K, g \otimes \overline{f})$, when $f \in \mathcal{S}(\mathbf{R}^{m_1})$ and $g \in \mathcal{S}(\mathbf{R}^{m_2})$.

Proof. Assume that T extends to continuous map from $M_{(\omega_1)}^1$ to $M_{(\omega_2)}^\infty$. It follows from the kernel theorem of Schwartz that (37) holds for some $K \in \mathcal{S}'(\mathbf{R}^m)$. We shall prove that K belongs to $M_{(\omega)}^\infty$.

From the assumptions and Proposition 2.1 (3) it follows that

$$|(K, g \otimes \overline{f})_{L^2}| \le C \|f\|_{M_{(\omega_1)}^1} \|g\|_{M_{(1/\omega_2)}^1} \tag{38}$$

holds for some constant C which is independent of $f \in \mathcal{S}(\mathbf{R}^{m_1})$ and $g \in \mathcal{S}(\mathbf{R}^{m_2})$. By letting $\chi = \overline{g} \otimes f$ be fixed, and replacing f and g with

$$f_{y,\eta} = e^{-i\langle \cdot, \eta \rangle} f(\cdot - y) \quad \text{and} \quad g_{x,\xi} = e^{i\langle \cdot, \xi \rangle} f(\cdot - x),$$

it follows that (38) takes the form

$$|\mathscr{F}(K\tau_{(x,y)}\chi)(\xi,\eta)| \leq C\|f_{y,\eta}\|_{M^1_{(\omega_1)}}\|g_{x,\xi}\|_{M^1_{(1/\omega_2)}}. \tag{38'}$$

We have to analyze the right-hand side of (38′). If v is chosen such that ω_1 is v-moderate, and $\chi_1 \in \mathscr{S}(\mathbf{R}^{m_1}) \setminus 0$ is a fixed function, then we obtain

$$\|f_{y,\eta}\|_{M^1_{(\omega_1)}} = \iint |\mathscr{F}(f\tau_{z-y}\chi_1)(\zeta+\eta)\omega_1(z,\zeta)|\,dzd\zeta$$

$$= \iint |\mathscr{F}(f\tau_z\chi_1)(\zeta)\omega_1(z+y,\zeta-\eta)|\,dzd\zeta$$

$$\leq C\omega_1(y,-\eta)\|f\|_{M^1_{(v)}} = C'\omega_1(y,-\eta).$$

In the same way we get

$$\|g_{x,\xi}\|_{M^1_{(1/\omega_2)}} \leq C\omega_2(x,\xi)^{-1}.$$

If these estimates are inserted into (38′), we obtain

$$|\mathscr{F}(K\tau_{(x,y)}\chi)(\xi,\eta)\omega(x,y,\xi,\eta)| \leq C,$$

for some constant C which is independent of x, y, ξ and η. By taking the supremum of the left-hand side it follows that $\|K\|_{M^\infty_{(\omega)}} < \infty$. Hence $K \in M^\infty_{(\omega)}$, and the necessity follows.

The sufficiency follows by straight-forward computations. The details are left for the reader. (See also the proof of Theorem 4.1 in [40].) The proof is complete. $\square$

Proposition 5.3 *Assume that $a \in \mathscr{S}'(\mathbf{R}^{2m})$, and that $K \in \mathscr{S}'(\mathbf{R}^{2m})$ is the distribution kernel for the Weyl operator $a^w(x,D)$. Also assume that $p \in [1,\infty]$, and that $\omega,\omega_0 \in \mathscr{P}(\mathbf{R}^{2m} \oplus \mathbf{R}^{2m})$ such that*

$$\omega(x,\xi,y,\eta) = \omega_0(x-y/2, x+y/2, \xi+\eta/2, -\xi+\eta/2).$$

Then $a \in \widetilde{M}^p_{(\omega)}(\mathbf{R}^{2m})$ if and only if $K \in M^p_{(\omega_0)}(\mathbf{R}^{2m})$. Moreover, if $\chi \in \mathscr{S}(\mathbf{R}^{2m})$ and

$$\psi(x,y) = \int \overline{\chi((x+y)/2,\xi)}e^{i\langle y-x,\xi\rangle}\,d\xi,$$

then $\|a\|_{\widetilde{M}^{p,\chi}_{(\omega)}} = \|K\|_{M^{p,\psi}_{(\omega_0)}}.$

Proof. By Fourier's inversion formula, it follows by straight-forward computations that

$$|\mathscr{F}(K\,\tau_{(x-y/2,x+y/2)}\psi)(\xi+\eta/2,-\xi+\eta/2)| = |\widetilde{\mathscr{F}}(a\tau_{(x,\xi)}\chi)(y,\eta)|.$$

The result now follows by applying the $L^p_{(\omega)}$-norm on these expressions. The proof is complete.   □

Proof of Theorem 10. In the case $t = 1/2$, the result follows immediately by combining Theorem 8, Proposition 5.2 and Proposition 5.3. For general t the result is now a consequence of (14) and Proposition 2.4. The proof is complete.   □

Remark 11 *Proposition 5.2 was proved in the case that ω_1 and $1/\omega_2$ are moderate functions in [21]. It was proved in the non-weighted case (i.e. $\omega_1 = \omega_2 = 1$) already in [9].*

Next we discuss embedding properties between Schatten–von Neumann classes in pseudo-differential calculus and modulation spaces. As a first step we consider Hilbert–Schmidt properties for operators acting on modulation spaces of Hilbert type.

Proposition 5.4 *Assume that $\omega_1, \omega_2 \in \mathscr{P}(\mathbf{R}^{2m})$ and $\omega \in \mathscr{P}(\mathbf{R}^{2m} \oplus \mathbf{R}^{2m})$ are such that (36) holds, and that T is a linear and continuous operator from $\mathscr{S}(\mathbf{R}^m)$ to $\mathscr{S}'(\mathbf{R}^m)$ with distribution kernel $K \in \mathscr{S}'(\mathbf{R}^{2m})$. Then $T \in \mathscr{I}_2(M^2_{(\omega_1)}, M^2_{(\omega_2)})$, if and only if $K \in M^2_{(\omega)}(\mathbf{R}^{2m})$, and then*

$$\|T\|_{\mathscr{I}_2} = \|K\|_{M^2_{(\omega)}}. \tag{40}$$

Proof. Let $(f_j) \in \mathrm{ON}_{\omega_1}$ and $(h_k) \in \mathrm{ON}_{\omega_2}$ be orthonormal basis for $M^2_{(\omega_1)}(\mathbf{R}^m)$ and $M^2_{(\omega_2)}(\mathbf{R}^m)$ respectively. Then

$$\|T\|^2_{\mathscr{I}_2} = \sum_{j,k} |(Tf_j, h_k)_{M^2_{(\omega_2)}}|^2 = \sum_{j,k} |(K, h_k \otimes \overline{f_j})_{M^2_{(\omega_2)} \otimes L^2}|^2 \tag{41}$$

Next we consider the operator $T'_{\omega_0} = I_{M^2_{(\omega_2)}} \otimes \mathscr{R}_{1/\omega_0}$, where $\omega_0(x, \xi) = \omega_1(x, -\xi)$, which acts from $M^2_{(\omega_2)} \otimes M^2_{(1/\omega_0)}$ to $M^2_{(\omega_2)} \otimes M^2_{(\omega_0)}$ (Hilbert tensor products). Then (41) gives

$$\|T\|^2_{\mathscr{I}_2} = \sum_{j,k} |(T'_{\omega_0} K, h_k \otimes \overline{f_j})_{M^2_{(\omega_2)} \otimes M^2_{(\omega_1)}}|^2$$

$$= \|T'_{\omega_0} K\|^2_{M^2_{(\omega_2)} \otimes M^2_{(\omega_0)}} = \|K\|^2_{M^2_{(\omega_2)} \otimes M^2_{(1/\omega_0)}} = \|K\|^2_{M^2_{(\omega)}}$$

Hence (40) holds, and the proof is complete.   □

The following result is now an immediate consequence of Proposition 5.4 and Proposition 5.3 for $p = 2$.

Proposition 5.5 *Assume that* $a \in \mathscr{S}'(\mathbf{R}^{2m})$, $\omega_1, \omega_2 \in \mathscr{P}(\mathbf{R}^{2m})$ *and that* $\omega \in \mathscr{P}(\mathbf{R}^{2m} \oplus \mathbf{R}^{2m})$ *are such that equality is attained in* (35) *for* $t = 1/2$ *and some constant* C. *Then* $a^w(x, D) \in \mathscr{I}_2(M^2_{(\omega_1)}, M^2_{(\omega_2)})$, *if and only if* $a \in \widetilde{M}^2_{(\omega)}(\mathbf{R}^{2m})$. *Moreover, for some constant* $C > 0$ *it holds*

$$C^{-1}\|a\|_{\widetilde{M}^2_{(\omega)}} \leq \|a^w(x, D)\|_{\mathscr{I}_2} \leq C\|a\|_{\widetilde{M}^2_{(\omega)}}.$$

For general Schatten–von Neumann classes, we have the following generalization of Proposition 5.5.

Theorem 12 *Assume that* $t \in \mathbf{R}$ *and* $p, q, p_j, q_j \in [1, \infty]$ *for* $j = 1, 2$, *satisfy*

$$p_1 \leq p \leq p_2, \quad q_1 \leq \min(p, p') \quad and \quad q_2 \geq \max(p, p'). \tag{42}$$

Also assume that $\omega \in \mathscr{P}(\mathbf{R}^{2m} \oplus \mathbf{R}^{2m})$ *and* $\omega_1, \omega_2 \in \mathscr{P}(\mathbf{R}^{2m})$ *are such that equality is attained in* (35), *for some constant* C. *Then*

$$\widetilde{M}^{p_1, q_1}_{(\omega)}(\mathbf{R}^{2m}) \hookrightarrow s_{t,p}(\omega_1, \omega_2) \hookrightarrow \widetilde{M}^{p_2, q_2}_{(\omega)}(\mathbf{R}^{2m}) \tag{43}$$

For the proof we need the following lemma.

Lemma 5.1. *Assume that*

$$(x_{j_1})_{j_1 \in I_1}, \quad (\xi_{j_2})_{j_2 \in I_2}, \quad (y_{k_1})_{k_1 \in I_1} \quad and \quad (\eta_{k_2})_{k_2 \in I_2}$$

are lattices in $\mathbf{R}^m$. *Also assume that* $\varphi(x) = e^{-|x|^2/2}$ *where* $x \in \mathbf{R}^m$, $\omega_1, \omega_2 \in \mathscr{P}(\mathbf{R}^{2m})$, $(f_l) \in \mathrm{ON}_{\omega_1}$, $(h_l) \in \mathrm{ON}_{\omega_2}$, *and set* $\kappa_l = \mathcal{R}_{\omega_2} h_l$ *and*

$$\theta_1(j, k, l) = \mathscr{F}(f_l \tau_{x_{j_1} + y_{k_1}/2}\varphi)(\xi_{j_2} - \eta_{k_2}/2)\omega_1(x_{j_1} + y_{k_1}/2, \xi_{j_2} - \eta_{k_2}/2),$$

$$\theta_2(j, k, l) = \mathscr{F}(\kappa_l \tau_{x_{j_1} - y_{k_1}/2}\varphi)(\xi_{j_2} + \eta_{k_2}/2)/\omega_2(x_{j_1} - y_{k_1}/2, \xi_{j_2} + \eta_{k_2}/2),$$

where $j = (j_1, j_2) \in I_1 \times I_2 \equiv I$, $k = (k_1, k_2) \in I$. *Then for some constant* C *and integer* $N \geq 0$ *it holds*

$$i = 1, 2, \tag{44}$$

$$\sum_{k \in I} |\theta_1(j, k, l)|^2 \leq C\|f_l\|^2_{M^2_{(\omega_1)}} \leq C < \infty, \tag{45}$$

$$\sum_{k \in I} |\theta_2(j, k, l)|^2 \leq C\|h_l\|^2_{M^2_{(\omega_2)}} \leq C < \infty. \tag{46}$$

Proof. Let $N \geq 0$ be chosen such that ω_1 and ω_2 are σ_N-moderate. Then (44) in the case $i = 1$ follows if we prove that

$$\sum_l |\mathscr{F}(f_l \tau_z \varphi)(\zeta)|^2 \leq C\|\varphi\|^2_{M^2_N}/\omega_1(z, \zeta)^2$$

for some constant C which is independent of (z, ζ). Since

$$\mathscr{F}(f_l \tau_z \varphi)(\zeta) = (f_l, e^{i\langle \cdot, \zeta \rangle} \tau_z \varphi)_{L^2} = (f_l, \mathcal{R}_{1/\omega_1}(e^{i\langle \cdot, \zeta \rangle} \tau_z \varphi))_{M^2_{(\omega_1)}}$$

and $(f_l) \in \mathrm{ON}_{\omega_1}$, we obtain for some $\chi \in \mathscr{S} \setminus 0$ that

$$\sum_l |\mathscr{F}(f_l \tau_z \varphi)(\zeta)|^2 = \sum_l |(f_l, \mathcal{R}_{1/\omega_1}(e^{i\langle \cdot, \zeta \rangle} \tau_z \varphi))_{M^2_{(\omega_1)}}|^2$$

$$\leq \|\mathcal{R}_{1/\omega_1}(e^{i\langle \cdot, \zeta \rangle} \tau_z \varphi)\|^2_{M^2_{(\omega_1)}} \leq C \|e^{i\langle \cdot, \zeta \rangle} \tau_z \varphi\|^2_{M^2_{(1/\omega_1)}}$$

$$= C \iint |\mathscr{F}(e^{i\langle \cdot, \zeta \rangle} \tau_z \varphi \tau_x \chi)(\xi)/\omega_1(x, \xi)|^2 \, dx d\xi$$

$$= C \iint |\mathscr{F}(\varphi \tau_{x-z} \chi)(\xi - \zeta)/\omega_1(x, \xi)|^2 \, dx d\xi$$

$$= C \iint |\mathscr{F}(\varphi \tau_x \chi)(\xi)/\omega_1(x + z, \xi + \zeta)|^2 \, dx d\xi$$

$$\leq C' \iint |\mathscr{F}(\varphi \tau_x \chi)(\xi) \sigma_N(x, \xi)|^2 \, dx d\xi / \omega_1(z, \zeta)^2$$

$$= C' \|\varphi\|^2_{M^2_N} / \omega_1(z, \zeta)^2.$$

This proves the assertion.

The case $i = 2$ in (44) follows from similar arguments together with the fact that

$$\mathscr{F}(\kappa_l \tau_z \varphi)(\zeta) = (\kappa_l, e^{i\langle \cdot, \zeta \rangle} \tau_z \varphi)_{L^2} = (h_l, e^{i\langle \cdot, \zeta \rangle} \tau_z \varphi)_{M^2_{(\omega_2)}}.$$

Next we prove (45). For some lattices $(z_{k_1})_{k_1 \in I_1}$ and $(\zeta_{k_2})_{k_2 \in I_2}$ we have

$$\sum_k |\theta_1(j, k, l)|^2 \leq \sum_{k_1, k_2} |\mathscr{F}(f_l \tau_{z_{k_1}} \varphi)(\zeta_{k_2}) \omega_1(z_{k_1}, \zeta_{k_2})|^2 \leq C \|f_l\|^2_{M^2_{(\omega_1)}}$$

for some constant C, where the last inequality follows from Proposition 2.2. This proves (45). By replacing the lattices (z_{k_1}) and (ζ_{k_2}) with other ones, if necessary, we obtain

$$\sum_k |\theta_2(j, k, l)|^2 \leq \sum_{k_1, k_2} |\mathscr{F}(\kappa_l \tau_{z_{k_1}} \varphi)(\zeta_{k_2}) \omega_2(z_{k_1}, \zeta_{k_2})|^2$$

$$\leq C \|\kappa_l\|^2_{M^2_{(1/\omega_2)}} \leq C' \|h_l\|^2_{M^2_{(\omega_2)}}$$

for some constants C and C', and (46) follows. The proof is complete. $\quad\square$

Proof of Theorem 12. We use the same notations as in Lemma 5.1. In view of Proposition 2.4 it is no restriction to assume that $t = 1/2$, and that equalities are attained in (42). Then the result is an immediate consequence of Proposition 5.5 in the case $p = q = 2$. Next we consider the case $q = 1$. Let $X_j = (x_{j_1}, \xi_{j_2})$ and $Y_k = (y_{k_1}, \eta_{k_2})$. Then it follows that $(X_j)_{j \in I}$ and $(Y_k)_{k \in I}$ are lattices in $\mathbf{R}^{2m}$. Assume that $a \in \widetilde{M}_{(\omega)}^{p,1}$. By Proposition 2.2 it follows that if the lattices here above are chosen sufficiently dense, then

$$a(X) = \sum_{j,k \in I} c_{j,k} e^{i\langle X, Y_k \rangle} (\tau_{X_j} W_{\varphi,\varphi})(X),$$

for some sequence $(c_{j,k})_{j,k \in I}$ which satisfies

$$C^{-1} \sum_{j \in I} \Big(\sum_{k \in I} \lambda_{j,k}^p \Big)^{1/p} \le C \|a\|_{\widetilde{M}_{(\omega)}^{p,1}} \le C \sum_{j \in I} \Big(\sum_{k \in I} \lambda_{j,k}^p \Big)^{1/p}, \tag{47}$$

$$\lambda_{j,k} = |c_{j,k} \omega(x_{j_1}, \xi_{j_2} y_{k_1}, \eta_{k_2})|.$$

Here $X = (x, \xi) \in \mathbf{R}^{2m}$ and $\langle X, Y_k \rangle = \langle x, \eta_{k_2} \rangle + \langle y_{k_1}, \xi \rangle$. Hence if $(f_l) \in \mathrm{ON}_{\omega_1}$ and $(\kappa_l) \in \mathrm{ON}_{\omega_2}^*$, then

$$|(a^w(x, D)f_l, \kappa_l)| \le \sum_{j,k} |c_{j,k}(e^{i\langle \cdot, Y_k \rangle} \tau_{X_j} W_{\varphi,\varphi}, W_{\kappa_l, f_l})|.$$

By straight-forward computations it follows that

$$e^{i\langle X, Y_k \rangle} (\tau_{X_j} W_{\varphi,\varphi})(X) = e^{i\Phi(Y_k, X_j)} W_{\varphi_{j,k}^1, \varphi_{j,k}^2},$$

where Φ is a real-valued quadratic form on $\mathbf{R}^{2m} \oplus \mathbf{R}^{2m}$, and

$$\varphi_{j,k}^1(x) = e^{i\langle x, \xi_{j_2} + \eta_{k_2}/2 \rangle} \varphi(x - x_{j_1} + y_{k_1}/2),$$

$$\varphi_{j,k}^2(x) = e^{i\langle x, \xi_{j_2} - \eta_{k_2}/2 \rangle} \varphi(x - x_{j_1} - y_{k_1}/2).$$

This gives

$$|(e^{i\langle \cdot, Y_k \rangle} \tau_{X_j} W_{\varphi,\varphi}, W_{\kappa_l, f_l})| = |(W_{\varphi_{j,k}^1, \varphi_{j,k}^2}, W_{\kappa_l, f_l})| = |(\varphi_{j,k}^1, \kappa_l)(\varphi_{j,k}^2, f_l)|$$

$$= |\theta_1(j, k, l)\theta_2(j, k, l)| \frac{\omega_2(x_{j_1} - y_{k_1}/2, \xi_{j_2} + \eta_j/2)}{\omega_1(x_{j_1} + y_{k_1}/2, \xi_{j_2} - \eta_{k_2}/2)}$$

$$\le C|\theta_1(j, k, l)\theta_2(j, k, l)| \omega(x_{j_1}, \xi_{j_2}, y_{k_1}, \eta_{k_2}),$$

for some constant C. Hence

$$|c_{j,k}(e^{i\langle \cdot, Y_k \rangle} \tau_{X_j} W_{\varphi,\varphi}, W_{\kappa_l, f_l})| \le C\lambda_{j,k} |\theta_1(j, k, l)\theta_2(j, k, l)|.$$

From these inequalities we obtain

$$\left(\sum_l |(a^w(x,D)f_l,\kappa_l)|^p\right)^{1/p} \le C\sum_{j,k,l} \lambda_{j,k}|\theta_1(j,k,l)\theta_2(j,k,l)|$$

$$\le C(J_1 + J_2)/2,$$

where

$$J_i = \left(\sum_l \left(\sum_{j,k} \lambda_{j,k}|(\theta_i(j,k,l)|^2\right)^p\right)^{1/p}, \quad j = 1,2.$$

We have to estimate J_1 and J_2. By Minkowski's and Hölder's inequalities we get

$$J_1 \le C\sum_j \left(\sum_l \left(\sum_k \lambda_{j,k}|\theta_1(j,k,l)|^2\right)^p\right)^{1/p}$$

$$= C\sum_j \left(\sum_l \left(\sum_k (\lambda_{j,k}|\theta_1(j,k,l)|^{2/p})|\theta_1(j,k,l)|^{2/p'}\right)^p\right)^{1/p}$$

$$\le C\sum_j \left(\sum_l \left(\sum_k \lambda_{j,k}^p|\theta_1(j,k,l)|^2\right)\left(\sum_k |\theta_1(j,k,l)|^2\right)^{p/p'}\right)^{1/p}.$$

Now (44)–(47) give

$$J_1 \le C_1 \sum_j \left(\left(\sum_{k,l} \lambda_{j,k}^p|\theta_1(j,k,l)|^2\right)\sup_l \left(\sum_k |\theta_1(j,k,l)|^2\right)^{p/p'}\right)^{1/p}$$

$$\le C_2\left(\sup_l \|f_l\|_{M^2_{(\omega_1)}}^{2/p'}\right)\sum_j \left(\sum_k \lambda_{j,k}^p\left(\sum_l |\theta_1(j,k,l)|^2\right)\right)^{1/p}$$

$$\le C_3\|\varphi\|_{M^2_N}^{2/p}\sum_{j\in I}\left(\sum_{k\in I} \lambda_{j,k}^p\right)^{1/p} \le C_4\|a\|_{\widetilde{M}^{p,1}_{(\omega)}},$$

for some constants $C_1,\ldots,C_4$. In the same way we get $J_2 \le C\|a\|_{\widetilde{M}^{p,1}_{(\omega)}}$ for some constant C. Hence it follows from these estimates that

$$\left(\sum_l |(a^w(x,D)f_l,\kappa_l)|^p\right)^{1/p} \le C\|a\|_{M^{p,1}_{(\omega)}}.$$

For some constant which is independent of the choice of sequences (f_l) and (κ_l). The result now follows by taking the supremum of the left-hand side with respect to all sequences (f_l) and (κ_l).

The first embedding in (43) now follows by interpolation of the case $q = 1$ and the case $p = q = 2$, using Proposition 2.3 and Proposition 3.6. The second embedding in (43) now follows from the first embedding and duality, using Theorem 6. The proof is complete.   □

Next we present some consequences of Theorem 12 when the weight functions ω_1 and ω_2 are the same and satisfy certain properties which are common in the applications. It is for example common that the moderate function $v \in \mathscr{P}(\mathbf{R}^{2m})$ satisfies the symmetry condition

$$v(x, \xi) = v(-x, \xi) = v(x, -\xi) = v(-x, -\xi), \tag{48}$$

for every $(x, \xi) \in \mathbf{R}^{2m}$.

Corollary 13 *Assume that $t \in \mathbf{R}$, $p, q, p_j, q_j \in [1, \infty]$ for $j = 1, 2$ are the same as in Theorem 12, and that $\omega_0 \in \mathscr{P}(\mathbf{R}^{2m})$ is v-moderate for some $v \in \mathscr{P}(\mathbf{R}^{2m})$ which satisfies (48). Also let $v_0(x, \xi, y, \eta) = v(y, \eta)$. Then*

$$\widetilde{M}^{p_1,q_1}_{(v_0)} \subseteq s_{t,p}(\omega_0) \subseteq \widetilde{M}^{p_2,q_2}_{(1/v_0)}.$$

Proof. It follows from the assumptions that

$$C^{-1} v(y, \eta)^{-1} \leq \frac{\omega_0(x - ty, \xi + (1 - t)\eta)}{\omega_0(x + (1 - t)y, \xi - t\eta)} \leq C v(y, \eta).$$

The result is therefore a consequence of Proposition 2.1 (2) and Theorem 12. The proof is complete.   □

Example 5.6 *Assume that $t \in \mathbf{R}$ and that $p, q, p_j, q_j \in [1, \infty]$ for $j = 1, 2$ are the same as in Theorem 12. Then it follows from the Corollary 13 that the following are true:*

(1) *If $s \in \mathbf{R}$, then*

$$\widetilde{M}^{p_1,q_1}_{|s|,0} \subseteq s_{t,p}(\sigma_s) \subseteq \widetilde{M}^{p_2,q_2}_{-|s|,0};$$

(2) *If $s_1, s_2 \in \mathbf{R}$ and $v(x, \xi, y, \eta) = \sigma_{|s_1|,|s_2|}(y, \eta)$, then*

$$\widetilde{M}^{p_1,q_1}_{(v)} \subseteq s_{t,p}(\sigma_{s_1,s_2}) \subseteq \widetilde{M}^{p_2,q_2}_{(1/v)}.$$

Remark 14 *By using embedding properties in [41] between modulation spaces and Besov spaces, the embeddings in Theorem 12, Corollary 13 and Example 5.6 give rise to embeddings between the $s_{t,p}$ spaces and Besov spaces (cf. [42]).*

Acknowledgements I am grateful to A. Holst for a careful reading and valuable comments, leading to improvement of the language and the content. I also thank P. Boggiatto, N. Kruglyak and I. Asekritova for fruitful discussions. I also thank L. Rodino and M. W. Wong for all support concerning the CIME meeting in Cetraro June 2006.

References

1. J. Bergh and J. Löfström *Interpolation Spaces, An Introduction*, Springer-Verlag, Berlin Heidelberg NewYork, 1976.

2. Birman, Solomyak · *Estimates for the singular numbers of integral operators (Russian)*, Usbehi Mat. Nauk. **32**, (1977), 17–84.

3. P. Boggiatto, E. Cordero, K. Gröchenig *Generalized Anti-Wick Operators with Symbols in Distributional Sobolev spaces*, Integr. equ. oper. theory (4), textbf48 (2004), 427–442.

4. P. Boggiatto, J. Toft *Embeddings and compactness for generalized Sobolev-Shubin spaces and modulation spaces*, Appl. Anal. (3) **84** (2005), 269–282.

5. A. Boulkemair *Remarks on a Wiener type pseudodifferential algebra and Fourier integral operators*, Math. Res. L. **4** (1997), 53–67.

6. _____ *L^2 estimates for Weyl quantization*, J. Funct. Anal. **165** (1999), 173–204.

7. E. Cordero, K. Gröchenig *Time–Frequency Analysis of Localization Operators*, J. Funct. Anal. (1) **205** (2003), 107–131.

8. M. Dimassi, J. Sjöstrand *Spectral Asymptotics in the Semi-Classical Limit*, vol 268, London Math. Soc. Lecture Note Series, Cambridge University Press, Cambridge, New York, Melbourne, Madrid, 1999.

9. H. G. Feichtinger *Un espace de Banach de distributions tempérés sur les groupes localement compacts abéliens (French)*, C. R. Acad. Sci. Paris Sér. A-B 290 **17** (1980), A791–A794.

10. H. G. Feichtinger *Banach spaces of distributions of Wiener's type and interpolation*, in: Ed. P. Butzer, B. Sz. Nagy and E. Görlich (Eds), Proc. Conf. Oberwolfach, Functional Analysis and Approximation, August 1980, Int. Ser. Num. Math. **69** Birkhäuser Verlag, Basel, Boston, Stuttgart, 1981, pp. 153–165.

11. _____ *Banach convolution algebras of Wiener's type*, in: Proc. Functions, Series, Operators in Budapest, Colloquia Math. Soc. J. Bolyai, North Holland Publ. Co., Amsterdam Oxford NewYork, 1980.

12. _____ *Modulation spaces on locally compact abelian groups. Technical report*, University of Vienna, Vienna, 1983; also in: M. Krishna, R. Radha, S. Thangavelu (Eds) Wavelets and their applications, Allied Publishers Private Limited, NewDehli Mumbai Kolkata Chennai Hagpur Ahmedabad Bangalore Hyderbad Lucknow, 2003, pp.99–140.

13. _____ *Atomic characterizations of modulation spaces through Gabor-type representations*, in: Proc. Conf. on Constructive Function Theory, Rocky Mountain J. Math. **19** (1989), 113–126.

14. H. G. Feichtinger and P. Gröbner *Banach Spaces of Distributions Defined by Decomposition Methods, I*, Math. Nachr. **123** (1985), 97–120.

15. H. G. Feichtinger and K. H. Gröchenig *Banach spaces related to integrable group representations and their atomic decompositions, I*, J. Funct. Anal. **86** (1989), 307–340.

16. _____ *Banach spaces related to integrable group representations and their atomic decompositions, II*, Monatsh. Math. **108** (1989), 129–148.

17. _____ *Gabor frames and time–frequency analysis of distributions*, J. Functional Anal. (2) **146** (1997), 464–495.

18. G. B. Folland *Harmonic analysis in phase space*, Princeton U. P., Princeton, 1989.

19. I. C. Gohberg, M. G. Krein *Introduction to the theory of linear non-selfadjoint operators in Hilbert space (Russian)*, Izdat. Nauka, Moscow, 1965.

20. K. H. Gröchenig *Describing functions: atomic decompositions versus frames*, Monatsh. Math.**112** (1991), 1–42.

21. _____ *Foundations of Time–Frequency Analysis*, Birkhäuser, Boston, 2001.

22. K. H. Gröchenig and C. Heil *Modulation spaces and pseudo-differential operators*, Integral Equations Operator Theory (4) **34** (1999), 439–457.

23. _____ *Modulation spaces as symbol classes for pseudodifferential operators* in: M. Krishna, R. Radha, S. Thangavelu (Eds) Wavelets and their applications, Allied Publishers Private Limited, NewDehli Mumbai Kolkata Chennai Hagpur Ahmedabad Bangalore Hyderbad Lucknow, 2003, pp. 151–170.

24. L. Hörmander *The Analysis of Linear Partial Differential Operators*, vol I, III, Springer-Verlag, Berlin Heidelberg NewYork Tokyo, 1983, 1985.

25. S. Pilipović, N. Teofanov *Wilson Bases and Ultramodulation Spaces*, Math. Nachr. **242** (2002), 179–196.

26. _____ *On a symbol class of Elliptic Pseudodifferential Operators*, Bull. Acad. Serbe Sci. Arts **27** (2002), 57–68.

27. M. Reed, B. Simon *Methods of modern mathematical physics*, Academic Press, London New York, 1979.

28. R. Schatten *Norm ideals of completely continuous operators*, Springer, Berlin, 1960.

29. B. W. Schulze, N. N. Tarkhanov *Pseudodifferential operators with operator-valued symbols*. Israel Math. Conf. Proc. **16**, 2003.

30. B. Simon *Trace ideals and their applications* I, London Math. Soc. Lecture Note Series, Cambridge University Press, Cambridge London New York Melbourne, 1979.

31. J. Sjöstrand *An algebra of pseudodifferential operators*, Math. Res. L. **1** (1994), 185–192.

32. _____ *Wiener type algebras of pseudodifferential operators*, Séminaire Equations aux Dérivées Partielles, Ecole Polytechnique, 1994/1995, Exposé n° IV.

33. K. Tachizawa *The boundedness of pseudo-differential operators on modulation spaces*, Math. Nachr. **168** (1994), 263–277.

34. N. Teofanov *Ultramodulation spaces and pseudodifferential operators*, Endowment Andrejević, Beograd, 2003.

35. J. Toft *Continuity and Positivity Problems in Pseudo-Differential Calculus, Thesis*, Department of Mathematics, University of Lund, Lund, 1996.

36. _____ *Subalgebras to a Wiener type Algebra of Pseudo-Differential operators*, Ann. Inst. Fourier (5) **51** (2001), 1347–1383.

37. _____ *Continuity properties for non-commutative convolution algebras with applications in pseudo-differential calculus*, Bull. Sci. Math. (2) **126** (2002), 115–142.

38. _____ *Modulation spaces and pseudo-differentianl operators*, Research Report 2002:05, Blekinge Institute of Technology, Karlskrona, 2002.

39. _____ *Continuity properties for modulation spaces with applications to pseudo-differential calculus, I*, J. Funct. Anal. (2), **207** (2004), 399–429.

40. _____ *Continuity properties for modulation spaces with applications to pseudo-differential calculus, II*, Ann. Global Anal. Geom., **26** (2004), 73–106.

41. _____ *Convolution and embeddings for weighted modulation spaces* in: P. Boggiatto, R. Ashino, M. W. Wong (eds) *Advances in Pseudo-Differential Operators,* Operator Theory: Advances and Applications **155**, Birkhäuser Verlag, Basel 2004, pp. 165–186.

42. _____ *Continuity and Schatten–von Neumann properties for Pseudo-Differential Operators on modulation spaces* in: J. Toft, M. W. Wong, H. Zhu (eds) *Modern Trends in Pseudo-Differential Operators,* Operator Theory: Advances and Applications, Birkhäuser Verlag, Basel 2007.

List of Participants

1. Boggiatto Paolo
   paolo.boggiatto@unito.it
   Univ. Torino, Italy

2. Bordeaux Montrieux William
   bordeaux@math.polytechnique.fr
   Ecole Polytechnique, France

3. Buzano Ernesto
   buzano@dm.unito.it
   Univ. Torino, Italy

4. Camperi Igor
   camperi@dm.unito.it
   Univ. Torino, Italy

5. Cappiello Marco
   marco.cappiello@unife.it
   Univ. Ferrara, Italy

6. Catania Davide
   catania@mail.dm.unipi.it
   Univ. Pisa, Italy

7. Ciraolo Giulio
   ciraolo@math.unifi.it
   Univ. Firenze, Italy

8. Concetti Francesco
   concetti@dm.unito.it
   Univ. Torino, Italy

9. Cordero Elena
   elena.cordero@unito.it
   Univ. Torino, Italy

10. Dasgupta Aparajita
    adgupta@yorku.ca
    York Univ., Canada

11. De Donno Giuseppe
    dedonno@dm.unito.it
    Univ. Torino, Italy

12. Feichtinger Hans
    hans.feichtinger@univie.ac.at
    Univ. Wien, Austria (**lecturer**)

13. Fernandez Rosell Carmen
    fernand@uv.es
    Univ. Valencia, Spain

14. Galbis Verdu Antonio
    antonio.galbis@uv.es
    Univ. Valencia, Spain

15. Garello Gianluca
    gianluca.garello@unito.it
    Univ. Torino, Italy

16. Helffer Bernard
    bernard.helffer@math.u-psud.fr
    Univ. Paris-Sud, France
    (**lecturer**)

17. Khmelynitskaya Alena
    Alena.Khmelynitskaya@ksu.ru
    Kazan State Univ., Russia

18. Labate Demetrio
    dlabate@math.ncsu.edu
    North Carolina State Univ., USA

19. Lamoureux Michael
    mikel@math.ucalgary.ca
    Univ. Calgary, Canada (**lecturer**)

20. Lerner Nicolas
    nicolas.lerner@univ-rennes1.fr
    Univ. Rennes, France (**lecturer**)

21. Luef Franz
    franz.luef@univie.ac.at
    Univ. Wien, Austria

22. Martinet Jerome
    jerome.martinet@math.u-psud.fr
    Univ. Orsay Paris-Sud 11, France

23. Obukhovskii Andrei
    avo-ob@mail.ru
    Voronezh St. For. Academy
    Russia

24. Oliaro Alessandro
    oliaro@dm.unito.it
    Univ. Torino, Italy

25. Rodino Luigi
    luigi.rodino@unito.it
    Univ. Torino, Italy (**editor**)

26. Teofanov Nenad
    tnenad@im.ns.ac.yu
    Univ. Novi Sad, Serbia

27. Toft Joachim
    joachim.toft@vxu.se
    Univ. Växjö, Sweden (**seminars**)

28. Wong Man Wah
    mwwong@mathstat.yorku.ca
    York Univ., Canada (**editor**)

29. Yashagin Eugene
    Evgene.Yashagin@ksu.ru
    Kazan State Univ., Russia

30. Yu Liu
    hanliu@yorku.ca
    York Univ., Canada

31. Zakharova Anastasia
    nastjka@list.ru
    Moscow State Univ., Russia

32. Zanelli Lorenzo
    lzanelli@math.unipd.it
    Univ. Padov., Italy

LIST OF C.I.M.E. SEMINARS

Published by C.I.M.E

Published by Ed. Cremonese, Firenze

Published by Ed. Liguori, Napoli

Published by Ed. Liguori, Napoli & Birkhäuser

Published by Springer-Verlag

Lecture Notes in Mathematics

For information about earlier volumes
please contact your bookseller or Springer
LNM Online archive: springerlink.com

Vol. 1812: L. Ambrosio, K. Deckelnick, G. Dziuk, M. Mimura, V. A. Solonnikov, H. M. Soner, Mathematical Aspects of Evolving Interfaces. Madeira, Funchal, Portugal 2000. Editors: P. Colli, J. F. Rodrigues (2003)

Vol. 1813: L. Ambrosio, L. A. Caffarelli, Y. Brenier, G. Buttazzo, C. Villani, Optimal Transportation and its Applications. Martina Franca, Italy 2001. Editors: L. A. Caffarelli, S. Salsa (2003)

Vol. 1814: P. Bank, F. Baudoin, H. Föllmer, L.C.G. Rogers, M. Soner, N. Touzi, Paris-Princeton Lectures on Mathematical Finance 2002 (2003)

Vol. 1815: A. M. Vershik (Ed.), Asymptotic Combinatorics with Applications to Mathematical Physics. St. Petersburg, Russia 2001 (2003)

Vol. 1816: S. Albeverio, W. Schachermayer, M. Talagrand, Lectures on Probability Theory and Statistics. Ecole d'Eté de Probabilités de Saint-Flour XXX-2000. Editor: P. Bernard (2003)

Vol. 1817: E. Koelink, W. Van Assche (Eds.), Orthogonal Polynomials and Special Functions. Leuven 2002 (2003)

Vol. 1818: M. Bildhauer, Convex Variational Problems with Linear, nearly Linear and/or Anisotropic Growth Conditions (2003)

Vol. 1819: D. Masser, Yu. V. Nesterenko, H. P. Schlickewei, W. M. Schmidt, M. Waldschmidt, Diophantine Approximation. Cetraro, Italy 2000. Editors: F. Amoroso, U. Zannier (2003)

Vol. 1820: F. Hiai, H. Kosaki, Means of Hilbert Space Operators (2003)

Vol. 1821: S. Teufel, Adiabatic Perturbation Theory in Quantum Dynamics (2003)

Vol. 1822: S.-N. Chow, R. Conti, R. Johnson, J. Mallet-Paret, R. Nussbaum, Dynamical Systems. Cetraro, Italy 2000. Editors: J. W. Macki, P. Zecca (2003)

Vol. 1823: A. M. Anile, W. Allegretto, C. Ringhofer, Mathematical Problems in Semiconductor Physics. Cetraro, Italy 1998. Editor: A. M. Anile (2003)

Vol. 1824: J. A. Navarro González, J. B. Sancho de Salas, $\mathscr{C}^\infty$ – Differentiable Spaces (2003)

Vol. 1825: J. H. Bramble, A. Cohen, W. Dahmen, Multiscale Problems and Methods in Numerical Simulations, Martina Franca, Italy 2001. Editor: C. Canuto (2003)

Vol. 1826: K. Dohmen, Improved Bonferroni Inequalities via Abstract Tubes. Inequalities and Identities of Inclusion-Exclusion Type. VIII, 113 p, 2003.

Vol. 1827: K. M. Pilgrim, Combinations of Complex Dynamical Systems. IX, 118 p, 2003.

Vol. 1828: D. J. Green, Gröbner Bases and the Computation of Group Cohomology. XII, 138 p, 2003.

Vol. 1829: E. Altman, B. Gaujal, A. Hordijk, Discrete-Event Control of Stochastic Networks: Multimodularity and Regularity. XIV, 313 p, 2003.

Vol. 1830: M. I. Gil', Operator Functions and Localization of Spectra. XIV, 256 p, 2003.

Vol. 1831: A. Connes, J. Cuntz, E. Guentner, N. Higson, J. E. Kaminker, Noncommutative Geometry, Martina Franca, Italy 2002. Editors: S. Doplicher, L. Longo (2004)

Vol. 1832: J. Azéma, M. Émery, M. Ledoux, M. Yor (Eds.), Séminaire de Probabilités XXXVII (2003)

Vol. 1833: D.-Q. Jiang, M. Qian, M.-P. Qian, Mathematical Theory of Nonequilibrium Steady States. On the Frontier of Probability and Dynamical Systems. IX, 280 p, 2004.

Vol. 1834: Yo. Yomdin, G. Comte, Tame Geometry with Application in Smooth Analysis. VIII, 186 p, 2004.

Vol. 1835: O.T. Izhboldin, B. Kahn, N.A. Karpenko, A. Vishik, Geometric Methods in the Algebraic Theory of Quadratic Forms. Summer School, Lens, 2000. Editor: J.-P. Tignol (2004)

Vol. 1836: C. Năstăsescu, F. Van Oystaeyen, Methods of Graded Rings. XIII, 304 p, 2004.

Vol. 1837: S. Tavaré, O. Zeitouni, Lectures on Probability Theory and Statistics. Ecole d'Eté de Probabilités de Saint-Flour XXXI-2001. Editor: J. Picard (2004)

Vol. 1838: A.J. Ganesh, N.W. O'Connell, D.J. Wischik, Big Queues. XII, 254 p, 2004.

Vol. 1839: R. Gohm, Noncommutative Stationary Processes. VIII, 170 p, 2004.

Vol. 1840: B. Tsirelson, W. Werner, Lectures on Probability Theory and Statistics. Ecole d'Eté de Probabilités de Saint-Flour XXXII-2002. Editor: J. Picard (2004)

Vol. 1841: W. Reichel, Uniqueness Theorems for Variational Problems by the Method of Transformation Groups (2004)

Vol. 1842: T. Johnsen, A. L. Knutsen, K_3 Projective Models in Scrolls (2004)

Vol. 1843: B. Jefferies, Spectral Properties of Noncommuting Operators (2004)

Vol. 1844: K.F. Siburg, The Principle of Least Action in Geometry and Dynamics (2004)

Vol. 1845: Min Ho Lee, Mixed Automorphic Forms, Torus Bundles, and Jacobi Forms (2004)

Vol. 1846: H. Ammari, H. Kang, Reconstruction of Small Inhomogeneities from Boundary Measurements (2004)

Vol. 1847: T.R. Bielecki, T. Björk, M. Jeanblanc, M. Rutkowski, J.A. Scheinkman, W. Xiong, Paris-Princeton Lectures on Mathematical Finance 2003 (2004)

Vol. 1848: M. Abate, J. E. Fornaess, X. Huang, J. P. Rosay, A. Tumanov, Real Methods in Complex and CR Geometry, Martina Franca, Italy 2002. Editors: D. Zaitsev, G. Zampieri (2004)

Vol. 1849: Martin L. Brown, Heegner Modules and Elliptic Curves (2004)

Vol. 1850: V. D. Milman, G. Schechtman (Eds.), Geometric Aspects of Functional Analysis. Israel Seminar 2002-2003 (2004)

Vol. 1851: O. Catoni, Statistical Learning Theory and Stochastic Optimization (2004)

Vol. 1852: A.S. Kechris, B.D. Miller, Topics in Orbit Equivalence (2004)

Vol. 1853: Ch. Favre, M. Jonsson, The Valuative Tree (2004)

Vol. 1854: O. Saeki, Topology of Singular Fibers of Differential Maps (2004)

Vol. 1855: G. Da Prato, P.C. Kunstmann, I. Lasiecka, A. Lunardi, R. Schnaubelt, L. Weis, Functional Analytic Methods for Evolution Equations. Editors: M. Iannelli, R. Nagel, S. Piazzera (2004)

Vol. 1856: K. Back, T.R. Bielecki, C. Hipp, S. Peng, W. Schachermayer, Stochastic Methods in Finance, Bressanone/Brixen, Italy, 2003. Editors: M. Fritelli, W. Runggaldier (2004)

Vol. 1857: M. Émery, M. Ledoux, M. Yor (Eds.), Séminaire de Probabilités XXXVIII (2005)

Vol. 1858: A.S. Cherny, H.-J. Engelbert, Singular Stochastic Differential Equations (2005)

Vol. 1859: E. Letellier, Fourier Transforms of Invariant Functions on Finite Reductive Lie Algebras (2005)

Vol. 1860: A. Borisyuk, G.B. Ermentrout, A. Friedman, D. Terman, Tutorials in Mathematical Biosciences I. Mathematical Neurosciences (2005)

Vol. 1861: G. Benettin, J. Henrard, S. Kuksin, Hamiltonian Dynamics – Theory and Applications, Cetraro, Italy, 1999. Editor: A. Giorgilli (2005)

Vol. 1862: B. Helffer, F. Nier, Hypoelliptic Estimates and Spectral Theory for Fokker-Planck Operators and Witten Laplacians (2005)

Vol. 1863: H. Führ, Abstract Harmonic Analysis of Continuous Wavelet Transforms (2005)

Vol. 1864: K. Efstathiou, Metamorphoses of Hamiltonian Systems with Symmetries (2005)

Vol. 1865: D. Applebaum, B.V. R. Bhat, J. Kustermans, J. M. Lindsay, Quantum Independent Increment Processes I. From Classical Probability to Quantum Stochastic Calculus. Editors: M. Schürmann, U. Franz (2005)

Vol. 1866: O.E. Barndorff-Nielsen, U. Franz, R. Gohm, B. Kümmerer, S. Thorbjønsen, Quantum Independent Increment Processes II. Structure of Quantum Lévy Processes, Classical Probability, and Physics. Editors: M. Schürmann, U. Franz, (2005)

Vol. 1867: J. Sneyd (Ed.), Tutorials in Mathematical Biosciences II. Mathematical Modeling of Calcium Dynamics and Signal Transduction. (2005)

Vol. 1868: J. Jorgenson, S. Lang, $Pos_n(R)$ and Eisenstein Series. (2005)

Vol. 1869: A. Dembo, T. Funaki, Lectures on Probability Theory and Statistics. Ecole d'Eté de Probabilités de Saint-Flour XXXIII-2003. Editor: J. Picard (2005)

Vol. 1870: V.I. Gurariy, W. Lusky, Geometry of Müntz Spaces and Related Questions. (2005)

Vol. 1871: P. Constantin, G. Gallavotti, A.V. Kazhikhov, Y. Meyer, S. Ukai, Mathematical Foundation of Turbulent Viscous Flows, Martina Franca, Italy, 2003. Editors: M. Cannone, T. Miyakawa (2006)

Vol. 1872: A. Friedman (Ed.), Tutorials in Mathematical Biosciences III. Cell Cycle, Proliferation, and Cancer (2006)

Vol. 1873: R. Mansuy, M. Yor, Random Times and Enlargements of Filtrations in a Brownian Setting (2006)

Vol. 1874: M. Yor, M. Émery (Eds.), In Memoriam Paul-André Meyer - Séminaire de Probabilités XXXIX (2006)

Vol. 1875: J. Pitman, Combinatorial Stochastic Processes. Ecole d'Eté de Probabilités de Saint-Flour XXXII-2002. Editor: J. Picard (2006)

Vol. 1876: H. Herrlich, Axiom of Choice (2006)

Vol. 1877: J. Steuding, Value Distributions of L-Functions (2007)

Vol. 1878: R. Cerf, The Wulff Crystal in Ising and Percolation Models, Ecole d'Eté de Probabilités de Saint-Flour XXXIV-2004. Editor: Jean Picard (2006)

Vol. 1879: G. Slade, The Lace Expansion and its Applications, Ecole d'Eté de Probabilités de Saint-Flour XXXIV-2004. Editor: Jean Picard (2006)

Vol. 1880: S. Attal, A. Joye, C.-A. Pillet, Open Quantum Systems I, The Hamiltonian Approach (2006)

Vol. 1881: S. Attal, A. Joye, C.-A. Pillet, Open Quantum Systems II, The Markovian Approach (2006)

Vol. 1882: S. Attal, A. Joye, C.-A. Pillet, Open Quantum Systems III, Recent Developments (2006)

Vol. 1883: W. Van Assche, F. Marcellàn (Eds.), Orthogonal Polynomials and Special Functions, Computation and Application (2006)

Vol. 1884: N. Hayashi, E.I. Kaikina, P.I. Naumkin, I.A. Shishmarev, Asymptotics for Dissipative Nonlinear Equations (2006)

Vol. 1885: A. Telcs, The Art of Random Walks (2006)

Vol. 1886: S. Takamura, Splitting Deformations of Degenerations of Complex Curves (2006)

Vol. 1887: K. Habermann, L. Habermann, Introduction to Symplectic Dirac Operators (2006)

Vol. 1888: J. van der Hoeven, Transseries and Real Differential Algebra (2006)

Vol. 1889: G. Osipenko, Dynamical Systems, Graphs, and Algorithms (2006)

Vol. 1890: M. Bunge, J. Funk, Singular Coverings of Toposes (2006)

Vol. 1891: J.B. Friedlander, D.R. Heath-Brown, H. Iwaniec, J. Kaczorowski, Analytic Number Theory, Cetraro, Italy, 2002. Editors: A. Perelli, C. Viola (2006)

Vol. 1892: A. Baddeley, I. Bárány, R. Schneider, W. Weil, Stochastic Geometry, Martina Franca, Italy, 2004. Editor: W. Weil (2007)

Vol. 1893: H. Hanßmann, Local and Semi-Local Bifurcations in Hamiltonian Dynamical Systems, Results and Examples (2007)

Vol. 1894: C.W. Groetsch, Stable Approximate Evaluation of Unbounded Operators (2007)

Vol. 1895: L. Molnár, Selected Preserver Problems on Algebraic Structures of Linear Operators and on Function Spaces (2007)

Vol. 1896: P. Massart, Concentration Inequalities and Model Selection, Ecole d'Été de Probabilités de Saint-Flour XXXIII-2003. Editor: J. Picard (2007)

Vol. 1897: R. Doney, Fluctuation Theory for Lévy Processes, Ecole d'Été de Probabilités de Saint-Flour XXXV-2005. Editor: J. Picard (2007)

Vol. 1898: H.R. Beyer, Beyond Partial Differential Equations, On linear and Quasi-Linear Abstract Hyperbolic Evolution Equations (2007)

Vol. 1899: Séminaire de Probabilités XL. Editors: C. Donati-Martin, M. Émery, A. Rouault, C. Stricker (2007)

Vol. 1900: E. Bolthausen, A. Bovier (Eds.), Spin Glasses (2007)

Vol. 1901: O. Wittenberg, Intersections de deux quadriques et pinceaux de courbes de genre 1, Intersections of Two Quadrics and Pencils of Curves of Genus 1 (2007)

Vol. 1902: A. Isaev, Lectures on the Automorphism Groups of Kobayashi-Hyperbolic Manifolds (2007)

Vol. 1903: G. Kresin, V. Maz'ya, Sharp Real-Part Theorems (2007)

Vol. 1904: P. Giesl, Construction of Global Lyapunov Functions Using Radial Basis Functions (2007)

Vol. 1905: C. Prévôt, M. Röckner, A Concise Course on Stochastic Partial Differential Equations (2007)

Vol. 1906: T. Schuster, The Method of Approximate Inverse: Theory and Applications (2007)

Vol. 1907: M. Rasmussen, Attractivity and Bifurcation for Nonautonomous Dynamical Systems (2007)

Vol. 1908: T.J. Lyons, M. Caruana, T. Lévy, Differential Equations Driven by Rough Paths, Ecole d'Été de Probabilités de Saint-Flour XXXIV-2004 (2007)

Vol. 1909: H. Akiyoshi, M. Sakuma, M. Wada, Y. Yamashita, Punctured Torus Groups and 2-Bridge Knot Groups (I) (2007)

Vol. 1910: V.D. Milman, G. Schechtman (Eds.), Geometric Aspects of Functional Analysis. Israel Seminar 2004-2005 (2007)

Vol. 1911: A. Bressan, D. Serre, M. Williams, K. Zumbrun, Hyperbolic Systems of Balance Laws. Cetraro, Italy 2003. Editor: P. Marcati (2007)

Vol. 1912: V. Berinde, Iterative Approximation of Fixed Points (2007)

Recent Reprints and New Editions

LECTURE NOTES IN MATHEMATICS   ◲ Springer

Edited by J.-M. Morel, F. Takens, B. Teissier, P.K. Maini

Editorial Policy (for Multi-Author Publications: Summer Schools/Intensive Courses)

1. Lecture Notes aim to report new developments in all areas of mathematics and their applications - quickly, informally and at a high level. Mathematical texts analysing new developments in modelling and numerical simulation are welcome. Manuscripts should be reasonably self-contained and rounded off. Thus they may, and often will, present not only results of the author but also related work by other people. They should provide sufficient motivation, examples and applications. There should also be an introduction making the text comprehensible to a wider audience. This clearly distinguishes Lecture Notes from journal articles or technical reports which normally are very concise. Articles intended for a journal but too long to be accepted by most journals, usually do not have this "lecture notes" character.

2. In general SUMMER SCHOOLS and other similar INTENSIVE COURSES are held to present mathematical topics that are close to the frontiers of recent research to an audience at the beginning or intermediate graduate level, who may want to continue with this area of work, for a thesis or later. This makes demands on the didactic aspects of the presentation. Because the subjects of such schools are advanced, there often exists no textbook, and so ideally, the publication resulting from such a school could be a first approximation to such a textbook. Usually several authors are involved in the writing, so it is not always simple to obtain a unified approach to the presentation.

   For prospective publication in LNM, the resulting manuscript should not be just a collection of course notes, each of which has been developed by an individual author with little or no co-ordination with the others, and with little or no common concept. The subject matter should dictate the structure of the book, and the authorship of each part or chapter should take secondary importance. Of course the choice of authors is crucial to the quality of the material at the school and in the book, and the intention here is not to belittle their impact, but simply to say that the book should be planned to be written by these authors jointly, and not just assembled as a result of what these authors happen to submit.

   This represents considerable preparatory work (as it is imperative to ensure that the authors know these criteria before they invest work on a manuscript), and also considerable editing work afterwards, to get the book into final shape. Still it is the form that holds the most promise of a successful book that will be used by its intended audience, rather than yet another volume of proceedings for the library shelf.

3. Manuscripts should be submitted either to Springer's mathematics editorial in Heidelberg, or to one of the series editors. Volume editors are expected to arrange for the refereeing, to the usual scientific standards, of the individual contributions. If the resulting reports can be forwarded to us (series editors or Springer) this is very helpful. If no reports are forwarded or if other questions remain unclear in respect of homogeneity etc, the series editors may wish to consult external referees for an overall evaluation of the volume. A final decision to publish can be made only on the basis of the complete manuscript; however a preliminary decision can be based on a pre-final or incomplete manuscript. The strict minimum amount of material that will be considered should include a detailed outline describing the planned contents of each chapter.

   Volume editors and authors should be aware that incomplete or insufficiently close to final manuscripts almost always result in longer evaluation times. They should also be aware that parallel submission of their manuscript to another publisher while under consideration for LNM will in general lead to immediate rejection.

4. Manuscripts should in general be submitted in English. Final manuscripts should contain at least 100 pages of mathematical text and should always include
   - a general table of contents;
   - an informative introduction, with adequate motivation and perhaps some historical remarks: it should be accessible to a reader not intimately familiar with the topic treated;
   - a global subject index: as a rule this is genuinely helpful for the reader.

   Lecture Notes volumes are, as a rule, printed digitally from the authors' files. We strongly recommend that all contributions in a volume be written in LaTeX2e. To ensure best results, authors are asked to use the LaTeX2e style files available from Springer's web-server at

   ftp://ftp.springer.de/pub/tex/latex/svmultt1/ (for summer schools/tutorials).

   Additional technical instructions are available on request from: lnm@springer.com.
5. Careful preparation of the manuscripts will help keep production time short besides ensuring satisfactory appearance of the finished book in print and online. After acceptance of the manuscript authors will be asked to prepare the final LaTeX source files (and also the corresponding dvi-, pdf- or zipped ps-file) together with the final printout made from these files. The LaTeX source files are essential for producing the full-text online version of the book. For the existing online volumes of LNM see: www.springerlink.com/content/110312

   The actual production of a Lecture Notes volume takes approximately 12 weeks.
6. Volume editors receive a total of 50 free copies of their volume to be shared with the authors, but no royalties. They and the authors are entitled to a discount of 33.3% on the price of Springer books purchased for their personal use, if ordering directly from Springer.
7. Commitment to publish is made by letter of intent rather than by signing a formal contract. Springer-Verlag secures the copyright for each volume. Authors are free to reuse material contained in their LNM volumes in later publications: a brief written (or e-mail) request for formal permission is sufficient.

Addresses:

Professor J.-M. Morel, CMLA,
École Normale Supérieure de Cachan,
61 Avenue du Président Wilson,
94235 Cachan Cedex, France
E-mail: Jean-Michel.Morel@cmla.ens-cachan.fr

Professor F. Takens, Mathematisch Instituut,
Rijksuniversiteit Groningen, Postbus 800,
9700 AV Groningen, The Netherlands
E-mail: F.Takens@math.rug.nl

Professor B. Teissier,
Institut Mathématique de Jussieu,
UMR 7586 du CNRS,
Équipe "Géométrie et Dynamique",
175 rue du Chevaleret
75013 Paris, France
E-mail: teissier@math.jussieu.fr

For the "Mathematical Biosciences Subseries" of LNM:

Professor P.K. Maini, Center for Mathematical Biology
Mathematical Institute, 24-29 St Giles,
Oxford OX1 3LP, UK
E-mail: maini@maths.ox.ac.uk

Springer, Mathematics Editorial I, Tiergartenstr. 17,
69121 Heidelberg, Germany,
Tel.: +49 (6221) 487-8410
Fax: +49 (6221) 4876-8259
E-mail: lnm@springer.com